This book explains the theory and methods by which gas molecules can be polarized by light, a subject of considerable importance for what it tells us about the electronic structure of molecules and the properties of chemical reactions. Starting with a brief review of molecular angular momentum, the text goes on to consider resonant absorption, fluorescence, photodissociation and photoionization, as well as the effects of collisions and static fields. A variety of macroscopic effects are considered, among them angular distribution and the polarization of emitted light, ground state depopulation, laser-induced dichroism, the effects of collisions, and external magnetic and electric field effects. Most examples in the book are for diatomic molecules, but symmetric-top polyatomic molecules are also included. The book concludes with a short appendix of essential formulae, tables for vector calculus, spherical functions, Wigner rotation matrices, Clebsch–Gordan coefficients and methods for expansion over irreducible tensors.

This book will be of interest to graduate students and researchers in atomic and molecular physics, chemical physics, quantum optics and spectroscopy.

T0172105

Cambridge Monographs on Atomic, Molecular, and Chemical Physics 4

General Editors: A. Dalgarno, P. L. Knight, F. H. Read, R. N. Zare

Optical Polarization of Molecules

Cambridge Monographs on Atomic, Molecular and Chemical Physics

Optical Polarization of Molecules

Marcis Auzinsh

and

Ruvin Ferber

University of Latvia

CAMBRIDGE
UNIVERSITY PRESS

CAMBRIDGE UNIVERSITY PRESS
Cambridge, New York, Melbourne, Madrid, Cape Town, Singapore, São Paulo

Cambridge University Press
The Edinburgh Building, Cambridge CB2 2RU, UK

Published in the United States of America by Cambridge University Press, New York

www.cambridge.org
Information on this title: www.cambridge.org/9780521443463

First published 1995
This digitally printed first paperback version 2005

A catalogue record for this publication is available from the British Library

Library of Congress Cataloguing in Publication data

Auzinsh, Marcis.
Optical polarization of molecules / Marcis Auzinsh, Ruvin Ferber.
   p.   cm. – (Cambridge monographs on atomic, molecular, and
chemical physics)
Includes bibliographical references.
ISBN 0-521-44346-6
1. Molecules–Optical properties.   2. Molecular structure.   3.
Gases–Optical properties.   4. Angular momentum (Nuclear
physics).   5. Polarization (Light).   I. Ferber, Ruvin   II. Title.
III. Series.
QC165.3.A99   1995
539′.6–dc20   94-6315   CIP

ISBN-13 978-0-521-44346-3 hardback
ISBN-10 0-521-44346-6 hardback

ISBN-13 978-0-521-67344-0 paperback
ISBN-10 0-521-67344-5 paperback

To
Dr Ojārs Šmits,
who taught and inspired us to this work,
and to our families

Contents

ix

Preface

From the very beginning of our involvement in the investigation of light induced polarization of angular momenta of molecules we were fascinated by the variety of information about the properties of molecules which they bear. At the same time the description and interpretation of these phenomena appeared to us to be extremely complicated and unclear. In fact, at times it seemed as if our computers understood the problem better than we did.

This book is an attempt to clarify the processes during the course of which polarized (ordered) angular momenta distribution is created in an ensemble of molecules in the gas phase by the effect of light. We discuss the effect of static external magnetic and electric fields on the angular momenta distribution. In particular, we wish to emphasize the 'geometric' meaning and interpretation of the phenomena. This may, we believe, be a further step in attempts to simplify the theoretical description, thus making it more accessible to a wide range of users, both physicists and chemists.

The fundamental basis for optical polarization (alignment, orientation) of angular momenta is the law of conservation of angular momenta in photon–molecule interaction. In this book we examine a variety of macroscopic manifestations of spatial anisotropy of angular momenta, such as angular distribution and polarization of emitted light, including changes under non-linear absorption, and the influence of collisions and external fields. Quantum angular momentum theory, in particular that which is based on irreducible tensorial set representation, presents a well-developed approach that is widely used in subatomic, atomic and molecular physics. At the same time, there exists a considerable gap between this rather complicated theory and the possibility of imagining the physical essence of the phenomena, as well as performing the explicit calculations required in the everyday work of the practical researcher in physics and chemistry.

In writing this book we aimed to close this gap by taking the reader all the way from general definitions through to the detailed treatment required in specific experimental situations. In the course of this aim we arrived at a kind of 'three-dimensional space' for the book, with pictorial illustration, strict theory and experimental examples as its 'eigenvectors'.

A distinctive feature of a molecule is the large, as a rule, value of angular momentum. We have therefore tried, while by no means ignoring quantum mechanical description, to benefit from classical approach. The book begins gently with a tangible description of how a light wave of given polarization is able to excite a Hertzian dipole in a rotating diatomic molecule, producing an ordering of angular momenta in the familiar P-, Q- and R-transition branches. We then demonstrate how the classical approach enables us to imagine visually the shape of the distribution of the angular momenta, in particular by isometric projections, in order to clarify symmetry properties by expansion over spherical functions, and to connect all of this with the observables.

Arriving subsequently at rigorous quantum mechanical descriptions, we have assumed that the reader has some preliminary knowledge of basic quantum mechanical formalism. We consider it methodologically important to illustrate the correspondence principle between quantum and classical concepts, in particular between the concept of coherence of the wave functions of magnetic sublevels, and the symmetry properties of spatial angular momenta distribution.

This is a good place to mention that in our efforts to preserve a unified approach we have restricted ourselves to the approximation that all molecules are affected by light independently of their velocities ('broad line approximation'). This makes it possible to factorize, in a natural way, the calculations into two parts: the dynamic and the angular (geometric) parts, the latter forming the subject of the present book.

Far from presenting even a part of the tremendous amount of experimental material, we have restricted ourselves to giving examples of some studies on diatomic molecules. However, a number of more complicated cases have been included in order to demonstrate that it is easy to extend the considerations discussed here to polyatomic and, in particular, to symmetric top molecules.

The authors are hopeful that the book will serve as a practical manual for the researcher, both theorist as well as experimentalist, in atomic, molecular or chemical physics. With this in mind, we have paid special attention to the comprehensive information necessary to follow all calculations through to their conclusions. We also hope that the material in the appendices will take the place of a short handbook with formulae on vector calculus, spherical functions, Wigner D-functions, Clebsch–Gordan coefficient tables, etc. We fully realize that one may find in various papers a multitude of ways of introducing expansion over irreducible tensors which differ in normalization, phase convention and transformation algebra under coordinate rotation. While keeping to one particular type, we have appended an overview on other existing methods and have included the necessary conversion formulae.

The birth of this book was to no small extent fostered and assisted by a considerable number of people, to whom the authors have the pleasure and honour to be deeply indebted.

We are grateful to our colleagues from the University of Latvia M.Ya. Tamanis, I.Ya. Pirags, Ya.A. Harya, I.P. Klintsare, as well as to A.V. Stolyarov (Moscow State University), who at some time have all either worked with us, or else have at various times been engaged in research connected to the theme of the present book. We also wish to express our thanks to A.I. Okunevich of the A.F. Ioffe Physical-Technical Institute, St Petersburg, whose influence drew the interest of the authors to the method of polarization moments.

We are grateful to W. Hanle, the discoverer of 'Hanle effect', to E.K. Kraulinya and M.L. Janson, founders of the Riga school of experimental atomic physics, as well as to the founders of the St Petersburg school of the study of interference of atomic states M.P. Chaika, N.I. Kaliteyevskii, E.B. Aleksandrov. Encouragement and approval from all these scientists gave us the confidence necessary to proceed with this work.

We are especially grateful to R.N. Zare, the author of pioneer ideas and research in polarization of molecules for his valuable comments and suggestions.

We are indebted to I. Dabolinya, M. Petrovska and P. Kapostinsh for their technical assistance. We greatly appreciate the considerable support given by E.G. Rapoport during all stages of our work.

Last, but not least, we fully realize the contribution of J. Eiduss, without whose enormous amount of work the English version of this book would never have materialized.

Riga

Marcis Auzinsh, Ruvin Ferber

1

Angular momentum and transition dipole moment

1.1 Anisotropy of angular momenta: ideas and methods

The fundamental idea forming the basis of early quantum theory was the quantization of the *angular momentum* of the atom, which, at a later stage, assumed the form of the Bohr–Sommerfeld quantization rule [360], namely $\oint p\,dq = n\hbar$, where p and q are the generalized momentum and coordinate. This concept has retained its physical meaning in modern-day quantum mechanics [371, 372]. In the course of the further development of quantum mechanical concepts it became increasingly clear that in the theory of atoms and molecules the angular momentum is a 'dominant' notion, the fundamental significance of which is in no small degree connected with its dimension of action, which coincides with that of Planck's constant $\hbar$.

In the interaction between particles and a directed beam of photons the angular momentum of the system is conserved, thus forming the basis for the optical method of producing spatial anisotropy of the distribution of the angular momenta of the ensemble of particles, i.e. its *optical polarization*. In terms of quantum concepts, spatial anisotropy means that in an ensemble of particles a non-equilibrium population of magnetic sublevels m_j of angular momenta is created. We make a distinction between the *alignment* of momenta, for instance in the absorption of linear polarized light, when there is no difference between the population of $\pm m_j$-sublevels, and *orientation*, when such a difference exists. Anisotropic distribution of angular momenta in an excited state, as produced in absorption, manifests itself directly in the *polarization of fluorescence*.

It is rather likely that understanding of the anisotropic distribution of the angular momenta of atoms was enhanced by Hanle's [184] discovery in 1924. Hanle observed the effect of the disappearance of linear polarization in fluorescence in a weak magnetic field directed at right angles to the **E**-

1

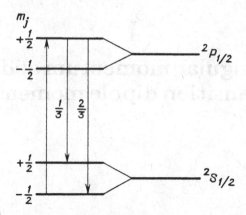

Fig. 1.1. Scheme of optical pumping (orientation) of the $^2S_{1/2}$-state; the figures show the relative probabilities of radiational transition.

vector of linear polarized exciting light (the *Hanle effect*). It may be worth pointing out that this effect was rather convincingly explained by the author of the discovery himself, who ascribed it to precession of the distribution of the angular momenta in the external magnetic field.

Up until 1950 such an anisotropy of angular momenta was assumed for excited states only. Then, however, it was suggested, for the first time by Kastler [224], to use circularly polarized light to produce anisotropic angular momenta distribution of atoms in the electronic ground state as a result of absorption of resonance frequency light, thus achieving macroscopic polarization (orientation) of the ensemble. Following quantum concepts, it may be said that one manages to transfer atoms which were primarily evenly distributed over m_j on to one magnetic sublevel in this way. Such a process was named 'optical pumping' (*pompage optique*) by Kastler, thus stressing the decisive role of the optical radiation as a 'pump'. Fig. 1.1 shows a, now classical, scheme and illustrates the simple, beautiful idea of the method. A hypothetical alkali atom without nuclear spin is used as an example. As may be seen, since in the case of lefthanded polarized irradiation no absorption may take place from level $m_j = 1/2$, and spontaneous emission goes to both levels $m_j = \pm 1/2$, the atoms finally find themselves 'pumped over' on to the magnetic sublevel with spin projection $1/2$, i.e. in *spinpolarized state*. It is, of course, necessary in practice for this process to compete successfully with non-radiational mixing of states with $m_j = \pm 1/2$. Note that circularly polarized light orients the spin of the electron along the light beam, which determines the direction of quantization. As a result we obtain an ensemble of atoms oriented through transfer of the angular momentum of the photons, and a macroscopic magnetic moment is created in the system. The effect can easily be observed from changes in absorption and from the appearance of dichroism and birefringence in fluorescence, as well as in magnetic resonance which causes induced transitions in the system of m_j-sublevels under the effect of a radiofrequency field.

The different variants of the method have been effected on paramagnetic states of atoms, in particular on atoms of the I–III groups of the periodic system of elements, on metastable levels of noble gases and on a number of ions. One may say that optical pumping has become a sufficiently standard instrument for investigating intervals of fine and hyperfine structure of relaxation processes. The method has made it possible to perform hypersensitive measurements of magnetic fields (quantum magnetometers) and establish frequency standards. The significance of this field of research is testified by the Nobel prize awarded to Kastler (1966), and the maturity of its development can be seen from the works of memoirs [185] and other publications in connection with various anniversaries [343]. International Kastler symposia are organized (their Proceedings include, e.g., [64, 344]). A large number of original papers, reviews and monographs have been published on these and on related problems. One might mention here [64, 65, 94, 106, 187, 316, 343, 344, 356] and references therein. A new wave of interest in optical pumping of atoms has sprung up recently in connection with superdeep cooling of spinpolarized atoms in traps to ultralow temperatures [101], and also in connection with ideas of obtaining 'squeezed' light with a reduced level of fluctuations [253]. One may now speak generally of optical pumping as of a fundamental instrument in laser spectroscopy, playing a central part in many phenomena of quantum optics, including orientation of atoms possessing hyperfine structure, in the trapping and cooling of particles, in laser 'shelving' and quantum jumps, in confirming non-conservation of parity in atoms, etc.

The object of our presentation is the optical 'ordering' of the angular momenta of gas phase molecules, using as examples mainly the most simple, diatomic ones. Historically, one of the first observations of the phenomenon was that of partial polarization of resonance fluorescence of diatomic molecules, discovered previously by Wood in 1908 [397]. Polarized emission from molecules was interpreted by Wawilow and Lewschin in 1923 [390] by means of the classical model of anisotropic distribution of radiating dipole oscillators (for details see the monograph by Feofilov [148]). But such a fundamental manifestation of alignment of angular momenta as the Hanle effect was not registered on molecules until 1969 (Zare, Demtröder *et al.*, on $Na_2(B^1\Pi_u)$ [290] and on $OH(A^2\Sigma)$ [165]). A classical and quantum description of the expected effect had been given by Zare shortly before [399]. It is rather significant that in the same year (1969), with the above authors participating, a number of interconnected experi-

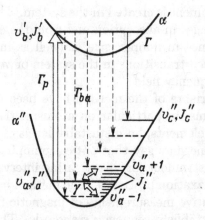

Fig. 1.2. Scheme of transitions of laser-induced fluorescence of molecules: Γ_p – rate of absorption, Γ; γ – rates of relaxation. Notation α, v, J denote the electronic–vibrational–rotational (rovibronic) state. The prime refers to the electronically excited state, the double prime refers to the ground state, whilst the indices a, b, c denote the states involved in an optical transition.

ments were performed [114, 124, 290] on the same test object, namely the Na$_2$ molecule. These experiments actually formed the basis of the method of *laser induced fluorescence* [114] (according to the scheme in Fig. 1.2) and of the Hanle method [290], and, finally, demonstrated [124] the possibility of ground state optical pumping by applying optical absorption to achieve anisotropic distribution of the angular momenta of not only the upper state but also of the initial (lower) state; see Fig. 1.2. This method, to some extent, can also be called *optical pumping* although the schematic for its implementation differs from the Kastler schematic; see Fig. 1.1. Indeed, whilst in Fig. 1.1 we are dealing with a *closed* absorption–emission cycle, the situation is different in the case presented in Fig. 1.2. Here the simplest approximation would be a *non-closed* (open) cycle, in which only an insignificant part of the molecules having absorbed light return spontaneously to the initial level v_a'', J_a'', owing to the large number of possible optical transitions on to various vibrational levels v_i''. As a result the initial level a shows a stationary population different from the equilibrium one (being smaller). However, more important to us is the fact that, owing to the angular dependence of absorption, an *anisotropic distribution of the angular momenta* (i.e. their optical polarization) emerges both in the upper (b) as well as in the lower (a) levels. Besides, emission of light on sufficiently high, and thus thermally unpopulated, levels v_i'' can also transfer the polarization of angular momentum. It is predominantly these phenomena and their various manifestations which form the principal theme of the present book.

A few words about terminology. In publications concerning the ground (or metastable) state of the atom the term 'optical pumping' [65, 124, 224, 343] is used in the above-mentioned sense. This is, however, not completely unambiguous, since, e.g., in the case of laser physics the term 'optical pumping' may be used simply in the sense of optical excitation. The term *'optical orientation'* would sound more precise in the case of Fig. 1.1. This term, however, does not cover the effect of linear polarized or directed non-polarized irradiation, when no preferred orientation of the angular momenta of the ensemble (orientation vector) arises. In this case a symmetry axis is created, along which (in both directions) there will be either more or fewer angular momenta than in the perpendicular plane. In this case we have *optical alignment*. In other words, we speak of *alignment* when the distribution of the angular momenta of the ensemble is symmetrical with respect to the reflection in the plane lying at right angles to the quantization axis z, and of *orientation* if such symmetry is absent. For clarity, it is best to imagine that alignment behaves as a double-headed arrow ($\Longleftrightarrow$), and orientation as a single-headed one ($\Longrightarrow$).

Thus one ought to speak of *'optical alignment and orientation'* in the general case. In attempting to find a single term generalizing the cases discussed, one might suggest the term *'optical polarization'* (which refers to the angular momenta of the ensemble of particles and not to the light beam), as is stated in the title of the present book. Also, not neglecting tradition, we reserve the right to use the term 'optical pumping' in the sense of Kastler and Zare. In other terms this phenomenon consists of the creation of a non-equilibrium population over magnetic sublevels. Here one must point out the fundamental role of the external field effects which define the axes of quantization.

The attentive reader ought to have realized that, in defining our problem of the interaction between light and molecules in such a way, basic attention must be paid to the so-called *angular part* of this interaction, i.e. to the so-called *geometric factor*. We wish to remind the reader that the power of the well-known Wigner–Eckart theorem [136] consists just in the circumstance that the transition matrix element between states with precisely definable angular momenta can be divided into two factors, of which one describes the *dynamic* part, and the second consists of the selection rules in the system of angular momenta at the given geometry (the *angular* part). It is just the angular part of the transition probability which, in quantum mechanical terms, includes the dependence of this probability on the magnetic quantum number M_J, and which describes the creation of the spatial distribution of angular momenta. It thus determines the *polarization of the radiation, coherence* phenomena and the effects of external *magnetic and electric fields*. In order to focus attention on the angular factor, we will limit ourselves to the approximation of a wide spectral range of the exciting light ('broad line approximation'). This permits us to exclude from consideration the dependence of absorption probability on the velocities and coordinates of the particles. Incidentally, such an approach appears rather more realistic in the case of 'usual' (i.e. without selection of a single mode) laser excitation, as compared to another limiting case, monochromatic excitation. For the peculiarities of a description based on velocity selected excitation; e.g., monographs [268, 320].

The most natural method of detecting the optical polarization produced in the excited state consists of using the polarization characteristics of molecular fluorescence from this state in some transition $b \rightarrow c$ which is convenient for observation; see Fig. 1.2. The same radiation, in the cycle $a \rightarrow b \rightarrow c$ may also provide information on the initial state a if the process of absorption is non-linear. A more direct way of probing the anisotropy of angular momenta distribution in the ground state consists of monitoring the absorption from this level by means of a test beam.

There also exist other methods of obtaining light-produced aligned or oriented molecules, such as fluorescence-caused population, photodissociation and photoionization, and collisions between optically aligned particles. It is important to underline that in all cases of direct light excitation the direction of the photon beam itself perturbs the isotropy of angular momenta distribution of the absorbing molecules in the bulk. If we are dealing with a molecular beam, the beam itself forms a selected direction. In this case anisotropy of both molecular axes, as well as that of the angular momenta, can obviously be produced without any action of light, for instance in collisions with the buffer gas, in collisions with the surface and also under the effect of an external electric or magnetic field. We will discuss these methods, however, in less detail.

Of the various methods of describing the phenomena, we have chosen the classical one, based on the probability density distribution of angular momenta, for the following reasons. First, we are often dealing with a state possessing a sufficiently large angular momentum, with a quantum number of the order of 10–100. Hence, a classical approach is fully justified. Second, this has the advantage of clarity, and in most cases it is also simpler. Finally, it is interesting to follow up how the transition takes place from quantum concepts and terms, such as the coherence between Zeeman sublevels, into classical ones; in other words, how the correspondence principle works. One might point out in this respect that in recent years interest in classical concepts has increased. As an example we refer to the treatment of the problem of superposition of Rydberg states in atoms by the authors of [1, 405], who confirm convincingly the classical orbital rotation of the Rydberg electron along an extended, as well as along a near circular orbit. In the case of molecules such interest is connected with the introduction of stationary quantum mechanical states with given maximal (accounting for the uncertainty principle) spatial localization of the angular momentum [128], or of the internuclear axis (the 'directed states') [157, 220].

The description used in the present book does not, however, confine itself to the limit of an infinitely large angular momentum. As is well known, the problem is in its simplest form in the case of either small momenta ($J \approx 1$), or in that of very large ones ($J \gg 1$). We will also cover the possibility of approaching the problem in the most 'unfavourable' case of an arbitrary angular momentum.

There is considerable value in information gained with participation of molecules with optically polarized angular momenta. The basic, the most obvious and rapidly progressing field of research will be that of the role of spatial orientation of molecules in the dynamics of elementary processes, e.g. in inelastic collisions leading to chemical reactions. Here the basis of approach will be the law of conservation of total angular momentum in an elementary process. The role of optical methods is reduced to the preparation of polarized reagents with selection over vibrational and rotational states, and to testing the angular distribution of the products. One might thus strive towards an 'ideal' stereodynamic experiment in which reagents have been created in selected states and predefined directions of angular momentum and (or) molecular axes with subsequent testing of the distribution over internal degrees of freedom, over scattering angles and over orientations of angular momenta or molecular axes of the products. This would be best in a real time scale which would also permit investigation of the stage of the intermediate reaction complex. An indication of the development of the above trend is the regular international conferences on dynamic stereochemistry, the first of which took place in Jerusalem in 1986 [66], in Bad Honef (FRG) in 1988 [341], in Santa Cruz (USA) in 1990 [67] and in Assisi (Peruggia, Italy) in 1992 [305].

Less popular (undeservedly, in our opinion) is the employment of optical polarization of angular momenta and of external field effects for the determination of structural parameters of molecules. In particular, unique information can be obtained from Landé factor measurements, the magnitude and sign of which form a singularly sensitive indicator of intramolecular interactions of various types [267, 294].

And, finally, one must keep in mind the fact that 'non-spherical' distribution of the angular momenta of a molecular gas can affect the results of quantitative spectroscopy. And if one uses laser sources, so popular today, non-linear effects, including non-linear optical polarization of ground state angular momenta, are by no means rare phenomena, and the art of their evaluation may become necessary in the everyday practical work of the spectroscopist.

1.2 The angular momentum of a molecule

The angular momentum of a molecule forms the basic object of our discussion. We will accordingly attempt to remind the reader briefly of which components it consists and how they combine. For details we refer the reader to well-known books, such as [194, 218, 294, 402].

The fundamental problem in the case of a rotating molecule (as a rule, for simplicity's sake, we will consider a *diatomic* or a *linear polyatomic* one) is that of the interaction between the electronic motion and rotation of the nuclei. For better clarity and in conformity with the style of our further presentation, we will apply the vector model approach.

Neglecting nuclear spin, the total *angular momentum* $\mathbf{J}$ of a molecule is composed of the total *orbital momentum* of the electrons $\mathbf{L}$, of their total *spin momentum* $\mathbf{S}$ and the *rotational momentum* $\mathbf{R}$ of the nuclei with respect to the mass center. For further comprehension it is important to range the various kinds of angular momenta coupling according to interaction energy. In terms of the vector model one usually speaks of corresponding precession frequencies of the momentum with respect to a certain axis. The latter forms the quantization axis, i.e. it is chosen in such a way as to conserve the projection on it of the momentum under consideration (at least, to a certain approximation). This may be the internuclear axis (or the figure axes) of the molecule, or the direction of one of the momenta. As a rule, the quantization axis itself also rotates (nutates) – see Fig. 1.3 – but much more slowly so that the momentum linked with it manages to become quantized on it. It is worth remembering that such an approach is similar to the adiabatic approximation largely applied in molecular physics, for instance in separating electronic motion from the, considerably slower, nuclear motion [374].

Let us first assume that $\mathbf{S} = 0$ and consider the behavior of the orbital momentum $\mathbf{L}$ in the inhomogeneous intramolecular electric field $\mathcal{E}_{mol}$ which possesses axial symmetry with respect to the internuclear axis. Under the effect of Stark interaction the vector $\mathbf{L}$ is in precessional motion around this axis. The interaction frequency (or energy) is strongly dependent on the orbitals forming the momentum $\mathbf{L}$. The molecule-fixed projection $|L_z| = \Lambda = 0, 1, 2, \ldots$ upon the internuclear axis is conserved, and the corresponding symbols of the molecular terms are $^1\Sigma, ^1\Pi, ^1\Delta, \ldots$, these terms, being singlet ones in this case. The energy difference between the terms with various Λ is a measure of the interaction between $\mathbf{L}$ and the internuclear axis. The interaction is strongly dependent on the one-electron orbits $\sigma, \pi, \delta, \ldots$ corresponding to the molecule-fixed projections, $|l_{iz}| = \lambda_i = 0, 1, 2, \ldots$, forming an electronic state with a total electronic angular momentum $\mathbf{L} = \sum_i \mathbf{l}_i$. The $^1\Sigma$ state with $\Lambda = 0$ emerges at $\mathbf{L} = 0$, or else **L** is in precession orthogonally around the internuclear axis. This is the ground state in the case of the large majority of linear molecules (with the exception of the O_2 group, as well as of such molecules as NO, NO_2, ClO_2, where the the number of valence electrons is odd). In this case the total momentum **J** coincides with the rotational momentum **R**, and the system forms a simple rotator. If $\Lambda \neq 0$, the **J** consists of the vectorial sum of mutually orthogonal Λ and **R**.

In the more general case $S \neq 0$ and the molecular angular momenta can be coupled in various ways. It is of primary importance to ascertain to what extent the interaction of the spin momentum **S** with the orbital momentum **L** is comparable to the rotation of the molecule, as well as to the interaction of each of the momenta **L** and **S** with the internuclear axis. An attempt to establish a hierarchy of interactions yields a number of possible, certainly idealized, coupling cases between angular momenta, first considered by Hund and known as *Hund's coupling cases*. Here we will discuss the three basic (out of five) cases of coupling of momenta in a linear molecule.

1 Hund's case (a) coupling

Let the orbital momentum **L** be coupled, due to Stark interaction, to the intermolecular axis considerably more strongly than to the spin momentum **S**. One may therefore neglect the interaction between **L** and **S**. Unlike **L**, the spin momentum is not directly affected, owing to its 'immateriality', by the electric field of the molecule. Nevertheless, interaction between **S** and the projection $\Lambda = -L, -L+1, \dots, L-1, L$ of momentum **L** upon the internuclear axis (assuming $\Lambda \neq 0$) may take place in the form of a magnetodipole interaction. More precisely, in this case we have an interaction between the respective magnetic moments μ_S and μ_Λ. Hence a projection of **S** upon the internuclear axis emerges which we denote by Σ, (Fig. 1.3(a)) (not to be confused with the symbol for the electronic state with $\Lambda = 0$), Σ covering the range of values $-S, -S+1, \dots, S-1, S$. Hund's case (a) coupling presumes that both **L** and **S** are strongly coupled to the internuclear axis, hence their total projection $\Omega = \Lambda + \Sigma$ is well defined; see Fig. 1.3(a). It is convenient to introduce Ω as a positive quantity which holds if we consider that the positive direction of the internuclear axis (the quantization axis) is defined by the largest of Λ and Σ. Hund's case (a) coupling scheme means that the rotation frequency of **L** and **S** around the internuclear axis considerably exceeds that of the relatively slow rotation (nutation) of momenta Ω and **R** around the total momentum $J = \Omega + R$. The latter ought to take place in such a way that the momentum **J** keeps its magnitude and direction in space. The

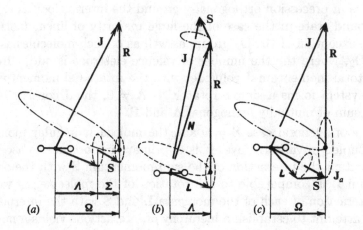

Fig. 1.3. Different cases of momentum coupling, after Hund.

frequency of this rotation may be represented as follows:

$$E_{rot} = hBR(R+1) = hB[J(J+1) - \Omega^2],   (1.1)$$

where $B = 1/(2I) = 1/(2\mu r_0^2)$ is the rotational constant, I being the moment of inertia, μ the reduced mass and r_0 the internuclear distance. Since $\mathbf{R}$ is orthogonal with respect to the internuclear axis, the projection of the total momentum $\mathbf{J}$ on the axis is obviously Ω. Such a sequence (hierarchy) of vector interaction in the case (a) coupling scheme may be symbolically represented in the form: $\{[(\mathbf{L}, \mathcal{E}_{mol}), \mathbf{S}], \mathbf{R}\}$. The quantum number of the total momentum assumes the following values:

$$J = \Omega, \Omega + 1, \Omega + 2, \ldots.   (1.2)$$

The classification of levels differs, in principle, only very little from the case in which we have zero spin. The notation of an electronic term usually includes the indication of projection Λ ($\Sigma, \Pi, \Delta, \Phi, \ldots$ for $|\Lambda| = 0, 1, 2, 3, \ldots$), multiplicities $2S + 1$ and the projections of Ω, such as, $^2\Pi_{1/2,3/2}$, $^3\Pi_{0,1,2}$, $^3\Delta_{1,2,3}$. The selection rules for an electric dipole transition are as follows: $\Delta\Sigma = 0$; $\Delta\Lambda = 0, \pm1$; $\Delta J = 0, \pm1$.

2   Hund's case (b) coupling

Our preceding assumption is an obvious fallacy if $\Lambda = 0$ but $\mathbf{S} \neq 0$, since in this case the spin cannot perform any precession around the internuclear axis in view of the absence of a magnetic moment directed along this axis. In the case of light molecules, when the rotation frequency is high, the coupling of $\mathbf{S}$ with the internuclear axis is very weak, even at $\Lambda \neq 0$. The quantum number Ω thus has no meaning. Under such circumstances we speak of Hund's case (*b*) coupling. Here the vector $\mathbf{\Lambda}$ (if it exists) is coupled with $\mathbf{R}$ (see Fig. 1.3(*b*)), forming the momentum $\mathbf{N} = \mathbf{R} + \mathbf{\Lambda}$, i.e. a 'total angular momentum without spin'. Subsequently, $\mathbf{N}$ is coupled with the spin $\mathbf{S}$, and they nutate, forming the total momentum $\mathbf{J}$, the quantum number of which assumes the values

$$J = N + S, N + S - 1, \ldots, |N - S|. \tag{1.3}$$

This type of interaction of momenta can be represented symbolically as $\{[(\mathbf{L}, \mathcal{E}_{mol}), \mathbf{R}], \mathbf{S}\}$. The notation of electronic terms is usually presented with an indication of the projection Λ and of the multiplicity, e.g. $^2\Sigma$, $^3\Sigma$, ... The selection rules are: $\Delta S = 0$, $\Delta N = \Delta \Lambda = \Delta J = 0, \pm 1$.

3 Hund's case (c) coupling

If the diatomic molecule consists of heavy atoms, one may naturally expect that, owing to the strong spin–orbit interaction, the coupling between $\mathbf{L}$ and $\mathbf{S}$ may prove stronger than the coupling between each of these momenta and the internuclear axis. Both Λ and Σ lose their meaning in this case; however, the total electronic angular momentum $\mathbf{J}_a = \mathbf{L} + \mathbf{S}$ is formed by the spin–orbit coupling; see Fig. 1.3(*c*). After this the momentum $\mathbf{J}_a$ is coupled to the molecular axis and forms a molecule-fixed projection of the total electronic angular momentum Ω. The nutation of Ω and $\mathbf{R}$, with the formation of the total angular momentum $\mathbf{J}$, takes place with the same frequency as in Hund's case (*a*) coupling. The rotational energy E_{rot} and the magnitude of the momentum $\mathbf{J}$ are also determined according to (1.1) and (1.2). Such a sequence of interactions can be represented symbolically as: $\{[(\mathbf{L}, \mathbf{S}), \mathcal{E}_{mol}], \mathbf{R}\}$. The notation of the terms is, as a rule, given after the value of the quantum number of the projection Ω (in numerical form).

It ought to be understood that all the above-mentioned basic coupling cases (there still exist cases (*d*) and (*e*), which, however, are rather rare) present idealized situations known as 'pure' coupling schemes. In reality one frequently ought to account for deviations from 'pure' coupling. Thus, e.g., interaction between $\mathbf{R}$ and $\mathbf{L}$ (or $\mathbf{J}_a$) may slightly 'break' the coupling of $\mathbf{L}$ with the molecular axis, leading to Λ- (or Ω)-type doubling. This manifests itself as the removal of degeneracy over two possible orientations of Λ with respect to the internuclear axis.

In our discussion which concerns chiefly diatomic or linear molecules we will often apply the same notation (J, or J', J'') for the total angular momentum and for the rotational quantum number because of the decisive contribution of rotational motion. Some remarks on accounting for nuclear spin can be found in Sections 1.4 and 5.3. The calculations of the rotational level energies and of the transition intensities have been

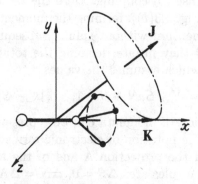

Fig. 1.4. Rotational motion and moments of inertia in symmetric top molecules (of the type of CH_3F or its halogensubstitutes).

considered in great detail in monographs [198, 244], and those for Hund's (*c*) coupling in [381].

It seems appropriate to make a few remarks concerning *polyatomic* (non-linear) molecules in which there is no axis with respect to which one may disregard rotation. Accordingly, one must generally account for rotation around several axes, applying the inertia tensor I_{ij}. Referring the inertia tensor to the main axes, we single out the main moments of inertia I_{xx}, I_{yy} and I_{zz}. If two of them are equal, e.g. $I_{xx} \neq I_{yy} = I_{zz}$, then the symmetry axis (figure axis) lies along the *x*-axis, and we have a *symmetric top* in this case; see Fig. 1.4. As examples of this type of molecule, one might mention NH_3, PH_3 and CH_3F and its halogensubstitutes, C_6H_6. The classical motion of such a molecule would be a combination of rotation around the symmetry axis with angular momentum **K**, and nutation of this momentum around the total momentum **J**. In this case, using a molecule-fixed coordinate system, the rotational state is characterized by two magnitudes: the quantum number of the total angular momentum **J**, and the quantum number of its projection K on the symmetry axis, the latter equalling $K = -J, -J+1, \ldots, J-1, J$. The energy levels of a symmetric top equal $hBJ(J+1) + h(C-B)K^2$, where B and C are the respective rotational constants of the momenta $I_{yy} = I_{zz}$ and I_{xx}.

The decisive contribution of the rotational motion in molecules frequently results in large total angular momentum values, $J \gg 1$. This circumstance leads to certain characteristic peculiarities of molecular magnetism which have to be taken into consideration in the subsequent presentation. First, the magnetic moment μ_Ω along the internuclear axis (if not equal to zero), being of the same order of magnitude as for atoms, is almost orthogonal to **J**. Hence the ratio between the **J**-projection of the magnetic moment μ_J and the **J**-magnitude $\sqrt{J(J+1)}$ is, as a rule, much smaller than Bohr's magneton μ_B. This means that the Landé factor of a rotating molecule g_J is much smaller than unity and, accordingly, high magnetic fields are required to observe precession of $\mathbf{J}$ over the external magnetic field $\mathbf{B}$. Second, the large J values correspond to a large number $2J + 1$ of Zeeman M_J components. This permits a transition to continuous projections of J upon the quantization axis (z-axis), and thus to the classical description which has largely been applied in our further discussion.

1.3   Transition dipole moment: basic concepts

In the classical description, the appearance of the polarization of angular momenta of molecules in absorption is determined, on the one hand, by the manner in which the oscillations of the light vector $\mathbf{E}$ and the absorbing dipole $\mathbf{d}$ are mutually positioned; on the other hand, it is determined by the way in which this dipole is *tied* to the molecular structure.

The basic aim of the present section is to analyze the orientation of $\mathbf{d}$ with respect to the space-fixed angular momentum $\mathbf{J}$ of the free molecule. Let us start by considering the mutual orientation of $\mathbf{J}$ and $\mathbf{d}$ in a diatomic molecule. In order to explain emission and absorption of light by molecules, classical electron theory assumes the following model. The electrons in ground state molecules possess certain equilibrium positions with respect to the nuclei. When displaced from this position, the electron behaves like a harmonic oscillator vibrating with eigenfrequency ω_0 which is determined by the *bond strength* keeping the electron near its equilibrium position, and by the mass of the electron. As a result of these oscillations an electromagnetic wave is emitted, according to the laws of electrodynamics.

In the process of scattering and absorption, the electric component of the incident wave excites the vibrations of the oscillator. Under the effect of this component the electron performs forced vibrations. If the eigenfrequency ω_0 of the oscillator coincides with that of the light wave ω_l, resonance absorption is observed. If these frequencies do not coincide, we have non-resonant scattering of light.

In analyzing the absorption and emission of light by molecules it is necessary to take into account the fact that, owing to the inner structure of molecules, oscillations of electrons may take place effectively only along a preferred direction with respect to the positioning of atoms in the molecule. Thus, in diatomic molecules dipole oscillations can be excited only along the internuclear axis in optical transitions between electronic states possessing equal values of the angular momentum projection Λ of

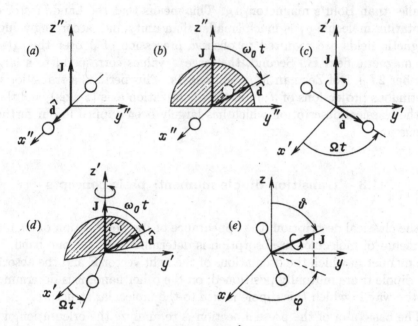

Fig. 1.5. Direction of the dipole moment $\hat{\mathbf{d}}$ of an optical transition in a diatomic molecule: (*a*), (*c*) *parallel transition*; (*b*), (*d*) *perpendicular transition*; (*e*) arbitrary orientation of the angular momentum **J**.

electronic motion on the internuclear axis (Σ–Σ, Π–Π and similar transitions); see Fig. 1.5(*a*). These are so-called *parallel transitions* [95]. Here we speak of the transition dipole moment which is created in the process of light absorption or emission by the molecule. In other words, we are dealing here with a transition between two stationary states, and not with a permanent electric dipole moment which may be present in polar heteronuclear molecules. The energy difference of the quantum states, between which the transition is taking place, determines the frequency of the emitted or absorbed light, i.e., classically speaking, it characterizes the eigenfrequency ω_0 of the dipole oscillator of the transition. If, in a diatomic molecule, the quantum number Λ changes by ± 1 (Σ–Π, Π–Δ, ...), then a corresponding classical circular oscillator – a rotator – may be excited. Here we have a rotational motion of the electron with frequency ω_0 in a plane which is perpendicular to the internuclear axis (see Fig. 1.5(*b*)). Such a transition is called *perpendicular* [95].

In addition, it ought to be kept in mind, considering the classical concept of light absorption by molecules, that, owing to the inertia of molecules, the light field is not capable of turning the molecule and thus excites the vibrations of the dipole oscillator only to the extent to which the **E**-vector of the wave possesses a component along the direction of possible vibrations of the oscillator (with the exception of the case of absorption from a state with zero angular momentum).

Simultaneously with the oscillations with respect to the internuclear axis, a diatomic molecule rotates as a whole with a frequency Ω (not to be confused with the quantum number of the projection of the total momentum, as in Section 1.2) around an axis which is perpendicular to the direction of the internuclear axis, exhibiting a total angular momentum **J**; see Fig. 1.3. Hence in the laboratory coordinate system, with respect to which the light wave possesses a certain polarization, electrons participate simultaneously in two types of motion: dipole oscillations with respect to the internuclear axis, and rotation of the molecule as a whole.

1.4   Absorption of light of different polarization

In the present work we are interested in the polarization of absorbed light. This is best characterized by expanding the unit $\hat{\mathbf{E}}$-vector (here and subsequently the symbol ˆ above the vector denotes the corresponding unit vector) over the cyclic unit vectors (see Appendix A). Such a choice of coordinates permits the most simple description of both *linear* and *circular polarized* radiation. Thus, e.g., an $\hat{\mathbf{E}}$-vector of light polarized along the z-axis possesses only one component – E^0. Righthand polarized light for which the $\hat{\mathbf{E}}$-vector rotates clockwise if viewed against the light beam moving along the z-axis, may be described by the E^{-1} component, the lefthand polarized light by the component E^{+1}. This definition of the circularity of light is traditionally used in optics [75], and we are going to use it in our subsequent discussion. Unfortunately it is not universally accepted. Thus, in quantum electrodynamics, in considering the connection between the spirality of a photon and righthanded, lefthanded or linear polarization of a light wave an agreement is used which is opposite to the traditional one in optics; it is sometimes called the natural one. Righthanded circular polarization in traditional terminology is called lefthanded in the natural one, and vice versa [73].

In the description of the absorption and emission of light it is convenient to expand the transition dipole moment $\hat{\mathbf{d}}$ of the rotating molecule over the same cyclic unit vectors. To this end we first express the vector $\hat{\mathbf{d}}$ in the Cartesian system of coordinates, which is rotating together with a molecule. We choose the rotating molecule-fixed frame with the z''-axis along the angular momentum **J**, and we draw the x''-axis along the internuclear axis, as shown in Fig. 1.5(a). In the case of parallel types of optical transition we have

$$\hat{d}^{mol} = \begin{pmatrix} e^{-i\omega_0 t} \\ 0 \\ 0 \end{pmatrix}. \tag{1.4}$$

In a non-rotating **J**-fixed coordinate system the z'-axis still coincides with the angular momentum **J** (see Fig. 1.5(c)), but the vector $\hat{d}$ has the following components:

$$\hat{d}^{J} = \begin{pmatrix} e^{-i\omega_0 t} \cos \Omega t \\ e^{-i\omega_0 t} \sin \Omega t \\ 0 \end{pmatrix}. \tag{1.5}$$

Now the vector $\hat{d}^{J}$, which describes the eigenvibrations of the dipole with frequency ω_0, as well as the rotation of the molecule with frequency Ω, may be expanded over cyclic unit vectors $e_{0,\pm 1}$; see A.1. As a result we obtain [95]:

$$\hat{d}^{J} = \begin{pmatrix} 1/2 \\ -i/2 \\ 0 \end{pmatrix} e^{-i(\omega_0 - \Omega)t} + \begin{pmatrix} 1/2 \\ i/2 \\ 0 \end{pmatrix} e^{-i(\omega_0 + \Omega)t}$$

$$= d^{-1} e_{-1} e^{-i(\omega_0 - \Omega)t} - d^{+1} e_{+1} e^{-i(\omega_0 + \Omega)t}. \tag{1.6}$$

The real part of both summands has the following physical meaning. The real part of the first term, namely

$$\mathrm{Re}[d^{-1} e_{-1} e^{-i(\omega_0 - \Omega)t}] = \frac{1}{2} d_x \cos(\omega_0 - \Omega)t - \frac{1}{2} d_y \sin(\omega_0 - \Omega)t \tag{1.7}$$

describes the dipole component which rotates in the $x'y'$ plane in the non-rotating **J**-fixed coordinate system with frequency $\omega_0 - \Omega$ (see Fig. 1.5(c)). The direction of its rotation is opposite to that of the molecule rotation. In the molecule-fixed x'', y'', z'' coordinate system this corresponds to *clockwise* rotation with frequency ω_0 if viewed from the end of the z''-axis [128, 129, 297] (Fig. 1.6).

The real part of the second term in (1.6), namely

$$\mathrm{Re}[d^{+1} e_{+1} e^{-i(\omega_0 + \Omega)t}] = -\frac{1}{2} d_x \cos(\omega_0 + \Omega)t - \frac{1}{2} d_y \sin(\omega_0 + \Omega)t \tag{1.8}$$

describes the dipole component which rotates in the $x'y'$ plane with frequency $\omega_0 + \Omega$. In this case the direction of rotation coincides with that of the molecule which corresponds to *counterclockwise* rotation with frequency ω_0 if viewed from the end of the z''-axis in the molecule-fixed x'', y'', z''-system; see Fig. 1.6.

The transition with dipole moment d^{+1}, which takes place at the summed electronic and rotational frequency $\omega_0 + \Omega$, is usually called an

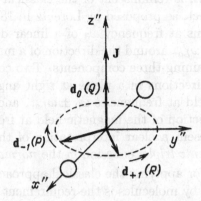

Fig. 1.6. Orientation of the transition dipole moment with respect to the angular momentum $\mathbf{J}$ for P-, Q-, R-type of a molecular transition; $d_Q = d^Q e_Q$, $Q = 0, \pm 1$.

R-type transition in spectroscopy. As a result of light absorption in this transition the difference $\Delta = J' - J''$ between the quantum numbers of the angular momentum in excited (J') and ground (J'') state equals $+1$, and the angular momentum of the molecule increases. The transition with transition dipole moment d^{-1} at frequency $\omega_0 - \Omega$ corresponds to a diminution in the angular momentum of molecular rotation, and we have $\Delta = J' - J'' = -1$. Such a transition is called a *P-type transition*.

The situation is somewhat different in transitions of the perpendicular type. In a rotating (molecular) coordinate system (see Fig. 1.5(b)) the vector $\hat{\mathbf{d}}$ has the following components:

$$\hat{\mathbf{d}}^{mol} = \begin{pmatrix} 0 \\ e^{-i\omega_0 t} \\ i e^{-i\omega_0 t} \end{pmatrix}. \tag{1.9}$$

Passing over to a space-fixed coordinate system with $\mathbf{z}' \parallel \mathbf{J}$ (Fig. 1.5(d)) and expanding the unit vector $\hat{\mathbf{d}}^J$ over cyclic unit vectors, we obtain three components. Two of these, namely d^{-1} and d^{+1}, similarly to (1.7) and (1.8), correspond to P- and R-types of transitions, with frequencies $\omega_0 - \Omega$ and $\omega_0 + \Omega$ respectively. In addition, a new component d^0 appears which corresponds to linear oscillations along the z-axis with frequency ω_0. This component of the vector $\hat{\mathbf{d}}$ is connected with a *Q-type transition*, as a result of which we obtain a change in the electronic state of the molecule, but no change in its rotational quantum number $\Delta = J' - J'' = 0$.

It is interesting to note that the procedure, as described above, of expansion over cyclic unit vectors of the transition dipole moment $\hat{\mathbf{d}}$, which has its own oscillation frequency ω_0, and rotates together with the molecule at frequency Ω, is reminiscent of the classical interpretation of the normal Zeeman effect, as proposed by Lorentz in 1896 [361]. He also expanded the oscillations at frequency ω_0 of a linear dipole which is in precession at frequency ω_{Lar} around the direction of a magnetic field, over cyclic unit vectors, obtaining three components. Two components rotate in mutually opposite directions in a plane at right angles to the direction of the magnetic field at frequencies $\omega_0 \pm \omega_{Lar}$, and one component oscillates along the direction of the magnetic field at frequency ω_0. This enabled Lorentz to present a clear interpretation of the frequency and polarization of the *Lorentz triplet* emission in the *normal Zeeman effect*.

The basic condition for applying the classical approach to the description of light absorption by molecules is the requirement that the angular momenta of molecular rotation must be considerably larger than Planck's constant $\hbar$. In this case we may assume approximately that the total angular momentum of the molecule $\mathbf{J}$ and the rotational momentum $\mathbf{R}$ coincide, and hence neither the total electron spin $\mathbf{S}$ of the molecule or its projection Σ, nor the projection of the orbital momentum on the internuclear axis Λ, nor the nuclear spin of the molecule $\mathbf{I}$ appear in the vector model; see Figs. 1.5, 1.6. Let us analyze what changes in the above discussion accounting for Λ and $\mathbf{I}$ may lead to. Fig. 1.7 shows the mutual orientation of the angular momenta $\mathbf{J}$, $\mathbf{R}$, of the projection Λ and of the dipole transition moment $\hat{\mathbf{d}}$ in the case of a parallel transition type, such as $^1\Pi$–$^1\Pi$, when $\Lambda = 1$. At $\Sigma = 0$ such a scheme satisfies practically all types of momentum coupling according to the (a), (b), (c) Hund cases (see Section 1.2, Fig. 1.3). In this case we obtain, in the molecule-fixed (rotating) frame x'', y'', z'', when the internuclear axis lies in the $x''z''$ plane, for the unit vector $\hat{\mathbf{d}}^{\mathbf{mol}}$,

$$\hat{\mathbf{d}}^{mol} = \begin{pmatrix} e^{-i\omega_0 t}\cos\delta \\ 0 \\ e^{-i\omega_0 t}\sin\delta \end{pmatrix}, \qquad (1.10)$$

where angle δ (Fig. 1.7) may be obtained from the condition $\sin\delta = \Lambda/[J(J+1)]^{1/2}$. In a space-fixed coordinate system with $\mathbf{z}' \parallel \mathbf{J}$ we have for the unit vector $\hat{\mathbf{d}}^J$

$$\hat{\mathbf{d}}^J = \begin{pmatrix} 1/2 \\ -i/2 \\ 0 \end{pmatrix} e^{-i(\omega_0-\Omega)t}\cos\delta + \begin{pmatrix} 1/2 \\ i/2 \\ 0 \end{pmatrix} e^{-i(\omega_0+\Omega)t}\cos\delta$$

$$+ \begin{pmatrix} 0 \\ 0 \\ 1 \end{pmatrix} e^{-i\omega_0 t}\sin\delta$$

$$= d^{-1}\mathbf{e}_{-1}e^{-i(\omega_0-\Omega)t} - d^{+1}\mathbf{e}_{+1}e^{-i(\omega_0+\Omega)t} + d^0\mathbf{e}_0 e^{-i\omega_0 t}. \quad (1.11)$$

Thus, accounting for non-zero value of Λ in the vector model of a molecule (Fig. 1.7) also leads to the appearance of a component d^0 of the dipole moment for a parallel transition. However, with increasing $\mathbf{J}$, owing to the decrease in $\sin \delta$, the value of this component falls rapidly, which agrees with well-known experimental facts [194].

In the case of perpendicular transitions at $\Lambda \neq 0$, accounting for the finite value of the angular momentum $\mathbf{J}$ does not lead to the appearance of new transitions, since all types of dipole transitions – P, Q and R – are already permitted for $\Lambda = 0$.

Up until now we have been concerned with the components of the transition dipole moment $\mathbf{d}$ with respect to the total angular momentum $\mathbf{J}$ of the molecule. The choice of $\mathbf{J}$ as the direction of reference is due to the circumstance that it is just the total angular momentum which is the vector keeping its orientation in space in a free molecule. This last assertion is correct if we neglect the effect of *nuclear spin*. If the atoms forming the molecule possess nuclear spin momentum, the molecule as a whole may also acquire a nuclear spin $\mathbf{I}$. The latter, together with the angular momentum $\mathbf{J}$, forms the total momentum $\mathbf{F}$ of the molecule. The transition dipole moment $\hat{\mathbf{d}}$ arising exclusively from electron motion in the molecule, remains, of course, linked with the momentum $\mathbf{J}$ and performs, together with it, precession around the total momentum $\mathbf{F}$. The frequency of this precession is determined by the energy of hyperfine interaction. Owing to this precession the components of vector $\hat{\mathbf{d}}$ are slightly 'smeared'. Thus, for instance, the component $\mathbf{d}_0 = d^0 \mathbf{e}_0$ moves along the generatrix of the cone produced by the precession of $\mathbf{J}$ around $\mathbf{F}$. For the purpose of accounting fully for the nuclear spin effect on the emission of light by the molecule we must consider, in addition to precession, the ratio between the quenching rate of the radiating dipole (the rate of radiational decay of the excited state of the molecule) and the precession frequency. It may be observed in a number of cases that, the dipole, in practice, does not manage to turn during its lifetime [290, 331], and, as a result, the emission of light takes place with the same orientation as the absorption. In our subsequent presentation the effect of nuclear spin $\mathbf{I}$ will not be taken into account in the overwhelming majority of cases.

Hence, the three components of $\hat{\mathbf{d}}$, i.e. d^{-1}, or the P-component ($\Delta = J' - J'' = -1$) at frequency $\omega_0 - \Omega$, d^0, or the Q-component ($\Delta = 0$) at frequency ω_0 and d^{+1}, or the R-component ($\Delta = +1$) at frequency $\omega_0 + \Omega$, cover all possible dipole transitions in the absorption or emission of light by diatomic molecules. In general form, the dipole transition moment may be represented as

$$\mathbf{d} = V d^\Delta \mathbf{e}_\Delta, \tag{1.12}$$

where V characterizes the *dynamic* part of the transition dipole moment,

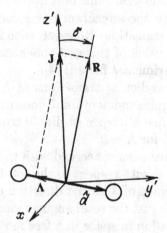

Fig. 1.7. Parallel type of molecular transition in the case where the projection Λ of the electron orbital momentum upon the internuclear axis differs from zero.

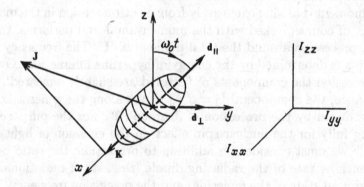

Fig. 1.8. Parallel ($d_\parallel$) and perpendicular ($d_\perp$) transitions in a symmetric top molecule.

and the component $d^\triangle$ represents its *orientation* with respect to the angular momentum $\mathbf{J}$ or $\mathbf{F}$ of the molecule.

The situation for the dipole transition moment remains unchanged in the case of *linear polyatomic molecules*.

The above results may be extended to electronic transitions in the case of *symmetric top molecules* [156, 374, 402], where, in molecule-fixed coordinates, the rotational state of the molecule is characterized by two values, namely the modulus (J) of the total angular momentum, and the projection ($\mathbf{K}$) of the momentum $\mathbf{J}$ on the symmetry axis of the molecule (the x-axis; see Fig. 1.4). In the case of an electronic transition where

$\Delta K = 0$ the transition dipole moment $\hat{\mathbf{d}}$ is parallel to the symmetry axis of the molecule. In the case of a transition where $\Delta K = \pm 1$ it rotates in a plane which is orthogonal to this direction [402] (see Fig. 1.8). If we compare this diagram with Fig. 1.7, we notice the analogy of the position of the two momenta, the only difference being that here the projection K (Fig. 1.8) acts in the place of the projection Λ (Fig. 1.7). Hence, for a parallel type of electronic transition ($\Delta K = 0$) in a symmetric top molecule the P ($\Delta = -1$) and R ($\Delta = +1$) transitions take place at any value of the angular momentum J, whilst the Q ($\Delta = 0$) transition appears only at $J \cong K$, vanishing at $J \gg K$. In a molecule-fixed coordinate system attached to the angular momentum ($\mathbf{z}' \parallel \mathbf{J}$) they are described by the corresponding d^{Δ}-component of the transition dipole moment. In the case of a perpendicular dipole transition ($\Delta K = \pm 1$) all three types of transition (P, Q and R) take place at all values of J, as in perpendicular transitions in diatomic molecules, and they possess the same orientations of $\hat{\mathbf{d}}$ with respect to $\mathbf{J}$.

We have thus elucidated the orientation of the dipole moment $\mathbf{d}$ of a molecular transition with respect to the angular momentum $\mathbf{J}$ of the molecule for various types of molecular transitions. We are now going to discuss how alignment and orientation of angular momenta $\mathbf{J}$ arise in the absorption of light by a transition dipole moment $\mathbf{d}$.

2

Excited state angular
momenta distribution

In the preceding chapter we ascertained the mutual orientation of the
dipole moment **d** and of the angular momentum **J** of the molecule. The
next step is to consider the probability of a molecule with arbitrary spatial orientation of the angular momentum **J** absorbing light of a given
polarization (in laboratory coordinates).

2.1   Absorption and emission probability

This probability is determined by the amount to which vector **E** is contained in vector **d**. This quantity equals the *hermitian product* of the **E**
and **d** [140]. The hermitian product can be written as the scalar product of **E** and **d***. As a result the probability W_{abs} of absorption of light,
which is characterized by its **E**-vector, by a classical dipole is propor-
tional to $|\mathbf{Ed}^*|^2$. This probability can be divided into a *dynamic* part Γ_p
(absorption rate, s^{-1}) and a dimensionless *angular* part $|\hat{\mathbf{E}}\hat{\mathbf{d}}^*|^2$:

$$W_{abs} = \text{Const} \cdot |\mathbf{Ed}^*|^2 = \Gamma_p|\hat{\mathbf{E}}\hat{\mathbf{d}}^*|^2. \qquad (2.1)$$

An analogue of such a difference in quantum approach is the application
of the Wigner–Eckart theorem [73, 136, 379, 402]. The latter permits us
to write the matrix element of a dipole transition between two rotational
states with definite angular momentum values as a product of the reduced matrix element, i.e. of the dynamic part, and the Clebsch–Gordan
coefficient, which is a quantity representing the geometry of interaction
and is based on the law of conservation of angular momentum in the
absorption (or emission) of a photon by a molecule. Similarly, the fac-
tor $|\hat{\mathbf{E}}\hat{\mathbf{d}}^*|^2$ expresses the law of conservation of the angular momentum
in the absorption or emission by a molecule of an electromagnetic wave
with polarization $\hat{\mathbf{E}}$. Conservation of the angular momentum constitutes

22

the basic factor in the optical alignment and orientation of the angular momenta of a molecular ensemble. It also determines the appearance of definite polarization in fluorescent light. Therefore we shall pay particular attention to the factor $|\hat{\mathbf{E}}\hat{\mathbf{d}}^*|^2$ in our subsequent discussion.

The explicit form of the factor $|\hat{\mathbf{E}}\hat{\mathbf{d}}^*|^2$ in laboratory coordinates x, y, z, in which the orientation of the angular momentum $\mathbf{J}$ of the molecule is characterized by angles θ and φ (see Fig. 1.5(e)), can be found by applying the rule of scalar multiplication of vectors in cyclic coordinates (see (A.9)):

$$|\hat{\mathbf{E}}\hat{\mathbf{d}}^*|^2 = \left| \sum_{Q=-1}^{1} (-1)^Q E^Q (\hat{\mathbf{d}}^*)^{-Q} \right|^2 = \left| \sum_Q E^Q d^{Q*} \right|^2, \qquad (2.2)$$

where E^Q, d^Q are the cyclic components of the respective vectors in a space-fixed coordinate system x, y, z. The sum over Q in the last part of Eq. (2.2) is the explicit form of the hermitian product of vectors $\hat{\mathbf{E}}$ and $\hat{\mathbf{d}}$. The relative phase of oscillations of the light wave and the dipole $\hat{\mathbf{d}}$ may be of a random nature. Therefore the magnitude (2.1) has to be averaged over the whole molecular ensemble. As a result, Eq. (2.2) does not contain any factor that is connected with the relative phase of these components.

In order to be able to apply Eq. (2.2) in practice, let us consider how we can determine the cyclic components E^Q of the vector $\hat{\mathbf{E}}$ for completely polarized light.

We will define light as *completely polarized* if there exists a polarization filter through which the light passes without loss, consisting of an ideal linear polarizer oriented in a definite way and an ideal quarter-wave plate. The state of polarization of such light may be fully described by means of only one vector $\hat{\mathbf{E}}$ [75]. Completely polarized light can be polarized linearly as well as elliptically. If a beam of completely polarized light propagates along the z-axis, then the unit polarization vector $\hat{\mathbf{E}}$ can be written as follows:

$$\hat{\mathbf{E}} = \mathbf{e}_x \cos v + \mathbf{e}_y e^{i\delta} \sin v, \qquad (2.3)$$

where δ is the phase shift of oscillations along the x- and y-axes, whilst the coefficients $\cos v$, $\sin v$ characterize the amplitude of these oscillations. At phase shift $\delta = 0$ we obtain light which is *linearly polarized* at angle v with respect to the x-axis. If the phase shift equals $\delta = \pm \pi/2$ we obtain the vector $\hat{\mathbf{E}} \propto (\mathbf{E}_x \pm i\mathbf{E}_y)$ which corresponds to lefthanded (plus sign) or to righthanded (minus sign) *elliptically polarized* light. The principal axes of the ellipse are positioned along the x- and y-axes, whilst the ratio between the lengths of the major and minor axes of the ellipse equals $\tan v$. In the general case the vector $\hat{\mathbf{E}}$ (2.3) can be resolved into two components E^{+1} and E^{-1}; see Eq. (A.1):

$$E^{+1} = (\hat{\mathbf{E}}\mathbf{e}^*_{+1}) = -(1/\sqrt{2})(\cos v - ie^{i\delta}\sin v),$$
$$E^{-1} = (\hat{\mathbf{E}}\mathbf{e}^*_{-1}) = (1/\sqrt{2})(\cos v + ie^{i\delta}\sin v). \qquad (2.4)$$

The component E^{+1} describes the light wave which transfers the angular momentum coinciding with the light beam direction, and E^{-1} describes that opposed to the former.

So far we have been analyzing the beam propagating along the z-axis. In the case of a beam propagating in an arbitrary direction described by the spherical angles θ_c, φ_c, it is necessary to turn the initial coordinate system in such a way as to have the light beam propagating in the new system of coordinates, in the desired direction, and possessing the desired orientation of the polarization ellipse. In the turned system of coordinates the cyclic components E^Q of the vector $\hat{\mathbf{E}}$ are expressed through components $E^{Q'}$ in the initial system by means of Wigner D-matrices (see [136, 379, 402], Table A.1 and (A.10)):

$$E^Q = \sum_{Q'} E^{Q'} D^{(1)}_{Q'Q}{}^*(\alpha, \beta, \gamma). \qquad (2.5)$$

The first turn by angle α is performed around the light beam and changes the direction of the principal axes of the polarization ellipse only, i.e. it changes the phases between the components E^{+1} and E^{-1}. The same result may be achieved by efficient selection of the parameters v and δ in Eq. (2.3). The second and third turns by angles β, γ transfer the direction of propagation of the beam into a state characterized by angles θ_c and φ_c. Simple geometric considerations will show that $\beta = \theta_c$, $\gamma = \pi - \varphi_c$. In this way, by applying parameters v and δ, as well as two turns by angles β and γ, it is possible to obtain any desired polarization of light propagating along the direction θ_c, φ_c. The cyclic components E^Q of such light may be found by using (2.4) and (2.5):

$$\begin{aligned} E^Q &= -(1/\sqrt{2})(\cos v - ie^{i\delta}\sin v)D^{(1)}_{1Q}{}^*(0, \theta_c, \pi - \varphi_c) \\ &\quad + (1/\sqrt{2})(\cos v + ie^{i\delta}\sin v)D^{(1)}_{-1Q}{}^*(0, \theta_c, \pi - \varphi_c) \\ &= -(-1)^Q/\sqrt{2}(\cos v - ie^{i\delta}\sin v)D^{(1)}_{1Q}(0, \theta_c, \varphi_c) \\ &\quad + (-1)^Q/\sqrt{2}(\cos v + ie^{i\delta}\sin v)D^{(1)}_{-1Q}(0, \theta_c, \varphi_c). \qquad (2.6) \end{aligned}$$

In writing Eq. (2.6) the symmetry properties of the D-functions have been used, as presented in Appendix A (see (A.11)). A detailed treatment of the properties of D-functions is considered in the quantum theory of angular momentum [379].

The second part of the problem of finding the angular part of the probability of emission or absorption of light in explicit form by a classical dipole (2.2) consists of determining the cyclic components d^Q of a transition dipole moment for a molecule possessing an arbitrarily oriented momentum $\mathbf{J}$ characterized by the spherical angles θ and φ (see Fig. 1.5(e)). As shown in Chapter 1, in a $\mathbf{J}$-fixed system of coordinates the transition dipole moment has only one non-zero cyclic component d^Δ in a molecular transition of a certain kind. Let us assume that the magnitude of this component (the amplitude of oscillations) equals unity. Then the components of dipole moment $\hat{\mathbf{d}}$ in laboratory-fixed coordinates may be obtained by a turn according to Appendix A (see (A.10)). As a result, the vector $\hat{\mathbf{d}}$ direction is described by the cyclic components:

$$d^Q = D^{(1)}_{\Delta Q}{}^*(\alpha, \beta, \gamma) = (-1)^Q D^{(1)}_{\Delta Q}(0, \theta, \varphi). \tag{2.7}$$

In this case we again have $\beta = \theta$ and $\gamma = \pi - \varphi$. Considering the fact that the phase of oscillations of component d^Q in Eq. (2.2) is of no significance, it is assumed that the first turning angle α around the vector $\mathbf{J}$ equals zero.

Substituting (2.7) into Eq. (2.2), we obtain the angular dependence $G(\theta, \varphi)$ of the *probability of absorption and emission* of light with polarization vector $\hat{\mathbf{E}}$:

$$G(\theta, \varphi) = \left|\hat{\mathbf{E}}\hat{\mathbf{d}}^*\right|^2 = \left|\sum_{Q=-1}^{1} (-1)^Q E^Q D^{(1)}_{\Delta Q}{}^*(0, \theta, \varphi)\right|^2. \tag{2.8}$$

Simultaneous application of Eqs. (2.6) and (2.8) makes it possible to find $G(\theta, \varphi)$ in explicit form for any case of interaction between completely polarized light and a transition dipole moment of a molecule.

As an example let us consider a righthanded circularly polarized light wave propagating along the y-axis and being absorbed in a molecular P-type transition (see Fig. 2.1). In beam-fixed coordinates the polarization vector of this wave has one non-zero component according to (2.4), namely $E^{-1} = 1$. This coordinate system may be superimposed on the laboratory one (see Fig. 2.1) by means of two turns corresponding to the Euler angles $\alpha = 0$, $\beta = \theta_c = \pi/2$, $\gamma = \varphi_c = \pi/2$. Then, in laboratory-fixed coordinates, the polarization vector $\hat{\mathbf{E}}$ has, according to (2.6), the following components: $E^{+1} = i/2$, $E^0 = 1/\sqrt{2}$, $E^{-1} = -i/2$. In P-absorption and at arbitrary orientation of the angular momentum of the molecule $\mathbf{J}(\theta, \varphi)$ the transition dipole moment $\hat{\mathbf{d}}$ has, according to (2.7), the following components: $d^{+1} = -e^{-i\varphi}(1 - \cos\theta)/2$, $d^0 = (\sin\theta)/\sqrt{2}$, $d^{-1} = -e^{i\varphi}(1 + \cos\theta)/2$. As a result we obtain, according to (2.8), the

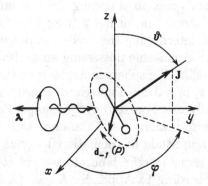

Fig. 2.1. Scheme of absorption of righthanded circular polarized light on a molecular P-type transition. .

explicit form of the angular dependence of absorption probability:

$$G(\theta, \varphi) = (1 + \sin\theta\sin\varphi)^2/4. \qquad (2.9)$$

The latter expression presents a very clear physical picture. When the direction of the angular momentum $\mathbf{J}$ coincides with that of the propagation of light, and when in Fig. 2.1 we have $\theta = \pi/2$, $\varphi = \pi/2$, then the direction of rotation of the $\hat{\mathbf{E}}$-vector of light and that of the dipole moment $\hat{\mathbf{d}}$ coincide. In this case the absorption probability is $G(\pi/2, \pi/2) = 1$. When the vector $\mathbf{J}$ is opposed to the light beam ($\theta = \pi/2$, $\varphi = -\pi/2$, see Fig. 2.1), then $\hat{\mathbf{E}}$ and $\hat{\mathbf{d}}$ rotate in opposite directions, and we have $G(\pi/2, -\pi/2) = 0$. This result is also understandable from the point of view of conservation of angular momentum in the absorption of light. The righthanded circular polarized light transfers the angular momentum $\boldsymbol{\lambda}$ which is directed in opposition to the light beam ($E^{-1} = 1$). In a situation where $\mathbf{J}$ and $\boldsymbol{\lambda}$ are antiparallel, the angular momentum of the molecule in the absorption process must decrease. A P-type transition in which $\Delta = J' - J'' = -1$ may occur (Section 1.4). In a situation where $\mathbf{J}$ and $\boldsymbol{\lambda}$ are parallel, the absorption of light must lead to an increase in angular momentum, which means that the P-type transition cannot take place.

2.2   Probability density of angular momenta distribution

The probability $G(\theta, \varphi)$ of absorption of light of a given polarization by a molecule possessing definite orientation of angular momentum $\mathbf{J}(\theta, \varphi)$ is of decisive importance in the alignment and orientation of molecules. Let us assume that the exciting light is sufficiently weak, i.e. it does not disturb the isotropic, spherically symmetric distribution of the angular

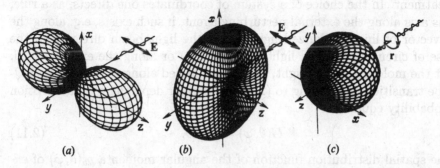

Fig. 2.2. Isometric projections of probability density $\rho_b(\theta, \varphi)$ of angular momentum distribution in an excited molecular state: (a) denotes a Q-transition (linear polarized excitation); (b) denotes P- or R-transition (linear polarized excitation); (c) denotes R-transition (excitation by righthanded circular polarized light).

momenta $\mathbf{J}_a$ of the ground (initial) level. In other words, the process of absorption does not produce optical pumping (in the sense of optical polarization of angular momenta) of the ground state. In this case the angular distribution of momenta $\mathbf{J}_b$ in the excited state is completely determined by the *absorption probability* $G(\theta, \varphi)$. This distribution is described by the *probability density* function $\rho_b(\theta, \varphi)$, which, in this case, is proportional to $G(\theta, \varphi)$ with proportionality factor $\Gamma_p/(4\pi\Gamma)$, where Γ is the rate of decay of level b. Knowledge of the function $\rho_b(\theta, \varphi)$ makes it possible to calculate the probability of detecting the angular momentum of the molecule $\mathbf{J}_b$ within the range of spherical angles between $\theta \div \theta + d\theta$ and $\varphi \div \varphi + d\varphi$. This probability equals $\rho_b(\theta, \varphi) \sin \theta \, d\theta \, d\varphi$. Knowledge of the form of the function $\rho(\theta, \varphi)$ yields all the information on the polarization of the angular momentum. The total probability W of finding the molecule in state b determines the normalization of the probability density

$$\int_0^{2\pi} \int_0^{\pi} \rho_b(\theta, \varphi) \sin \theta \, d\theta \, d\varphi = W. \qquad (2.10)$$

In each case the probability density may be represented in spherical coordinates in the form of a three-dimensional diagram (see Fig. 2.2) in which the angles θ, φ correspond to the arguments of the function $\rho_b(\theta, \varphi)$, whilst the value of the function is plotted along the radius-vector $\mathbf{r}(\theta, \varphi)$.

Let us now consider a few examples of probability density $\rho_b(\theta, \varphi)$. In cyclic coordinates, as used by us, the z-axis is a selected one; see (A.1). It corresponds to the quantization axis in quantum mechanical treatment. In the choice of a system of coordinates one directs, as a rule, this axis along the external perturbing agent, if such exists, e.g. along the $\hat{\mathbf{E}}$-vector of linear polarized light, along the light beam direction in the case of circular polarized light (see Fig. 2.2) or along the external field. Let the molecule absorb light, linearly polarized along the z-axis, in a Q-type transition. According to (2.8) the angular dependence of absorption probability equals

$$G(\theta, \varphi) = \cos^2 \theta. \tag{2.11}$$

The spatial distribution function of the angular momenta $\rho_b(\theta, \varphi)$ of excited molecules thus emerging is shown in Fig. 2.2(a). In this case the angular momenta of the molecules are mainly directed along the z-axis. Therefore alignment along the z-axis is called *longitudinal alignment*. As can be seen in Fig. 2.2(a), the number of molecular angular momenta along the axis of alignment (the z-axis) is larger than in the plane perpendicular to it, and such an alignment is called *positive*. If the $\hat{\mathbf{E}}$-vector of linear polarized light is directed at right angles with respect to the z-axis, i.e. is positioned in the xy plane, then we have the so-called *transversal alignment*.

A different situation arises in the absorption of linear polarized light causing P- or R-type molecular transition. If the light is polarized along the z-axis, we obtain, according to (2.8), the probability of light absorption as

$$G(\theta, \varphi) = (\sin^2 \theta)/2. \tag{2.12}$$

The probability density for angular momentum $\mathbf{J}_b$ distribution in the excited state thus obtained is shown in Fig. 2.2(b). In this case we also have longitudinal alignment. However, since there are fewer molecules with angular momenta directed along the z-axis of alignment than in the plane perpendicular to the latter, we have *negative* alignment here.

Finally, if the righthanded circular polarized light propagating along the z-axis ($E^{-1} = 1$) produces an R-type transition in the molecule, then the probability of absorption is

$$G(\theta, \varphi) = (1 - \cos \theta)^2/4. \tag{2.13}$$

In this case the *orientation* of the angular momenta of the excited molecules appears (see Fig. 2.2(c)). Since the angular momenta are largely directed against the z-axis, this orientation is called *negative*. If the angular momenta of the molecules are directed mainly along (or opposite, as is the case in the given example) the 'axis of quantization' (z-axis), then the orientation is called *longitudinal*; if orthogonally it is called *transversal*.

In the investigation of the stereodynamics of chemical reactions, as dependent on the mutual orientation of the reagents, an important part, in addition to the polarization of the angular momenta of the molecules, belongs to the polarization of the *internuclear axes*; see e.g., [66, 86, 158, 271]. It is, of course, understandable that both orientation and alignment of the angular momenta of rotating molecules lead only to alignment of molecular axes and not to orientation. Thus, for instance, in Q-absorption of light polarized linearly along the z-axis, when the probability density of angular momenta distribution **J** is proportional to $\cos^2\theta$ (see (2.11)), then the probability density of the distribution of molecular axes is proportional to $(1/2)\sin^2\theta_r$ (see (6.3)), where θ_r characterizes the direction of the internuclear axes **r**. In other words, the internuclear axes are *aligned*, the alignment however being of opposite sign to that of the angular momenta and, accordingly, *negative*.

In the case of righthanded circular polarized R-type excitation, when *orientation* of angular momenta takes place and the probability density of the angular momenta distribution is proportional to $(1 - \cos\theta)^2/2$ (see (2.13)), only *alignment* of internuclear axes occurs, described by the probability density, which is proportional to $(1/2)[1 + (\sin^2\theta_r)/2]$.

2.3 Expansion of probability density over multipoles. Coherence

It is convenient, both from the point of view of calculations, as well as from the point of view of interpretation of results, to expand the continuous probability density function $\rho(\theta, \varphi)$ which depends on the spherical angles θ, φ over certain basic functions forming a complete orthogonal set. After such an expansion the spatial distribution $\rho(\theta, \varphi)$ can be characterized by the *discrete* coefficients of this expansion. In the general case, nothing can be gained by it, of course, since instead of a continuous function we obtain an infinite quantity of expansion coefficients. However, considering a certain symmetry of absorption probability, and with an appropriate choice of basic functions, one may arrive at a situation in which only a small number of expansion coefficients will differ from zero. In this case the description of the angular distribution will be simplified considerably. If, in addition, a definite physical meaning can be ascribed to these coefficients, then the advantage becomes obvious.

In the case of absorption of light by molecules such an advantageous set of basic functions is represented by *spherical functions* $Y_{KQ}(\theta, \varphi)$ [379]. In Appendix B some such functions are given in explicit form. Making use of the latter we have

$$\rho(\theta, \varphi) = (4\pi)^{-1/2} \sum_{K=0}^{\infty} \sum_{Q=-K}^{K} (2K+1)^{1/2} \rho_Q^K (-1)^Q Y_{KQ}^*(\theta, \varphi). \quad (2.14)$$

The expansion coefficients ρ_Q^K are called *polarization moments* or *multipole moments*. The expansion (2.14) may also be carried out by slightly alternative methods which are presented in Appendix D and differ from the above one by the normalization and by the phase of the complex coefficients ρ_Q^K. The normalization used in (2.14) agrees with [19]. Considering the formula (B.2) from Appendix B of the complex conjugation for the spherical function $Y_{KQ}(\theta, \varphi)$, in order to obtain a real function $\rho(\theta, \varphi)$, the coefficients ρ_Q^K must satisfy the following symmetry property:

$$\rho_Q^K = (-1)^Q (\rho_{-Q}^K)^*. \tag{2.15}$$

The orthogonality properties of the functions $Y_{KQ}(\theta, \varphi)$ (see Eq. (B.3)), enable us to find the relationship which is reciprocal to (2.14):

$$\rho_Q^K = (-1)^Q \sqrt{\frac{4\pi}{2K+1}} \int_0^{2\pi} \int_0^\pi \rho(\theta, \varphi) Y_{KQ}(\theta, \varphi) \sin\theta \, d\theta \, d\varphi, \tag{2.16}$$

The latter expression reveals the physical meaning of the multipole moments. Thus, substituting the function $Y_{00}(\theta, \varphi)$ in explicit form (B.1) into (2.16), we obtain

$$\rho_0^0 = \int_0^{2\pi} \int_0^{2\pi} \rho(\theta, \varphi) \sin\theta \, d\theta \, d\varphi, \tag{2.17}$$

i.e. the multipole moment ρ_0^0 yields, according to (2.10), the *probability* of finding a particle with arbitrary orientation of the angular momentum **J** in the state under consideration. This is, of course, a scalar magnitude which remains unchanged at any turn of the coordinates. If only the multipole moment ρ_0^0 differs from zero in Eq. (2.14), then the distribution of angular momenta is *isotropic*.

Three components ($Q = -1, 0, +1$) of the multipole moment of rank $K = 1$ form the cyclic components of the vector. They are proportional to the mean value of the corresponding spherical functions (B.1) for angular momenta distribution in the state of the molecule as described by the probability density $\rho(\theta, \varphi)$. These components of the multipole moments enable us to find the cyclic components of the angular momentum of the molecule:

$$\langle J \rangle_Q = (-1)^Q |\mathbf{J}| \rho_Q^1 / \rho_0^0. \tag{2.18}$$

The connection between the covariant cyclic and cartesian coordinates of the vector **J** yields Eq. (A.6), whilst (A.5) makes it possible to form the vector itself out of the components $\langle J \rangle_Q$. As follows from (2.18), the components of the multipole moment ρ_Q^1 characterize the preferred *orientation* of the angular momentum **J** in the molecular ensemble. Fig. 2.3(a, b) shows the probability density $\rho(\theta, \varphi)$ in the case where only coefficients

ρ_0^0 and ρ_Q^1 are non-zero in Eq. (2.14). As can be seen, ρ_0^1 denotes the *longitudinal orientation*, whilst $\rho_{\pm 1}^1$ denotes the *transversal orientation*.

In the molecule the angular momentum $\mathbf{J}$ is always connected with a magnetic moment $\boldsymbol{\mu}$ parallel or antiparallel to the former, as will be discussed in greater detail in Sections 4.1 and 4.5. The cyclic components of the magnetic moment $\langle \mu \rangle_Q$, averaged over the ensemble (see (D.28)), can be found as

$$\langle \mu \rangle_Q = g\mu_B \langle \mathbf{J} \rangle_Q, \tag{2.19}$$

where g is the Landé factor, and μ_B the Bohr magneton.

Five components ($Q = -2, -1, 0, 1, 2$) of the multipole moment ρ_Q^2 of rank $K = 2$ form the tensor which characterizes *alignment*. The form of the probability density in the case where only ρ_0^0 and ρ_Q^2 are non-zero is presented in Fig. 2.3(c,d,e). The component ρ_0^2 characterizes *longitudinal alignment*, whilst components $\rho_{\pm 1}^2$, $\rho_{\pm 2}^2$ characterize *transversal alignment*. The method of transforming ρ_Q^K on turning the coordinate system is analyzed in Appendix D.

Let us analyze how to find the excited state multipole moments $_b\rho_Q^K$. As explained in the previous paragraph, at excitation by weak light the probability density $\rho_b(\theta, \varphi)$ of excited state angular momentum distribution is proportional to the absorption probability $G(\theta, \varphi)$. This means that the multipole moments $_b\rho_Q^K$ of an excited level b can be found as

$$_b\rho_Q^K = \frac{\Gamma_p}{\Gamma 4\pi\sqrt{2K+1}} \int_0^{2\pi} \int_0^{\pi} G(\theta, \varphi)(-1)^Q Y_{KQ}(\theta, \varphi) \sin\theta \, d\theta \, d\varphi. \tag{2.20}$$

Substituting function $G(\theta, \varphi)$ (see (2.8)) in its explicit form into (2.20) and using the relations (A.13) and (B.4), as well as the symmetry properties of the Clebsch–Gordan coefficients (C.3) and (C.4), we obtain

$$_b\rho_Q^K = (-1)^{\Delta+K} \frac{\Gamma_p}{\Gamma\sqrt{2K+1}} C_{1\Delta 1-\Delta}^{K0} \Phi_Q^K(\hat{\mathbf{E}}), \tag{2.21}$$

where $C_{1\Delta 1-\Delta}^{K0}$ is the Clebsch–Gordan coefficient. The quantity $\Phi_Q^K(\hat{\mathbf{E}})$ characterizes the polarization of the exciting light $\hat{\mathbf{E}}$ and is called the *Dyakonov tensor* [133]. Its components may be calculated as

$$\Phi_Q^K(\hat{\mathbf{E}}) = \frac{1}{\sqrt{2K+1}} \sum_{Q'Q''} (-1)^{Q''} E^{Q'} \left(E^{Q''}\right)^* C_{1-Q'1Q''}^{KQ}. \tag{2.22}$$

The components satisfy the relation $\Phi_Q^K = (-1)^Q (\Phi_{-Q}^K)^*$; their value can be found in Table 2.1 (p. 39). It follows from the properties of the Clebsch–Gordan coefficients ([379], (C.1)) that in Eq. (2.22) $K \leq 2$. This means that light of any polarization can be described by tensors Φ_Q^K of rank $K \leq 2$, i.e. by a total of *nine* components. As a result, excitation by weak light generally produces an excited state angular momentum distribution characterized by nine components of polarization moments $_b\rho_Q^K$ ($K \leq 2$). In each separate case the number of non-zero components may be even less.

Another advantage of the polarization moments is the possibility of describing relaxation processes in the most rational way. Thus, if the relaxation process is isotropic, then the various moments and their components relax independently; for details see Section 5.8. All multipole components of certain rank K relax at one and the same *rate constant* Γ_K.

If excitation takes place by intensive light, then, naturally, there may emerge in the molecular ensemble polarization moments of *rank higher than* 2. This may take place both in excited and in ground states. In this case the set ρ_Q^K of *odd* rank characterizes *orientation*, whilst that of *even* rank characterizes *alignment* of the angular momenta of the molecules.

The *multipole* (or *polarization*) moments introduced according to (2.14) present a classical analogue of *quantum mechanical polarization moments* [6, 73, 96, 133, 304]. They are obtained by expanding the quantum density matrix [73, 139] over irreducible tensor operators [136, 140, 379] and will be discussed in Chapters 3 and 5.

In the case of quantum mechanical description of particles an important concept is that of the *coherence* of an ensemble [73, 227]. If coherence, which is connected to synchronization of the phases of the wave functions of separate particles, is introduced into a system of non-interacting particles, then the ensemble reveals new properties. One way of achieving coherence in an ensemble of molecules consists of irradiating this ensemble by light polarized in such a way as to create at least two non-zero cyclic components $E^{Q'}$ and $E^{Q''}$ of the vector $\hat{\mathbf{E}}$. According to quantum mechanical treatment we then achieve synchronization of the phases of rotational wave functions $|JM\rangle$ showing a difference in magnetic quantum numbers $\Delta M = M' - M'' = Q' - Q'' = Q$. Such a state is described by *quantum polarization moments* f_Q^K with index value $Q = \Delta M$ [39]. The presence or absence of coherence in a molecular ensemble, and in any ensemble of quantum particles in general, depends to a certain extent on the choice of the z-axis (quantization axis) in the system of coordinates, with respect to which the components $E^{Q'}$, $E^{Q''}$ and the wave functions $|JM\rangle$ are referred. Thus, for instance, at excitation of a molecular ensemble by light which is linearly polarized along the y-axis, the polarization vector $\hat{\mathbf{E}}$ is characterized by two non-zero components $E^{+1} = i/\sqrt{2}$, $E^{-1} = i/\sqrt{2}$ (A.4). In this case coherence is created between two magnetic sublevels with $\Delta M = Q = 2$.

Let us now discuss the meaning of the coherence concept in the classical description of a molecular ensemble [39]. To this purpose we will use the

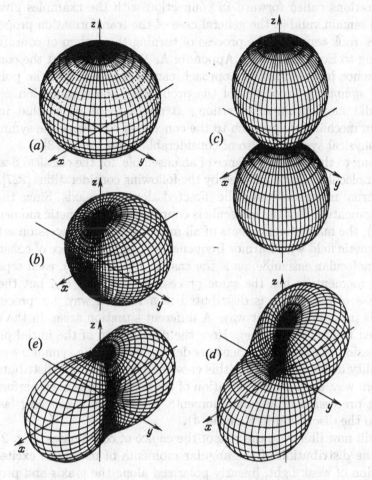

Fig. 2.3. Probability density at different numerical values of multipole moments
[25]; (a) $\rho_0^0 = 1$, $\rho_0^1 = 0,3$; (b) $\rho_0^0 = 1$, $\rho_{-1}^1 = -\rho_1^1 = 0,15$; (c) $\rho_0^0 = 1$, $\rho_0^2 = 0,3$;
(d) $\rho_0^0 = 1$, $\rho_{-1}^2 = -\rho_1^2 = 0,15$; (e) $\rho_0^0 = 1$, $\rho_{-2}^2 = \rho_2^2 = 0,15$.

expansion of the probability density over spherical functions and analyze
the meaning of the index Q in terms of classical multipole moments ρ_Q^K.
In Fig. 2.3 we have ρ_Q^K where $Q = 0$ in only two cases (a) and (c) This
corresponds to *complete rotational symmetry* of the probability density
of angular momenta distribution $\rho(\theta, \varphi)$ with respect to the z-axis. In
other words, there is no coherence in the ensemble in this case. In cases
(b) and (d), when we have non-zero ρ_Q^K values where $Q = 1$, the z-axis
is the axis of *first-order rotational symmetry*. Finally, in case (e) which
corresponds to a situation when non-zero ρ_Q^K where $Q = 2$ are present,
the z-axis presents an axis of *second-order rotational symmetry*. The

considerations called forward in connection with the examples given in Fig. 2.3 remain valid in the general case of the transformation properties of the K rank tensor in the process of turning the system of coordinates according to Eq. (A.12) from Appendix A. We thus see that the concept of coherence in the *classical* approach can be treated from the point of view of *symmetry properties* of the probability density function $\rho(\theta, \varphi)$ of angular momentum distribution. It ought to be noted that in the quantum mechanical approach to the concept of coherence the symmetry of the physical system is also of considerable importance [73].

Sticking to the term 'coherence of an ensemble' for the classical description of molecules can be justified by the following considerations [227]. Let the external magnetic field **B** be directed along the z-axis. Since the angular momentum **J** of the molecule is coupled to the magnetic momentum **μ** (2.19), the magnetic moments of all molecules are in precession around the magnetic field with Larmor frequency ω_J. In the absence of coherence in the molecular ensemble, as is the case in Fig. 2.3(a, c), each separate angular momentum has the same precession frequency ω_J, but the initial phase of precession is distributed in a random way, i.e. precession proceeds in a non-coherent way. A different situation arises in the cases presented in Fig. 2.3(b, d, e). Here the distribution of the initial phases of precession of angular momenta is determined by the symmetry of the probability density $\rho(\theta, \varphi)$. In this case the phases are not distributed in a random way, since synchronization of these phases has been performed, and the precession of angular momenta proceeds in a *coherent* fashion (see also the discussion in Chapter 4).

We will now illustrate the role of the choice of coordinates. Fig. 2.4(a) shows the distribution of the angular momenta of molecules excited by absorption of weak light, linearly polarized along the y-axis and producing Q-type transition. Coherence is created in the ensemble in this case. However, if we direct the z-axis along the polarization vector $\hat{\mathbf{E}}$, i.e. if we turn the system of coordinates in such a way that the new position of the z-axis coincides with the former position of the y-axis, then the distribution $\rho_b(\theta, \varphi)$ will show complete rotational symmetry with respect to the new z-axis. This is illustrated by the picture of $\rho_b(\theta, \varphi)$ as viewed from the end of vector $\hat{\mathbf{E}}$; see Fig. 2.4(b). As can be seen, in the turned system of coordinates the ensemble of angular momenta will not be coherent. In carrying out calculations it is always convenient to choose the coordinate system such that the molecular ensemble should be a non-coherent one, i.e. described only through the zero components of multipole moments ρ_0^K which are real magnitudes, as follows from (2.15). It is, however, not always possible to find such a system of coordinates. Fig. 2.4(c) shows the shape of function $\rho_b(\theta, \varphi)$ in the case where an external magnetic field **B** is applied along the z-axis (see Chapter 4 for a more detailed discus-

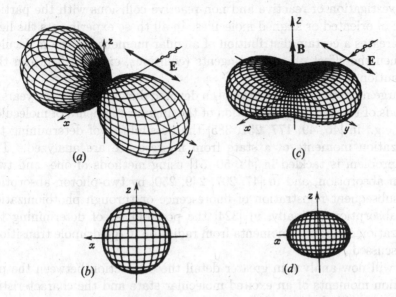

Fig. 2.4. Distribution $\rho_b(\theta, \varphi)$ at Q-excitation by weak light, linearly polarized along the y-axis: (a) isometric projection of the distribution in the absence of an external magnetic field; (b) the same distribution, but viewed from the end of the y-axis; (c) isometric projection of the distribution $\rho_b(\theta, \varphi)$ in the presence of an external magnetic field; (d) the same distribution as (c), but viewed from the end of the symmetry axis of $\rho_b(\theta, \varphi)$ in the xy plane.

sion on the magnetic field effect on angular momenta distribution). In this situation there does not exist any direction with respect to which the distribution $\rho_b(\theta, \varphi)$ could possess complete rotational symmetry. Thus, even if we view the distribution along the axis of its 'elongation' (see Fig. 2.4(d)), it will still not possess complete rotational symmetry. This means that with any choice of a coordinate system the ensemble of angular momenta of the molecules is described in a coherent way.

2.4 Intensity and polarization of fluorescence

The degree of alignment and orientation of angular momenta in an excited state of molecules directly determines the intensity, angular distribution and polarization of fluorescence appearing in the transition of the molecule from this state. In practice the reverse problem frequently arises, namely the determination of the multipole moments characterizing angular momenta distribution via examination of molecular fluorescence. One is faced with this problem in studying the depolarization of fluorescence in an external magnetic field (see Chapter 4), for instance, as well as in the investigation of reactive and non-reactive collisions with the partici-pation of oriented or aligned molecules. In all these experiments the light wave creates a certain distribution of angular momenta of the molecules, and then the effect of external agents (collisions, external field) on this distribution is studied.

A large number of papers have been devoted during the last ten years to methods of testing the polarization of the angular momenta of molecules. Thus, e.g., in [46, 49, 177, 251, 385] the possibilities of determining the polarization moments of a state from fluorescence are analyzed. The same problem is tackled in [48, 50, 51] using methods of one- and two-photon absorption, and in [47, 207, 249, 250] by two-photon absorption with subsequent registration of fluorescence or through photoionization after absorption. Finally, in [324] the possibilities of determining the polarization of angular momenta from radiation in quadrupole transitions are discussed.

We will now analyze in greater detail the connection between the po-larization moments of an excited molecular state and the characteristics of the emitted light.

In Section 2.1 we presented expressions for the probability of a molecule with given orientation θ, φ of the angular momentum $\mathbf{J}(\theta, \varphi)$ absorbing or emitting an electromagnetic wave with polarization vector $\hat{\mathbf{E}}$ at a dipole transition of a certain type (P, Q or R). Let us denote the probability of emitting light by $G'(\theta, \varphi)$ in order to make it differ from an analogous probability $G(\theta, \varphi)$ in absorption, and the polarization vector of emitted light by $\hat{\mathbf{E}}'$. The full intensity of light emission $I(\hat{\mathbf{E}}')$ can be found by averaging $G'(\theta, \varphi)$ over the molecular ensemble having probability density $\rho_b(\theta, \varphi)$ of angular momenta distribution:

$$I(\hat{\mathbf{E}}') = A \int_0^{2\pi} \int_0^{\pi} \rho_b(\theta, \varphi) G'(\theta, \varphi) \sin \theta \, d\theta \, d\varphi, \qquad (2.23)$$

where A is the coefficient of proportionality which includes the dynamic part of the probability of the molecular transition. The intensity $I(\hat{\mathbf{E}}')$ can be conveniently expressed through the excited state polarization moments $_b\rho_Q^K$. Substituting Eq. (2.14) into (2.23) and using the emission probability $G'(\theta, \varphi)$ in the explicit form of (2.8), we obtain

$$I(\hat{\mathbf{E}}') = A(-1)^{\Delta'} \sum_{K=0}^{2} \sqrt{2K+1} C_{1-\Delta'1\Delta'}^{K0}$$

$$\times \sum_{Q=-K}^{K} (-1)^Q {}_b\rho_Q^K \Phi_{-Q}^K(\hat{\mathbf{E}}'), \qquad (2.24)$$

where the difference in the quantum numbers of the angular momenta of the initial and final states of the radiational transition is denoted by

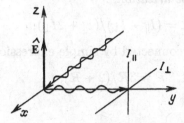

Fig. 2.5. The geometry adopted in calculating the degree of polarization $\mathcal{P}$ and the anisotropy of polarization of fluorescence $\mathcal{R}$ at excitation by linear polarized light beam directed along the x-axis with $\hat{\mathbf{E}} \parallel \mathbf{z}$.

$\Delta' = J' - J_1''$. Here the Dyakonov tensor Φ_Q^K characterizes the polarization of the registered light. In deriving Eq. (2.24) the properties of the D-functions were used, as previously in the case of Eq. (2.21) and according to (A.13) and (B.4). It may thus be seen from (2.24) that only the multipole moments $_b\rho_Q^K$ of rank $K \leq 2$ have any direct effect on the intensity and polarization of molecular fluorescence. This latter assertion also holds in the case where multipole moments of rank higher than $K = 2$ are created in the excited state (b) in the case of absorption of sufficiently intensive light.

There are two methods of experimental determination of the magnitude of the multipoles $_b\rho_Q^{K \leq 2}$ and of restoration of the shape of the spatial distribution $\rho_b(\theta, \varphi)$ of the angular momenta of the ensemble. One may measure either the intensity of radiation in a certain direction, but with different polarization characteristics (by changing the polarization devices in front of the light detector), or one may measure the intensity of radiation in different directions. The former method is technically more convenient and is therefore applied more frequently.

Let us consider the traditional geometry of excitation and observation, where the exciting light beam, its polarization vector $\hat{\mathbf{E}}$ and the direction of observation are mutually orthogonal. Let us place the z-axis along the vector $\hat{\mathbf{E}}$ (see Fig. 2.5). It is then convenient for registration to select two components of fluorescence intensity $I(\hat{\mathbf{E}}')$: one having the $\hat{\mathbf{E}}'$-vector parallel to $\hat{\mathbf{E}}$, which we denote $I_\parallel$, and one with the $\hat{\mathbf{E}}'$-vector at right angles to $\hat{\mathbf{E}}$, which we denote $I_\perp$ (see Fig. 2.5). The linear polarization of fluorescence in this geometry may be characterized by the *degree of polarization*:

$$\mathcal{P} = (I_\parallel - I_\perp)/(I_\parallel + I_\perp), \tag{2.25}$$

or by the *anisotropy of polarization:*

$$\mathcal{R} = (I_\parallel - I_\perp)/(I_\parallel + 2I_\perp), \qquad (2.26)$$

Both quantities are interconnected by simple expressions, namely,

$$\mathcal{P} = 3\mathcal{R}/(2 + \mathcal{R}), \qquad (2.27)$$

and, conversely,

$$\mathcal{R} = 2\mathcal{P}/(3 - \mathcal{P}). \qquad (2.28)$$

In order to ascertain, by means of (2.24), how $\mathcal{P}$ and $\mathcal{R}$ are connected with the polarization moments ${}_b\rho_Q^K$, it is necessary to find the Dyakonov tensor Φ_Q^K in explicit form for linear polarized light. In the more general case where the direction of the vector of light polarization is characterized by arbitrary angles θ, φ, this can be done by substituting the components of the polarization vector E^Q in the explicit form of (2.5) into (2.22). The expressions thus obtained for the components of the tensor Φ_Q^K, as well as their numerical values for various types of polarization, are presented in Table 2.1.

It is now possible to express the fluorescence intensities in Eqs. (2.25) and (2.26) through the multipole moments ${}_b\rho_Q^K$ of the excited level:

$$I_\parallel = A(-1)^{\Delta'}[-(1/\sqrt{3})C_{1-\Delta'1\Delta'b}^{00}\rho_0^0 + (2/\sqrt{6})C_{1-\Delta'1\Delta'b}^{20}\rho_0^2], \qquad (2.29)$$

$$I_\perp = A(-1)^{\Delta'}[-(1/\sqrt{3})C_{1\Delta'1\Delta'b}^{00}\rho_0^0 - (1/\sqrt{6})C_{1-\Delta'1\Delta'b}^{20}\rho_0^2]. \qquad (2.30)$$

The values of the Clebsch–Gordan coefficients for $\Delta' = 0, \pm 1$ are presented in Appendix C, Table C.1. It may be seen that in the passage from $I_\parallel$ (2.29) to $I_\perp$ (2.30) the contribution of the multipole moment ${}_b\rho_0^0$ to the light intensity does not change, whilst the contribution of ${}_b\rho_0^2$ changes not only its value, but also its sign. One may therefore assume that there exists some 'magic angle' θ_0 between the z-axis and the polarization vector $\hat{\mathbf{E}}'$ of the observed fluorescence, at which the contribution of ${}_b\rho_0^2$ to the signal passes through zero. The value of this angle can easily be determined. The ${}_b\rho_0^2$ contribution to the fluorescence intensity is proportional to Φ_0^2 (2.24) and, consequently, the angle θ_0 can be found from the equation

$$\Phi_0^2 = (3\cos^2\theta - 1)/\sqrt{30} = 0,$$

and equals

$$\theta_0 = \arccos(1/\sqrt{3}) \approx 54.7^\circ. \qquad (2.31)$$

Such an angle θ_0 is used in cases where it is necessary to measure a signal which is proportional to the state *population* of level ρ_0^0 only, and does

Table 2.1. General expressions and numerical values of the tensor Φ_Q^K characterizing the state of polarization of light at a certain beam direction. For linear polarized light the spherical angles θ, φ determine the direction of the $\hat{E}$-vector; in other cases, the direction of the light beam

Φ_Q^K	Linear polarization			
	$\theta=0$	$\theta=\pi/2,\ \varphi=0$	$\theta=\pi/2,\ \varphi=\pi/2$	Arbitrary θ,φ
Φ_0^0	$-1/\sqrt{3}$	$-1/\sqrt{3}$	$-1/\sqrt{3}$	$-1/\sqrt{3}$
Φ_0^1	0	0	0	0
Φ_1^1	0	0	0	0
Φ_{-1}^1	0	0	0	0
Φ_0^2	$2/\sqrt{30}$	$-1/\sqrt{30}$	$-1/\sqrt{30}$	$(3\cos^2\theta-1)/\sqrt{30}$
Φ_1^2	0	0	0	$e^{i\varphi}\sin\theta\cos\theta/\sqrt{5}$
Φ_{-1}^2	0	0	0	$-e^{-i\varphi}\sin\theta\cos\theta/\sqrt{5}$
Φ_2^2	0	$1/(2\sqrt{5})$	$-1/(2\sqrt{5})$	$e^{2i\varphi}\sin^2\theta/(2\sqrt{5})$
Φ_{-2}^2	0	$1/(2\sqrt{5})$	$-1/(2\sqrt{5})$	$e^{-2i\varphi}\sin^2\theta/(2\sqrt{5})$

Φ_Q^K	Lefthanded circularly polarized light			
	$\theta=0$	$\theta=\pi/2,\ \varphi=0$	$\theta=\pi/2,\ \varphi=\pi/2$	Arbitrary θ,φ
Φ_0^0	$-1/\sqrt{3}$	$-1/\sqrt{3}$	$-1/\sqrt{3}$	$-1/\sqrt{3}$
Φ_0^1	$1/\sqrt{6}$	0	0	$\cos\theta/\sqrt{6}$
Φ_1^1	0	$1/(2\sqrt{3})$	$i/(2\sqrt{3})$	$e^{i\varphi}\sin\theta/(2\sqrt{3})$
Φ_{-1}^1	0	$-1/(2\sqrt{3})$	$i/(2\sqrt{3})$	$-e^{-i\varphi}\sin\theta/(2\sqrt{3})$
Φ_0^2	$-1/\sqrt{30}$	$1/(2\sqrt{30})$	$1/(2\sqrt{30})$	$-(3\cos^2\theta-1)/(2\sqrt{30})$
Φ_1^2	0	0	0	$-e^{i\varphi}\sin\theta\cos\theta/(2\sqrt{5})$
Φ_{-1}^2	0	0	0	$e^{-i\varphi}\sin\theta\cos\theta/(2\sqrt{5})$
Φ_2^2	0	$-1/(4\sqrt{5})$	$1/(4\sqrt{5})$	$-e^{2i\varphi}\sin^2\theta/(4\sqrt{5})$
Φ_{-2}^2	0	$-1/(4\sqrt{5})$	$1/(4\sqrt{5})$	$-e^{-2i\varphi}\sin^2\theta/(4\sqrt{5})$

Φ_Q^K	Righthanded circularly polarized light			
	$\theta=0$	$\theta=\pi/2,\ \varphi=0$	$\theta=\pi/2,\ \varphi=\pi/2$	Arbitrary θ,φ
Φ_0^0	$-1/\sqrt{3}$	$-1/\sqrt{3}$	$-1/\sqrt{3}$	$-1/\sqrt{3}$
Φ_0^1	$-1/\sqrt{6}$	0	0	$-\cos\theta/\sqrt{6}$
Φ_1^1	0	$-1/(2\sqrt{3})$	$-i/(2\sqrt{3})$	$-e^{i\varphi}\sin\theta/(2\sqrt{3})$
Φ_{-1}^1	0	$1/(2\sqrt{3})$	$-i/(2\sqrt{3})$	$e^{-i\varphi}\sin\theta/(2\sqrt{3})$
Φ_0^2	$-1/\sqrt{30}$	$1/(2\sqrt{30})$	$1/(2\sqrt{30})$	$-(3\cos^2\theta-1)/(2\sqrt{30})$
Φ_1^2	0	0	0	$-e^{i\varphi}\sin\theta\cos\theta/(2\sqrt{5})$
Φ_{-1}^2	0	0	0	$e^{-i\varphi}\sin\theta\cos\theta/(2\sqrt{5})$
Φ_2^2	0	$-1/(4\sqrt{5})$	$1/(4\sqrt{5})$	$-e^{2i\varphi}\sin^2\theta/(4\sqrt{5})$
Φ_{-2}^2	0	$-1/(4\sqrt{5})$	$1/(4\sqrt{5})$	$-e^{-2i\varphi}\sin^2\theta/(4\sqrt{5})$

Contd on p. 40

not depend on the polarization of the angular momenta of the molecules. This is applied, for instance, in [111] and in many other works.

Let us now return to the polarization $\mathcal{P}$ and anisotropy $\mathcal{R}$ of fluorescence. Substituting (2.29) and (2.30) into (2.25) and (2.26), and using

Table 2.1. *continued*

Φ_Q^K	Elliptically polarized light
Φ_0^0	$-1/\sqrt{3}$
Φ_0^1	$(1/\sqrt{6})\sin 2v\cos\theta\sin\delta$
Φ_1^1	$[1/(2\sqrt{3})]e^{i\varphi}\sin 2v\sin\theta\sin\delta$
Φ_{-1}^1	$-[1/(2\sqrt{3})]e^{-i\varphi}\sin 2v\sin\theta\sin\delta$
Φ_0^2	$-(1/\sqrt{30})(1-3\cos^2 v\sin^2\theta)$
Φ_1^2	$-[1/(2\sqrt{5})e^{i\varphi}(\cos^2 v\sin 2\theta + i\sin\theta\cos\delta\sin 2v)$
Φ_{-1}^2	$[1/(2\sqrt{5})]e^{-i\varphi}(\cos^2 v\sin 2\theta + i\sin\theta\cos\delta\sin 2v)$
Φ_2^2	$[1/(2\sqrt{5})]e^{2i\varphi}(\cos^2 v\cos^2\theta - \sin^2 v + i\sin 2v\cos\theta\cos\delta)$
Φ_{-2}^2	$[1/(2\sqrt{5})]e^{-2i\varphi}(\cos^2 v\cos^2\theta - \sin^2 v - i\sin 2v\cos\theta\cos\delta)$

Φ_Q^K	Non-polarized light			
	$\theta = 0$	$\theta=\pi/2,\ \varphi=0$	$\theta=\pi/2,\ \varphi=\pi/2$	Arbitrary θ,φ
Φ_0^0	$-1/\sqrt{3}$	$-1/\sqrt{3}$	$-1/\sqrt{3}$	$-1/\sqrt{3}$
Φ_0^1	0	0	0	0
Φ_1^1	0	0	0	0
Φ_{-1}^1	0	0	0	0
Φ_0^2	$-1/\sqrt{30}$	$1/(2\sqrt{30})$	$1/(2\sqrt{30})$	$-(3\cos^2\theta - 1)\ (2\sqrt{3})$
Φ_1^2	0	0	0	$-e^{i\varphi}\sin 2\theta/(4\sqrt{5})$
Φ_{-1}^2	0	0	0	$e^{-i\varphi}\sin 2\theta/(4\sqrt{5})$
Φ_2^2	0	$-1/(4\sqrt{5})$	$1/(4\sqrt{5})$	$-e^{2i\varphi}\sin^2\theta/(4\sqrt{5})$
Φ_{-2}^2	0	$-1/(4\sqrt{5})$	$1/(4\sqrt{5})$	$-e^{-2i\varphi}\sin^2\theta/(4\sqrt{5})$

the relation (C.5), we obtain:

$$\mathcal{P} = \frac{9C_{1-\Delta'1\Delta'b}^{20}\rho_0^2}{2\sqrt{6}(-1)^{\Delta'}{}_b\rho_0^0 + 3C_{1-\Delta'1\Delta'b}^{20}\rho_0^2}, \tag{2.32}$$

$$\mathcal{R} = (\sqrt{6}/2)(-1)^{\Delta'}C_{1-\Delta'1\Delta'b}^{20}\rho_0^2/{}_b\rho_0^0. \tag{2.33}$$

From the form of Eq. (2.33) it becomes understandable why the anisotropy of polarization $\mathcal{R}$ is sometimes called the *degree of alignment*. From the point of view of the determination of the magnitude of the polarization moments ${}_b\rho_0^2$ the measurement of $\mathcal{R}$ is preferable, as compared with that of $\mathcal{P}$, all the more so if one bears in mind that the population ${}_b\rho_0^0$ appears only as a normalizing factor for all other ${}_b\rho_Q^K$ and does not influence the shape of the probability density $\rho(\theta,\varphi)$, but determines its size only. The explicit form of the multipole moment dependence of $\mathcal{P}$ and $\mathcal{R}$ for various types of radiational transition ($\Delta' = 0, \pm 1$) can be obtained using the numerical values of the Clebsch–Gordan coefficient from Table C.1, Appendix C.

If the molecule, from the moment of absorption up to the moment of emission, is not subjected to external influence, e.g. to disorienting collisions or external fields, then the multipole moments ${}_b\rho_Q^K$ entering into

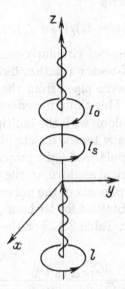

Fig. 2.6. Geometry for the calculation of the degree of circularity of fluorescence C at excitation by lefthanded circularly polarized light.

(2.32) and (2.33) can be calculated according to (2.21). In doing so we obtain the expression for anisotropy of polarization (degree of alignment) in the form

$$\mathcal{R} = (3/5)(-1)^{\Delta-\Delta'} C^{20}_{1\Delta1-\Delta} C^{20}_{1-\Delta'1\Delta'}, \qquad (2.34)$$

which permits us to calculate $\mathcal{R}$ and also, by applying (2.27), $\mathcal{P}$ for any type of molecular transition. The results of such calculations are presented in Table 3.6 of the next chapter (Section 3.2).

If the excited molecules undergo *disorienting collisions* before emitting light, then the ${}_b\rho^K_Q$ of various rank K relax at different rates Γ_K. The anisotropy of polarization $\mathcal{R}$ permits us to find the ratio between the *population relaxation rate* Γ_0 and that of the *alignment relaxation* Γ_2:

$$\mathcal{R} = (3\Gamma_0/5\Gamma_2)(-1)^{\Delta-\Delta'} C^{20}_{1\Delta1-\Delta} C^{20}_{1-\Delta'1\Delta'}. \qquad (2.35)$$

A similar result may be obtained from the degree of polarization; however, considering the connection between $\mathcal{P}$ and $\mathcal{R}$, Eq. (2.27), the dependence of $\mathcal{P}$ on Γ_0/Γ_2 will be more complex and less convenient for practical use.

Let us now consider the excitation of molecules by circularly polarized light. In this case one traditionally chooses the z-axis in the direction of propagation of the exciting beam (see Fig. 2.6). The registered signal indicates the *degree of circularity*:

$$\mathcal{C} = (I_s - I_o)/(I_s + I_o), \tag{2.36}$$

where I_s, I_o are the intensities of circularly polarized light coinciding (*same*) and *opposite* to the $\hat{\mathbf{E}}$-vector rotation direction of exciting radiation. Observation usually takes place from the end of the z-axis, i.e. 'meeting' the exciting beam. This time, in order to find the dependence of the degree of circularity $\mathcal{C}$, along with the multipole moments $_b\rho_Q^K$, it is necessary to know the Dyakonov tensor for circularly polarized light. In the general case of circularly polarized light propagating along the direction characterized by the spherical angles θ, φ, the tensor Φ_Q^K can be found by substituting the cyclic components of the polarization vector (2.6) into Eq. (2.22). The expressions obtained for lefthanded and righthanded circular polarization are given in Table 2.1. For the geometry presented in Fig. 2.6 we have

$$\mathcal{C} = \frac{3C_{1-\Delta'1\Delta'b}^{10}\rho_0^1}{(-1)^{\Delta'}\sqrt{2}\,_b\rho_0^0 - \sqrt{3}C_{1-\Delta'1\Delta'b}^{20}\rho_0^2}. \tag{2.37}$$

In the case of absence of disorienting processes we have

$$\mathcal{C} = \frac{15C_{1-\Delta1\Delta}^{10}C_{1-\Delta'1\Delta'}^{10}}{(-1)^{\Delta+\Delta'}10 + 3C_{1\Delta1-\Delta}^{20}C_{1-\Delta'1\Delta'}^{20}}. \tag{2.38}$$

In the presence of such processes the degree of circularity of fluorescence depends simultaneously on two ratios – Γ_0/Γ_1 and Γ_0/Γ_2:

$$\mathcal{C} = \frac{(\Gamma_0/\Gamma_1)15C_{1-\Delta1\Delta}^{10}C_{1-\Delta'1\Delta'}^{10}}{(-1)^{\Delta+\Delta'}10 + 3(\Gamma_0/\Gamma_2)C_{1\Delta1-\Delta}^{20}C_{1-\Delta'1\Delta'}^{20}}. \tag{2.39}$$

This fact complicates the measurement of the relaxation rate of the orientation Γ_1; for details see Section 4.2. The numerical values of the Clebsch–Gordan coefficients entering into (2.37)–(2.39) are given in Table C.1. $\mathcal{C}$ values for particular cases are presented in Table 3.6.

It is sometimes necessary to analyze the emerging multipole moments (2.21), or, in other cases, the emission intensity (2.24) of light possessing arbitrary polarization and propagating along the direction θ, φ. In this case the cyclic components E^Q, as obtained from (2.6), namely

$$\left.\begin{aligned}
E^{+1} &= (1/\sqrt{2})e^{-i\varphi}(\cos\upsilon\cos\theta - ie^{i\delta}\sin\upsilon), \\
E^0 &= \cos\upsilon\sin\theta, \\
E^{-1} &= (-1/\sqrt{2})e^{i\varphi}(\cos\upsilon\cos\theta + ie^{i\delta}\sin\upsilon),
\end{aligned}\right\} \tag{2.40}$$

must be substituted into the expression for the Dyakonov tensor (2.22). The explicit form of the components Φ_Q^K obtained is presented in Table 2.1.

So far we have been dealing with *completely polarized* light only. If the light is *partially polarized*, it can always be represented as the superposition of two non-coherent and, generally speaking, elliptically polarized waves [259]. The problem of partially polarized light lies outside the scope of the present book, and we will just present (Table 2.1) the expressions for Φ_Q^K taken from [96, 133] and pertaining to unpolarized light propagating along the direction θ, φ.

We will also mention that, turning the system of coordinates, the quantity Φ_Q^K transforms according to the formula

$$\Phi_{Q'}^K = \sum_Q \Phi_Q^K (-1)^{Q+Q'} D_{QQ'}^{(K)}. \tag{2.41}$$

2.5 Alignment and orientation studies (experimental)

One of the practical applications of the optical polarization of molecular angular momenta is the investigation of the stereochemical forces in the process of molecule–atom collisions. The most complete information on the dependence of atom–molecule interaction potential on the orientation of the molecule with respect to the relative collision velocity can be obtained by the method of *molecular beams*, and often in conjunction with inhomogeneous magnetic and electric fields which orient the molecules. Such investigations are undoubtedly very complex, and their realization rather costly. To convince oneself of it one might just peruse the *Proceedings of the First Workshop on Dynamic Stereochemistry in Jerusalem* in 1986 [66]; see also [67, 341].

At the same time it is possible to obtain sufficiently interesting information on the dynamics of collisions by means of technically simpler measurements of the polarization of laser-induced fluorescence at excitation of molecules in bulk. A short account of such investigations is given in [288].

In studies of molecular collisions in *thermal cells* it is important that the exciting light of a given polarization produces well-defined polarization of the excited molecules, the quantitative characteristics of which are easily determinable; see Section 2.4. If we now add into the cell a buffer gas as a partner in collisions, two problems can be solved. The first consists of the possibility of studying elastic collisions leaving the molecule on the initial level, but causing a change in the orientation of the angular momentum of the molecule, i.e. *disorienting collisions*. The second set of problems is connected with the transition of molecules on to different rovibronic levels of the same excited electronic state. Since all these levels are not populated in thermal equilibrium and in the absence of collisions, it becomes possible to judge, by the intensity of collisionally induced fluorescence from these levels, the efficiency of *collisional population transfer* and, by the polarization, the behavior of the angular momentum of the molecule in inelastic collisions, or *collisional polarization transfer*.

The simplest method consists of investigating the collisional depopulation of a laser excited rovibronic level, i.e. of measuring the rates and cross-sections of the collisional relaxation of its population $_b\rho_0^0$. The relaxation rate Γ_K of polarization moments $_b\rho_Q^K$ of various rank K may be represented, in the case of isotropic collisions, as follows:

$$\Gamma_K = \Gamma_{sp} + \sigma_K \langle v_{rel} \rangle N, \qquad (2.42)$$

where σ_K is the effective cross-section of the collisional relaxation of the polarization moment, $\langle v_{rel} \rangle$ is the average relative velocity of the partners in collision and N is the concentration of the particles, the collisions with which are under investigation. The rate Γ_{sp} describes the radiational decay of the excited molecular level and equals the reciprocal spontaneous lifetime τ_{sp}^{-1}. If collisions occur, then, obviously, the channel of collisional exit is opened, in addition to the radiative channel of decay of the excited level. As a result, the intensity of laser-induced fluorescence must decrease. This decrease permits us to judge the efficiency of the relaxation of the population $_b\rho_0^0$ in collisions. However, the registered fluorescence signal of certain direction and certain polarization is influenced not only by the population $_b\rho_0^0$ of the level under investigation, but also by its orientation $_b\rho_0^1$ and its alignment $_b\rho_0^2$, which may relax at different rates Γ_K (2.42) in collisions. One must therefore take special measures in registering only population $_b\rho_0^0$. Thus, if the experiment is carried out according to the geometry given in Fig. 2.5, then one may, as already pointed out with regard to (2.29) and (2.30), observe the fluorescence component I_{θ_0} which is linearly polarized at the 'magic' angle θ_0 (2.31) with respect to the z-axis. The second method also follows from Eqs. (2.29) and (2.30). As can be seen, in measuring $I_{\|} + 2I_{\perp}$, the signal will be proportional to the population $_b\rho_0^0$ only. Measuring by means of one of the above methods the dependence of the signal I_{θ_0} or $I_{\|} + 2I_{\perp}$ on concentration N, and taking into account that at stationary excitation we have

$$_b\rho_0^0(N) = {}_b\rho_0^0(0)\Gamma_0/(\Gamma_0 + \sigma_0 \langle v_{rel} \rangle N), \qquad (2.43)$$

it is easy to determine the value of the cross-section σ_0 from the slope of the straight line in the $1/I(\theta_0)$ or $1/(I_{\|} + 2I_{\perp})$ dependence on N. Such a dependence is called the Stern–Volmer graph [116]. This method was applied for determining the cross-sections of the collisional relaxation of the population in a number of molecules. Thus, for instance, σ_0 values of 115 Å^2 and 517 Å^2 were determined for $Na_2(B^1\Pi_u)$ in collisions with

Table 2.2. Cross-sections of population relaxation σ_0 in collisions with an atom A for NaK $D^1\Pi_u$, $(v' = 17, J' = 94)$ excited by 476.5 nm Ar^+ laser line [24] and for K_2, $B^1\Pi_u$, $(v' = 8, J' = 73)$ excited by 632.8 nm He-Ne laser line [312]

A $\sigma_0(\text{Å}^2)$	He	Ne	Ar	Kr	Xe	K
NaK	138 ± 4	213 ± 5	219 ± 6	257 ± 6	312 ± 8	900 ± 50
K_2	87 ± 7	126 ± 12	136 ± 13	176 ± 14	195 ± 33	642 ± 72

He and Na atoms [116]. Table 2.2 presents example results of [24, 312] in which the σ_0 of a given rovibronic level of NaK and K_2 molecules were determined from Stern–Volmer graphs. Such large cross-sections indicate that the partners of collisions interact fairly efficiently at comparatively large distances.

In a similar fashion one may examine fluorescence from the neighbouring rovibronic levels of an excited state populated in collisions, and thus determine the cross-sections of population transfer between the levels. Such measurements have been performed, for instance, for I_2 [255], Li_2 [138], Na_2 [60] and NaK [292] in collisions with various noble gases. Results have shown that in cases where the vibrational quantum number v' does not change after collision, $\Delta v' = 0$, then the cross-sections $\sigma_0(\Delta J')$ are of the order of $\sigma_0(1) \approx 10\text{Å}^2$ (for the states in which such transitions are permitted by selection rules by symmetry, e.g. for the Π-state), sometimes even attaining values close to 100 Å^2 [116]. With increasing $\Delta J'$ the cross-sections $\sigma_0(\Delta J')$ diminish relatively quickly, although fluorescence from levels with $\Delta J'$ of the order of several tens can still be registered reliably [91, 332]. Transfer of population in vibrational transitions can, in principle, be studied in a similar way. For instance, cross-sections of vibrational relaxation $\sigma_0(\Delta v' = \pm 1)$, summed over all final rotational states, for the Na_2 molecule in the state $B^1\Pi_u(v' = 6, J' = 43)$ in collisions with atoms of noble gases He, Ne, Ar, Kr and molecules H_2, D_2 and N_2, are of the order of $(1-5) \text{ Å}^2$ [60], which is an order of magnitude smaller than the cross-sections of rotational transfer.

So far we have discussed various studies on the relaxation of the population $_b\rho_0^0$ of excited state rovibrational levels in elastic and inelastic collisions. To this end the intensity of fluorescence was measured in one or the other way. If an analysis of the state of polarization of the radiation is performed, one may obtain information on the behavior of alignment and orientation of the molecular angular momenta in elastic and inelastic collisions. If we register, under collisional conditions, the polarization properties of a directly laser-induced rovibrational level of the molecule, then, according to (2.35) and (2.39), it is possible to determine the rates

Γ_2, Γ_1 and the cross-sections σ_2, σ_1 of the collisional relaxation of alignment and orientation.

It is frequently necessary, in studies of collisions, to single out the contribution of elastic processes or of the so-called 'purely disorienting' collisions. As a result of such collisions the rotational state of the molecule does not change, or, classically speaking, the modulus of the angular momentum $\mathbf{J}$ remains unchanged, but its spatial orientation changes. The cross-sections σ_1 and σ_2 obviously do not reflect to the full extent the role of such processes, since the multipole moments $_b\rho_0^1$, $_b\rho_0^2$ may decrease together with the decrease in the number of molecules in the state under investigation, due to relaxation of population. This may well be understood as a diminution of the size of the 'bodies' depicted in Fig. 2.2, their shape remaining unchanged. Indeed, in the absence of purely depolarizing collisions we have $\sigma_0 = \sigma_1 = \sigma_2$, hence the contribution of such collisions may reasonably be characterized by the differences $\sigma_1 - \sigma_0$ and $\sigma_2 - \sigma_0$.

Studies of the depolarization process in elastic collisions for excited states of a number of diatomic molecules, such as $H_2(B^1\Sigma_u^+)$ [311], $Li_2(A^1\Sigma_u^+)$ [147, 332], $NaK(D^1\Pi)$ [24, 291], $CdH(A^2\Pi_{1/2})$ [130], $Na_2(B^1\Pi_u)$ [237], $BaO(A^1\Sigma^+)$ [350], Se_2 [205] and Te_2 [154] have shown that, unlike the case of atoms, the efficiency of purely disorienting collisions is mostly very low in comparison with quenching collisions. The results of the works in which the differences $\sigma_2 - \sigma_0$, $\sigma_1 - \sigma_0$ have been measured directly are presented in Table 2.3. Comparison between Tables 2.2 and 2.3 show that usually the cross-sections of purely disorienting processes rarely differ from zero within the error range. Even if one considers these differences as being significant, the cross-sections of disorienting processes are at least two orders of magnitude smaller than the cross-sections of quenching collisions. An exception to this rule is represented by the CdH [130] molecule, for which collisions with the relatively heavy Ar atom proceed rather efficiently.

In the classical model the inefficiency of purely disorienting elastic collisions can easily be understood if one takes into account the fact that such disorientation is connected with a turn of the angular momentum $\mathbf{J}$ by a tangible angle (collisional randomization of $\mathbf{J}(\theta, \varphi)$). However, in this case one must also expect, with high probability, a change in the magnitude of $\mathbf{J}$, which *de facto* means a transition on to a different rotational level, i.e. the collision becomes an inelastic one.

Finally, studying the polarization properties of fluorescence emitted from rovibronic levels populated in the process of *inelastic collisions*, one may, according to (2.33) and (2.37), determine the degree of alignment $_b\rho_0^2/_b\rho_0^0$ and the degree of orientation $_b\rho_0^1/_b\rho_0^0$ of these levels. Compar-

Table 2.3.   Cross-sections of purely disorienting collisions

Molecule and state	Partner of collision	$\sigma_2 - \sigma_0$ (Å^2)	$\sigma_1 - \sigma_0$ (Å^2)	Source
Li$_2(A^1\Sigma_u^+)$				
$J' = 10$	He		$0,35 \pm 0,15$	[332]
	Ar		$0,65 \pm 0,3$	[332]
$J' = 6$	Xe		$2,4 \pm 0,45$	[147]
$J' = 20$	Xe		$0,74 \pm 0,22$	[147]
$J' = 32$	Xe		$0,42 \pm 0,12$	[147]
NaK $(D^1\Pi)$				
$J' = 30$	He		$0,1 \pm 0,2$	[291]
$J' = 94$	He	$\sigma_2/\sigma_0 = 0,99 \pm 0,01$		[24]
CdH $(A^2\Pi_{1/2})$				
$J' = 3,5$	He	0 ± 2		[130]
	Ar	7 ± 3		[130]
$J' = 7,5$	He	1 ± 2		[130]
	Ar	13 ± 4		[130]
$J' = 16,5$	He	1 ± 2		[130]
	Ar	32 ± 5		[130]
Na$_2(B^1\Pi_u)$				
$J' = 43$	He	$0,25 \pm 0,10$		[237]
$J' = 27$	He	$0,10 \pm 0,25$		[237]
$J' = 12$	He	$1,4 \pm 0,5$		[237]

ing these results with the degree of alignment and orientation of the directly optically populated level, it is possible to investigate the *transfer of polarization* of the molecular angular momenta in collisions. The most careful attention to the study of such problems is due to McCaffery and his co-workers. Inelastic collisions were studied in such molecules as I$_2$ [213, 214], Li$_2$ [215, 332] and, finally, NaK [292]. All these experiments revealed remarkable peculiarities of inelastic collisions, namely that, as a result of these collisions the projection of the angular momentum changes very insignificantly, which, in quantum terms, means conservation of the quantum number $M_{J'}$.

All the experiments described in this chapter were performed on molecules in gas cells, and, unlike in beam experiments, the results obtained are averaged over all possible orientations of the molecules with respect to the relative velocity vector, and over the Maxwellian distribution of the relative velocities of the colliding particles. Nevertheless, the experimental data stimulated interest in the development of theoretical models of elastic and inelastic collisions between alkali dimers and atoms; see, e.g., [229, 230, 265]. The general conclusion is that the reason for the conservation of $M_{J'}$ in elastic and inelastic collisions consists in the dynamics and kinematics of the collisional process and reflects the anisotropy of the intermolecular potential.

We also wish to point out the experimental works [357, 358] in which collisions of the molecule $Li_2(A^1\Sigma)$ with Ne, Ar and Xe were studied. The authors employed a narrow laser line, exciting, from Doppler velocity distribution, only molecules having a definite projection of velocity upon the direction of the laser beam. Within the range of kinetic energies corresponding to temperatures between 120 and 3500 K a pronounced dependence of cross-sections on the velocity of the collision partners was observed.

In concluding this chapter, we can say that the description presented thus permits us to connect the probability of absorption and emission of light of any polarization with the orientation of the angular momentum of the molecule. This may serve as a basis for more detailed analysis of the processes of creating an anisotropic distribution of angular momenta both on the upper, as well as on the lower, level of an optical transition (Chapter 3), including the effect of an external field (Chapters 4, 5)

3

Ground state angular momenta polarization

We know now how alignment and orientation of angular momenta arise in the excited state as a result of optical excitation from the primarily isotropic distribution of the ensemble of angular momenta. This phenomenon reflects, in essence, the anisotropy of absorption probability $G(\theta, \varphi)$. Since spatial structures of the type shown in Fig. 2.2 appear in the excited state the same structure must be 'eaten out' from the angular momentum distribution in the ground (initial) state. If the gap created is not closed quickly enough, the ensemble of the remaining angular momenta of the *absorbing* molecules in the ground state must necessarily also be polarized.

If, from another point of view, fluorescent decay leads to a considerable population of the ground state level c, (see, e.g., Fig. 1.2), then molecules on this level may exhibit angular momenta polarization, being particularly noticeable for thermally unpopulated, high lying levels.

In the present chapter we intend to discuss the conditions necessary for the creation of ground state angular momenta polarization, the possibilities of its experimental observation, and to develop the theoretical description of broad line radiation interaction with molecules further.

3.1   Angular momenta polarization via depopulation

For many years there existed the widespread opinion that in molecules we obtain an anisotropic distribution of angular momenta only in the excited state in the absorption process, whilst in the ground state isotropy is conserved [124]. This conjecture is based essentially on the assumption of the existence of a so-called 'weak' excitation. At first sight it may appear, for instance, from Eq. (2.20) that, as is usual in non-linear spectroscopy, such an admission is correct if condition $\Gamma_p/\Gamma \ll 1$ is valid. This may not be the case for molecular systems, since spontaneous emission does not

49

constitute the basic mechanism for restoring the population of the initial (lower) level. Indeed, as it follows from Fig. 1.2 and from the following discussion, only a very small number ($\sim \Gamma_{ba}/\Gamma$) of the excited particles return to the initial state a by way of spontaneous emission. The basic process competing with depopulation via absorption is the relaxation within the system of rotational and vibrational levels inside the lower electronic state α''. Consequently, if the rate of absorption Γ_p is comparable to the rate of relaxation of the anisotropy of the lower level, then the initial state (α'', v_a'', J_a'') will be *optically polarized* (aligned, oriented). We will now (see also Section 3.3) discuss the main relaxation mechanisms and estimate their importance, taking as an example a situation typical for many diatomic molecules in rarefied vapor or in molecular beams.

The following processes may take place:

  i collisional relaxation;

  ii relaxation due to the finite time which the molecule spends in the zone of action of the optical field;

  iii radiational transitions within the system of rovibrational levels.

The latter mechanism is excluded for homonuclear diatomic molecules (dimers) owing to the absence of a permanent dipole moment in these molecules. In the case of heteronuclear molecules, on the other hand, such transitions are well known and are widely employed, in particular for the determination of configurational and relaxational parameters by methods of infrared spectroscopy.

The second mechanism is connected to the exchange between 'polarized' molecules inside the light beam and unpolarized molecules from the remaining volume. Such an exchange takes place either as a result of chaotic thermal motion of the bulk molecules, or by directed motion of a molecular beam through the laser beam. The mechanism is always present and dominates at small concentrations, as well as in beams, when the role of collisions becomes less significant. Thus, the polarization of angular momenta of the ground state molecules, even in the absence of collisions, possesses an effective 'lifetime' T_0 of the order of the mean *transit time* through the light beam. In molecular beam experiments T_0^{-1} characterizes, as a rule, the dominant mechanism of the relaxation. As molecules possess spherically symmetric angular momenta distribution outside the light beam region, this 'fly-through' relaxation process supplies only the population to the depopulated level. In order to produce noticeable alignment or orientation of non-excited molecules under these conditions it is obviously sufficient to satisfy the condition $\Gamma_p T_0 \geq 1$.

The transit time is only the upper limit for the existence of light-produced optical anisotropy of angular momenta in the lower state. Under realistic conditions, particularly in the vapor bulk, one must take into consideration the collisional relaxation mechanism of level a (see Fig. 1.2). Without digressing too far from the main theme of our discussion, let us try to go into slightly greater detail regarding collision processes.

Let collisions AB–X between a certain molecule AB, in the electronic ground state α'', on a selected rovibrational level v'', J'', and at given spherical angles θ, φ determining angular momentum $\mathbf{J}_a$ orientation, and a particle X, tend to restore the equilibrium state population and the isotropic distribution of momenta $\mathbf{J}_a$ in a bimolecular reaction:

$$AB(\alpha'', v_i'', J_i'', \theta_i, \varphi_i) + X \longleftrightarrow AB(\alpha'', v_a'', J_a'', \theta_a, \varphi_a) + X + \Delta E. \quad (3.1)$$

Experiments show that the dominating mechanism in collisional relaxation in reaction (3.1) consists of rotational mixing. Since for most of the molecules in conditions $\Delta E_{ia} = E_i - E_a \ll kT$ the number of levels J_i'' involved in the relaxation process is sufficiently large (transitions $|J_i'' - J_a''|$ up to 20 and even up to 80 have been registered for Na_2, see [62, 216, 345]), they can serve as a kind of 'thermostat' of isotropic states. Such a thermal reservoir can, to a certain extent, be considered as unaffected by the cycle of optical pumping, and can be regarded as a source 'supplying' population to the level v_a, J_a. Experiments, such as those on Li_2 [306] may serve as a certain confirmation of a model. The situation is, in one way, the reverse of that in the excited state (Chapter 2). There we have a populated level surrounded by 'empty' ones, whilst here a partly emptied level of the ground state is surrounded by equilibrated populated ones (see Fig. 3.1). The diagram shows that we have an 'open' optical pumping cycle, and the angular momenta anisotropy is transferred from the ground state level v_a'', J_a'' into the thermostat. The attractive property of such an extremely lucid model consists of considerable simplification of the description of collisions in the ground state by means of only one rate constant γ_{col}.

It is important to produce a simultaneous description of collisional and fly-through relaxation. At first sight this appears to be a simple enough problem, namely, similarly to the excited state (2.42), one must introduce the total rate γ_Σ additively:

$$\gamma_\Sigma = \gamma_0 + \gamma_{col}. \quad (3.2)$$

A more detailed analysis (Section 3.6) reveals, however, the limits of applicability of (3.2).

It ought to be stressed that the above description of relaxation does not account for the velocity of translational motion of a separate particle. That is justified in the case where the spectral intensity of the exciting laser light is constant within the range of the Doppler contour of the optical transition, and the molecules can thus absorb light irrespective

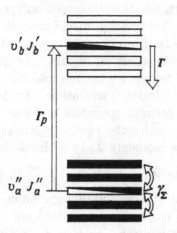

Fig. 3.1. Optical depopulation of the vibrational–rotational level v_a'', J_a''.

of their velocity. This implies the validity of broad line excitation [110, 127, 231], which means that the spectral width of the exciting line $\Delta\omega_{las}$ considerably exceeds both the homogeneous and the nonhomogeneously broadened excitation contour $\Delta\omega_D$.

Apart from the processes already discussed, other relaxation channels may also take part. Thus, if condition $\Gamma_p \ll \Gamma$ is not fulfilled, one must consider *stimulated transitions*. One may state in this case that an effect of saturation of the transition is taking place. The effect of saturation of transition on laser-induced fluorescence polarization characteristics has been discussed in detail in the classical approach by Atcorn and Zare; see [9] and the literature therein. We will include stimulated transitions in our subsequent discussion by presenting a more generalized description (see Chapter 5), which we are omitting now for better understanding of the main characteristics of optical polarization. The feasibility of such an approximation under the concrete circumstances may be judged from the examples given in Table 3.7.

Another radiational process consists of *spontaneous transitions* causing a return to the initial level with rate constant Γ_{ba}.

Let us sum up the above. A simple model of excitation and relaxation is discussed, which permits us (at least in the case of the stipulated objects and conditions) to single out the quantity

$$\chi = \Gamma_p/\gamma_\Sigma, \tag{3.3}$$

namely the ratio between the absorption rate Γ_p and the total relaxation rate γ_Σ of the lower level, as the basic dimensionless parameter governing the process of polarization of the angular momenta of this level via absorption. Hence, if $\chi \ll 1$, then one may assume that the angular momenta distribution of the absorbing level a remains isotropic and its population is not affected by absorption (linear absorption conditions). If, on the other hand, $\chi \geq 1$, the anisotropic angular momenta distribution is created in the absorbing state along with its depopulation (non-linear absorption conditions). The approximate nature of such a model is fully compensated by the fact that it becomes actually possible to pass over to a two-level model which allows understanding of the physical essence of the phenomena. Moreover, such an approach yields good agreement with experimental results in many cases, as will be shown.

1 Balance equations: classical description

Applying the latest model, the simplest equation for the probability density of angular momenta distribution in the absorbing state a may be written as follows:

$$\dot{\rho}_a(\theta, \varphi) = -\Gamma_p G(\theta, \varphi) \rho_a(\theta, \varphi) + \gamma_\Sigma [\rho_a^0 - \rho_a(\theta, \varphi)], \qquad (3.4)$$

where the first term on the righthand side describes the depopulation of state a in the process of light absorption, and the second one describes its population with the relaxation rate γ_Σ; ρ_a^0 is the isotropic population of the state.

The stationary solution of (3.4) is of an extremely simple form:

$$\rho_a(\theta, \varphi) = \frac{\rho_a^0}{1 + \chi G(\theta, \varphi)}. \qquad (3.5)$$

Using the absorption probability coefficients $G(\theta, \varphi)$, as described by Eq. (2.8), one may easily obtain the spatial distribution of the angular momenta $\mathbf{J}_a(\theta, \varphi)$ of the ground (initial) state molecules.

Examples of isometric projection of such a distribution are shown in Fig. 3.2 (lower part of the diagram). It is assumed that the light is linearly polarized along the z-axis. As can be seen, the light 'eats out' a certain amount of the initially spherical symmetric $\mathbf{J}_a(\theta, \varphi)$ distribution and transfers this amount to the excited state (upper part of the diagram). In both examples cylindrical symmetry with respect to the z-axis is, of course, present. The ρ_b scale on the diagram is enlarged by the factor Γ/γ_Σ with respect to that of ρ_a so that the sum of ρ_a and ρ_b again forms a sphere. It may be seen that in the case of a Q transition a negative alignment is created in the ground state along the z-axis. This is understandable from the discussion on dipole moments; see Chapter 1. Since we have $\hat{\mathbf{d}} \parallel \mathbf{J}_a$ for the Q transition, and classical transition dipole moments do not rotate together with the internuclear axis (see Fig. 1.6),

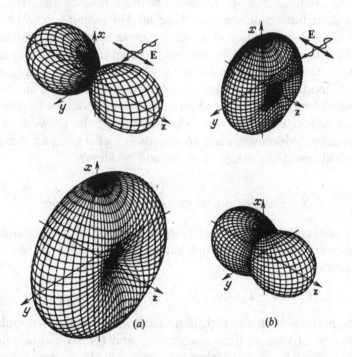

Fig. 3.2. Isometric projection of the probability density of angular momenta distribution of ground (lower part) and excited (upper part) states: (a) Q excitation; (b) (P, R) excitation by light with $\hat{\mathbf{E}} \parallel \mathbf{z}$.

then we have, owing to absorption a 'deficit' of ground state angular momentum $\mathbf{J}_a$ along the z-axis.

For (P, R)-type transitions the lower state angular momenta $\mathbf{J}_a$ in Fig. 3.2 show positive alignment along the z-axis, or along the $\hat{\mathbf{E}}$-vector. This follows directly from the orthogonality $\hat{\mathbf{d}} \perp \mathbf{J}_a$; see Fig. 1.5(a,c) and Fig. 1.6.

In the case of circular polarized excitation the distribution of initial state angular momenta $\mathbf{J}_a$ shows orientation, its sign being opposite to that of the excited state.

Following the general approach, as presented in Chapter 2, let us expand the solution (3.5) over spherical functions (2.14) in order to pass from $\rho_a(\theta, \varphi)$ to 'classic' ground state polarization moments $_a\rho_q^\kappa$ (2.16). It is important to stress that, since absorption is non-linear with respect to the exciting light, here, unlike in Section 2.3, we obtain polarization moments $_a\rho_q^\kappa$ of rank $\kappa > 2$ in the ground state. We will denote the rank and the projection by κ and q respectively (unlike K and Q for the excited state). We can, however, state that all the produced polarization moments $_a\rho_q^\kappa$ in the geometry of excitation presented in Figs. 2.5 and 2.6, and in the absence of external fields, will possess zero projection values $q = 0$.

Analytical formulae obtained for $_a\rho_0^\kappa(\chi)$ with $\kappa \leq 4$ from Eqs. (2.8), (2.16) and (3.5) are presented in Table 3.1 [40], both for the case of linearly polarized (geometry Fig. 2.5), and for the case of circularly polarized (geometry Fig. 2.6), excitation. Analytical expressions for the mutipole moments of higher ranks can be obtained in a similar way. Thus, for $_a\rho_0^6$ in the case of linear polarized Q-excitation we have

$$_a\rho_0^6 = \frac{1}{16\chi}\left[\frac{231}{5} + 238\chi^{-1} + 231\chi^{-2}\right.$$
$$\left. - (231\chi^{-5/2} + 315\chi^{-3/2} + 105\chi^{-1/2} + 5\chi^{1/2})\arctan\chi^{1/2}\right], \tag{3.6}$$

but for (P, R)-excitation

$$_a\rho_0^6 = -\frac{1}{8\chi}\left[\frac{231}{5} - 238a^2 + 231a^4\right.$$
$$\left. - (231a^5 - 315a^3 - 105a - 5a^{-1})\mathrm{Arctha}\right], \tag{3.7}$$

where $a^2 = 1 + 2\chi^{-1}$.

In order to render the expressions from Table 3.1 and Eq. (3.6) for $_a\rho_0^\kappa$ more clearly perceivable, in Table 3.2 we present numerical $_a\rho_0^\kappa$ values at the non-linear absorption parameter value $\chi = 1$ and at Q type transition. The multipole moments $_a\rho_0^\kappa$ may be seen to form a series of terms alternating in sign and diminishing in absolute value.

A graphic illustration of the relations in Table 3.1 and in (3.6) is presented in Figs. 3.3–3.5, showing the ratio $_a\rho_0^\kappa/_a\rho_0^0$ as dependent on the parameter χ. It is this relationship which is independent of population and thus ought to be chosen to characterize the angular momenta polarization of the level.

The limiting values of $_a\rho_0^\kappa/_a\rho_0^0$ for arbitrary κ and an infinitely large parameter $\chi \to \infty$ equal, in the case of linearly polarized Q-absorption of light:

$$\lim_{\chi\to\infty}\left(\frac{_a\rho_0^\kappa}{_a\rho_0^0}\right) = (-1)^{\kappa/2}\frac{1 \times 2 \times 3 \ldots (\kappa - 1)}{2 \times 4 \times 6 \ldots \kappa}$$
$$= (-1)^{\kappa/2}\pi^{-1/2}\frac{\Gamma(\kappa/2 + 1/2)}{\Gamma(\kappa/2 + 1)} \approx (-1)^{\kappa/2}\left(\frac{2}{\pi\kappa}\right)^{1/2}, \tag{3.8}$$

where $\Gamma(\kappa)$ is a gamma function. Limiting values of $_a\rho_0^\kappa/_a\rho_0^0$ for $\kappa \leq 4$ thus obtained are presented in Table 3.3. One must, however, keep in

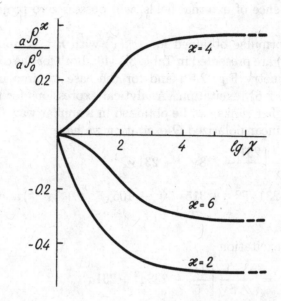

Fig. 3.3.   Relative polarization moment value in ground (initial) state at linearly polarized Q-absorption.

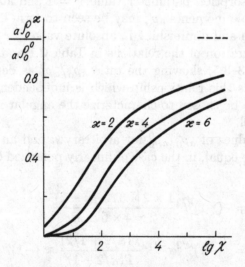

Fig. 3.4.   Relative polarization moment value in ground (initial) state at linearly polarized (P, R)-absorption.

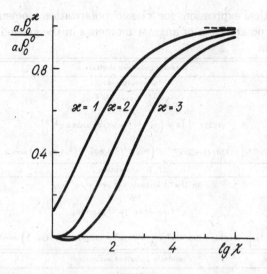

Fig. 3.5. Relative polarization moment value in ground (initial) state at circularly polarized (P,l)- or (R,r)-absorption.

mind the fact that the expressions make sense if the conditions for which the simplest balance equation (3.4) is obtained are obeyed.

Fig. 3.4 shows $_a\rho_0^\kappa/_a\rho_0^0$ dependences for linear polarized (P,R)-type excitation (in the same geometry as shown in Fig. 3.3). It must be noted that $_a\rho_0^\kappa/_a\rho_0^0$ is positive for all ranks κ; the limiting values at $\chi \to \infty$ are also presented in Table 3.3.

The case of excitation by circularly polarized light is illustrated by Fig. 3.5. As may be seen, in addition to orientation (odd rank moments) here we also have alignment ($\kappa = 2$) which changes its sign at $\chi \approx 1.871$. We wish to draw attention to the fact that the octupole moment $_a\rho_0^3$ also changes its sign at $\chi \approx 9.037$, having the same sign as that of $_a\rho_0^1$ at high χ value.

2   Balance equations: quantum mechanical treatment

We have so far limited ourselves to a classical description, the natural requirement for which is the condition $J', J'' \to \infty$. In order that the description is valid for any angular momentum value, it is necessary to employ the quantum mechanical approach. We presume that the reader is acquainted with the *density matrix* (or the *statistic operator*) introduced into quantum mechanics for finding the mean values of the observables averaged over the particle ensemble. Under the conditions and symmetry of excitation considered here one must simply pass from the *prob-*

Table 3.1. Analytical expressions for 'classic' polarization moments $_a\rho_0^\kappa$ ($\kappa \leq 4$) describing optical polarization of angular momenta in the ground (initial) state via light absorption

$_a\rho_0^\kappa$	Linearly polarized excitation
	Q-absorption
$_a\rho_0^0$	$\chi^{-1/2}\arctan\chi^{1/2}$
$_a\rho_0^1$	0
$_a\rho_0^2$	$(1/2)\chi^{-1}\left[3 - \left(3\chi^{-1/2} + \chi 1/2\right)\arctan\chi^{1/2}\right]$
$_a\rho_0^3$	0
$_a\rho_0^4$	$(1/16)\chi^{-1}\left[-(110/3) - 70\chi^{-1} + \left(70\chi^{-3/2} + 60\chi^{-1/2} + 6\chi 1/2\right)\arctan\chi^{1/2}\right]$
	(P, R)-absorption
$_a\rho_0^0$	$2\chi^{-1}a^{-1}\mathrm{Arctha}, \quad a^2 = 1 + 2\chi^{-1}$
$_a\rho_0^1$	0
$_a\rho_0^2$	$-\chi^{-1}\left[3 - (3a - a^{-1})\mathrm{Arctha}\right]$
$_a\rho_0^3$	0
$_a\rho_0^4$	$-\chi^{-1}\left[-(55/14) + (35/4)a^2 - \left((35/4)a^3 - (30/4)a + (3/4)a^{-1}\right)\mathrm{Arctha}\right]$

$_a\rho_0^\kappa$	Circularly polarized excitation
	Q-absorption
$_a\rho_0^0$	$2\chi^{-1}a^{-1}\mathrm{Arctha}, \quad a^2 = 1 + 2\chi^{-1}$
$_a\rho_0^1$	0
$_a\rho_0^2$	$-\chi^{-1}\left[3 - (3a - a^{-1})\mathrm{Arctha}\right]$
$_a\rho_0^3$	0
$_a\rho_0^4$	$-\chi^{-1}\left[-(55/14) + (35/4)a^2 - \left((35/4)a^3 - (30/4)a + (3/4)a^{-1}\right)\mathrm{Arctha}\right]$
	(P, R)-absorption*
$_a\rho_0^0$	$\chi^{-1/2}\arctan\chi^{1/2}$
$_a\rho_0^1$	$\pm\chi^{-1/2}\arctan\chi^{1/2} \mp \chi^{-1}\ln(1+\chi)$
$_a\rho_0^2$	$6\chi^{-1} + (\chi^{-1/2} - 6\chi^{-3/2})\arctan\chi^{1/2} - 3\chi^{-1}\ln(1+\chi)$
$_a\rho_0^3$	$\pm 20\chi^{-1} \pm (\chi^{-1/2} - 30\chi^{-3/2})\arctan\chi^{1/2} \pm (10\chi^{-2} - 6\chi^{-1})\ln(1+\chi)$
$_a\rho_0^4$	$-70\chi^{-2} + (130/3)\chi^{-1} + (70\chi^{-5/2} - 90\chi^{-3/2} + \chi^{-1/2}\arctan\chi^{1/2} + (70\chi^{-2} - 10\chi^{-1})\ln(1+\chi)$

* upper sign holds for (P, l), (R, r), transitions, the lower sign for (P, r), (R, l) absorptions

Table 3.2. Values of polarization moments $_a\rho_0^\kappa$ describing optical alignment in the ground (initial) state under linearly polarized Q-type absorption. It is assumed that the pumping parameter $\chi = 1$

$_a\rho_0^0$	$\pi/4$	$\approx 0.785\,398\,1$
$_a\rho_0^2$	$(3 - \pi)/2$	$\approx -0.070\,796\,3$
$_a\rho_0^4$	$-(20/3) + (17\pi)/8$	$\approx 0.009\,212\,1$
$_a\rho_0^6$	$(161/5) - (41\pi)/4$	$\approx -0.001\,297\,5$

ability density $\rho(\theta, \varphi)$, which can be considered as the '*classical density matrix*' as written in θ, φ representation of spherical angles characterizing the direction (but not the magnitude) of the vector $\mathbf{J}(\theta, \varphi)$, to the *quantum mechanical density matrix* $f_{MM'}$ written in $|J, M\rangle$ representation, where J is the quantum number of the angular momentum value

Table 3.3. Limiting values of the ratio $_a\rho_0^\kappa/_a\rho_0^0$ of various rank ground state multipole moments $_a\rho_0^\kappa$ and the population $_a\rho_0^0$ for an infinitely large parameter $\chi \to \infty$

	Linearly polarized excitation		Circularly polarized excitation	
$_a\rho_0^\kappa/_a\rho_0^0$	Q-abs.	(P,R)-abs.	Q-abs.	(P,R)-abs.
$_a\rho_1^\kappa/_a\rho_0^0$	0	0	0	$\pm 1^*$
$_a\rho_2^\kappa/_a\rho_0^0$	$-1/2$	1	1	1
$_a\rho_3^\kappa/_a\rho_0^0$	0	0	0	$\pm 1^*$
$_a\rho_4^\kappa/_a\rho_0^0$	3/8	1	1	1

* The plus sign holds for (P,l), (R,r), the sign minus for (P,r), (R,l) absorption.

$|\mathbf{J}| = \hbar\sqrt{J(J+1)}$, $M\hbar$ being the $\mathbf{J}$-projection upon the quantization axis (z-axis) and M being the magnetic quantum number. For closer acquaintance with quantum mechanical density matrix approach we recommend monographs and reviews [6, 73, 139, 140, 304]. We wish to point out that in the treatment of the present chapter only the diagonal elements f_{MM} of the density matrix, which characterize the distribution of population over the M sublevels (see Fig. 3.6), will differ from zero (this is equivalent to the condition $q = Q = 0$ in the classical polarization moments $_b\rho_Q^K$, $_a\rho_0^\kappa$). One says in such cases that there is an *absence of coherence* between the M-sublevels (for greater detail see Chapter 5, as well as Appendix D). The values of M change from $-J$ to J, and the sum $\sum_{-J}^{J} f_{MM}$ is the population of the level.

We will now introduce the following notation. Let the ground (initial) state of the molecule with rotational quantum number J'' be characterized by the ground state *density matrix* $\varphi_{\mu\mu'}$ (reserving notation $f_{MM'}$ for the excited state), where the indices μ cover the set M'' of magnetic quantum numbers. The quantum mechanical equivalent of (3.4) for the situations presented in Figs. 2.5 and 2.6 is

$$\dot{\varphi}_{\mu\mu} = -\Gamma_p \left(C_{J''\mu 1\eta}^{J'\mu+\eta}\right)^2 \varphi_{\mu\mu} + \gamma_\Sigma \left(\varphi_{\mu\mu}^0 - \varphi_{\mu\mu}\right), \qquad (3.9)$$

where $\varphi_{\mu\mu}^0$ is the 'isotropic' population of the magnetic sublevel μ, in thermal equilibrium; index $\eta = M' - M'' = M - \mu$ characterizes the change in the angular momentum projection at absorption of light. Note that η has zero value in the case of linear polarized excitation, or equals ± 1 for circular polarized excitation, since η is the projection upon the z-axis of the angular momentum of the photon; $C_{a\alpha b\beta}^{c\gamma}$ in (3.9) is the Clebsch–Gordan coefficient, the square of which is an analogue to the classical absorption probability $G(\theta, \varphi)$ in (3.4). The values of $C_{J''\mu 1\eta}^{J'\mu+\eta}$ for $J' - J'' = \Delta = 0, \pm 1$ are presented in Appendix C; see Table C.2. Thus, for instance, in a Q-transition in a geometry according to Fig. 3.2, where $\eta = 0$, $M' = M''$, we have from Table C.2

$$\left(C^{J'\mu}_{J''\mu 10}\right)^2 = \frac{(M'')^2}{J''(J''+1)}. \tag{3.10}$$

It follows that on switching to a vector model of the quantum angular momentum we obtain $G(\theta, \varphi) = (J_z/|\mathbf{J}|)^2 = \cos^2 \theta$.

The population of the magnetic sublevel $M'' \equiv \mu$ emerges as a stationary solution of Eq. (3.9) and equals

$$\varphi_{\mu\mu} = \frac{\varphi^0_{\mu\mu}}{1 + \chi \left(C^{J'\mu+\eta}_{J''\mu 1\eta}\right)^2}. \tag{3.11}$$

The stationary population of sublevels $M'' \equiv \mu$ for linear polarized Q-excitation, as obtained by substitution of (3.10) into (3.11), is demonstrated in Fig. 3.6(a), lower drawings, for $J'' = 3$. The length of the corresponding horizontal line is proportional to $\varphi_{\mu\mu}$. As may be easily understood, the population, which adds up to an isotropic one (indicated by a rectangle inside the dotted line), is transferred to the excited state (upper part of the diagram). Fig. 3.7 demonstrates, in the vector model, the spatial distribution of the ground state angular momenta with quantum number $J'' = 3$ and its z-projections M'', as calculated for the same conditions used in Fig. 3.6. The length of the section along the angular momentum $\mathbf{J}_a$ from the origin of coordinates to the mark (open circle in Fig. 3.7) characterizes the probability of having a direction of $\mathbf{J}_a$, which corresponds to the projection $M'' \equiv \mu$. Such a probability is obviously proportional to $\varphi_{\mu\mu}$. The solid line connecting the open circles in Fig. 3.7 corresponds to the classic limit of continuous angular momentum distribution, i.e. to the section of the body in Fig. 3.2(a) (bottom) by the yz plane.

The expressions for $\varphi_{\mu\mu}$ in the case of linear or circular polarized P- or R-absorption may also be easily obtained by substitution of the respective $C^{c\gamma}_{a\alpha b\beta}$ (presented in the Table C.2 of Appendix C) into (3.11) (the explicit form of $(C^{c\gamma}_{a\alpha b\beta})^2$ can be found in Table 3.6 as coefficients multiplied by χ). The cases of linear and circular polarized P-absorption are demonstrated in Fig. 3.6(b) and 3.6(c).

Comparing Fig. 3.6(a) and 3.6(b) we see that in both situations anisotropy with respect to M'' is created in the form of alignment, since $\varphi_{\mu''\mu''} = \varphi_{-\mu''-\mu''}$. Therefore, the mean value of the projection of the angular momentum upon the quantization axis equals zero. There is, however, a substantial difference. For a Q-transition, mainly the sublevels in the vicinity of $M'' = 0$ are populated, see Fig. 3.6(a) (bottom), whilst the level $M'' = 0$ itself is not depopulated, as follows from (3.10).

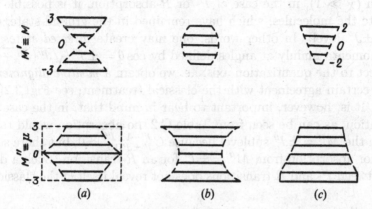

Fig. 3.6. Distribution of the population f_{MM} (top) and $\varphi_{\mu\mu}$ (bottom) over magnetic sublevels at $J = 3$, $\chi = 2$: (a) linear polarized Q-excitation; (b) linear polarized P-excitation; (c) P-excitation by righthanded circular polarized light; the lengths of the horizontal lines are proportional to f_{MM} (top) and to $\varphi_{\mu\mu}$ (bottom). The f_{MM} scale accounts for the Γ/γ_{Σ} factor.

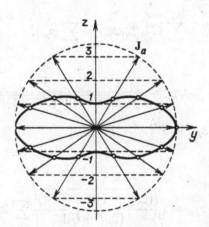

Fig. 3.7. Spatial quantization of the angular momentum $\mathbf{J}_a$ with quantum number $J'' = 3$ at Q-absorbtion ($\chi = 2$) of linearly polarized light with $\hat{\mathbf{E}} \parallel \mathbf{z}$. The figures show the values of $M'' \equiv \mu$. The solid line indicates the spatial distribution of $\mathbf{J}_a$ in the classic limit ($J \to \infty$).

Hence, at $\chi \gg 1$, the angular momenta are positioned in a plane which is orthogonal to the z-axis, i.e. the alignment is *negative*. At the same time, for P-transitions (as well as for R-transitions) there remain mainly populated ground state sublevels in the vicinity of $M'' = \pm J''$; see Fig. 3.6(b) (bottom). This leads to an important conclusion: at very strong depopulation ($\chi \gg 1$), in the case of P- or R-absorption, it is possible to concentrate the molecules, which have remained in the ground state, on a $M'' = \pm J''$ level. In other words, one may create *directed states* of angular momenta mainly at angles defined by $\cos \theta = \pm J''/\sqrt{J''(J'' + 1)}$ with respect to the quantization axis, i.e. we obtain a *positive alignment*. This is in certain agreement with the classical treatment; see Fig. 3.2(b) (bottom). It is, however, important to bear in mind that, in the case of a P-transition, as can be seen from Table C.2, no absorption would take place from the $M'' = \pm J''$ sublevel because $C^{J''-1\pm J''}_{J''\pm J''10} = 0$, but there still exists minor absorption from $M'' = \pm J''$ for an R-transition. Such a difference between P- and R-transitions does not reveal itself in the classical model.

The quantitative characteristic of the alignment created is given, as already stated, by multipole moments of even rank. A more rigorous treatment of the expansion of the quantum mechanical density matrix over irreducible tensorial operators will be performed later, in Chapter 5 and in Appendix D. As an example we will write the zero, second and fourth rank polarization moments φ_0^0, φ_0^2 and φ_0^4 as expressed through $\varphi_{\mu\mu}$:

$$\varphi_0^0 = \sum_\mu C^{J''\mu}_{J''\mu 00}\varphi_{\mu\mu} = \sum_\mu \varphi_{\mu\mu}, \qquad (3.12)$$

$$\varphi_0^2 = \sum_\mu C^{J''\mu}_{J''\mu 20}\varphi_{\mu\mu} = \frac{1}{\sqrt{(2J'' - 1)J''(J'' + 1)(2J'' + 3)}}$$
$$\times \sum_\mu \left[3\mu^2 - J''(J'' + 1)\right]\varphi_{\mu\mu}, \qquad (3.13)$$

and

$$\varphi_0^4 = \sum_\mu C^{J''\mu}_{J''\mu 40}\varphi_{\mu\mu} = \sqrt{\frac{(2J'' + 1)(2J'' - 4)!}{(2J'' + 5)!}} \sum_\mu 2[3(J''^2 + 2J''$$
$$-5\mu^2)(J''^2 - 5\mu^2 - 1) - 10\mu^2(4\mu^2 - 1)]\varphi_{\mu\mu}. \qquad (3.14)$$

Fig. 3.6(c) presents an example of P-excitation by righthanded circularly polarized light. In the case of excitation by circularly polarized light we obtain $\varphi_{\mu\mu} \neq \varphi_{-\mu-\mu}$, i.e. the system is optically oriented. As can be seen, orientation has been created in the positive direction of the z-axis. The *polarization moment* φ_0^1 emerging in the system (Appendix D) may be written using averaged $M'' \equiv \mu = 0$ values:

$$\varphi_0^1 = \sum_\mu C^{J''\mu}_{J''\mu 10}\varphi_{\mu\mu} = \frac{1}{\sqrt{J''(J'' + 1)}} \sum_\mu \mu\varphi_{\mu\mu}. \qquad (3.15)$$

Table 3.4. Limiting values of ratios φ_0^2/φ_0^0 and φ_0^4/φ_0^0 for various J at very strong depopulation ($\chi \to \infty$) in Q-transitions

J	φ_0^2/φ_0^0	φ_0^4/φ_0^0
1	$-\sqrt{2/5}$	—
2	$-\sqrt{2/7}$	$\sqrt{2/7}$
3	$-\sqrt{4/15}$	$\sqrt{2/11}$
5	$-\sqrt{10/39}$	$\sqrt{30/195}$
10	$-\sqrt{110/437}$	$9\sqrt{66/37145}$
∞	$-1/2$	$3/8$

At Q-absorption and under very strong depopulation ($\chi \to \infty$) only one term remains in Eqs. (3.13) and (3.14), where $\mu = M'' = 0$. One may therefore obtain limiting values of $\varphi_0^\kappa/\varphi_0^0$, similar in meaning to the 'classical' ones presented in Table 3.3, but with arbitrary values of J''. The φ_0^2/φ_0^0 and φ_0^4/φ_0^0 values, as calculated in this way, are given in Table 3.4.

Thus, in this section we have described the manner in which absorption of light by a molecule leads to polarization of the angular momenta of the *absorbing level*. We have also shown how to calculate the multipole moments created on the lower level. It is important to stress that the adopted model of description enables us to obtain precise analytical expressions for the multipole moments, including both cases, namely those for arbitrary values of angular momenta and those for the classic limit $J \to \infty$. Our subsequent discussion will concern problems connected with the manifestation of ground state angular momenta anisotropy in experimentally observable quantities.

3.2 Ground state effects in fluorescence

The discussion in Chapter 2 shows that the basic method of registering the anisotropy of angular momenta in the excited state consists of measuring the degree of fluorescence polarization. It may seem, at first glance, that once one speaks of the ground state of an optical transition, it ought to be convenient to register the absorption in transition from this level as dependent on the state of polarization of either the pumping beam or the probe beam. Research experience shows, however, that problems arise in connection with the registration of small relative intensity changes against the background of a strong transmitted light. This may lead, at least within the optical frequency range, to a poor S/N ratio.

(Some possibilities of overcoming these problems are considered further; see Section 3.5.)

As has been proved in earlier experiments it is much easier to apply the laser-induced fluorescence method of detection. Indeed, if we observe, for instance, fluorescence from an excited level in the cycle $a \to b \to c$ (see Fig. 1.2), then the anisotropy of angular momenta J_a distribution will manifest itself in $b \to c$ emission.

1   Classical approach

In order to describe a signal by this method we will first use the classical approach. At the beginning we will ascertain how either probability density $\rho_b(\theta, \varphi)$ or multipole moments ${}_b\rho_Q^K$ of the excited state b, entering into the fluorescence intensity expressions (2.23) or (2.24), are connected to the corresponding magnitudes $\rho_a(\theta, \varphi)$ and ${}_a\rho_q^\kappa$ of the ground state a. The respective kinetic balance equation for probability density and its stationary solution, assuming that the conditions supposed to hold in Eq. (3.4) are in force, is very simple indeed:

$$\dot{\rho}_b(\theta, \varphi) = -\Gamma\rho_b(\theta, \varphi) + \Gamma_p G(\theta, \varphi)\rho_a(\theta, \varphi), \qquad (3.16)$$

yielding the stationary solution

$$\rho_b(\theta, \varphi) = \frac{\Gamma_p}{\Gamma}\rho_a(\theta, \varphi)G(\theta, \varphi). \qquad (3.17)$$

If we substitute Eq. (3.5) for $\rho_a(\theta, \varphi)$ into (3.17) we obtain

$$\rho_b(\theta, \varphi) = \frac{\Gamma_p}{\Gamma}\frac{\rho_a^0 G(\theta, \varphi)}{[1 + \chi G_1(\theta, \varphi)]}. \qquad (3.18)$$

The coefficients $G(\theta, \varphi)$ and $G_1(\theta, \varphi)$ coincide if we do not use a separate probe beam and if the upper level b is excited by the same light beam by which optical polarization of the lower level angular momenta is created. Substitution of (3.18) into (2.23) permits us to connect the intensity and the polarization of radiation with the pumping parameter χ, as was first performed in [124]. We will reach this in a slightly different way, making use of the advantage offered by the technique of multipole expansion. Expanding $\rho_a(\theta, \varphi)$, $\rho_b(\theta, \varphi)$ over spherical harmonics according to (2.16), and expressing $G(\theta, \varphi)$ according to (2.8), we obtain an analogue of (3.17) for multipole moments, namely:

$$\begin{aligned}
{}_b\rho_Q^K &= \frac{\Gamma_p(-1)^\Delta}{\Gamma_K\sqrt{2K+1}}\sum_{X\kappa}\sqrt{(2X+1)(2\kappa+1)}C_{1\Delta1-\Delta}^{X0}C_{X0K0}^{\kappa0} \\
&\times \sum_q C_{XQ-q\kappa q}^{KQ}\Phi_{Q-q}^X(\hat{\mathbf{E}}){}_a\rho_q^\kappa = \frac{\Gamma_p}{\Gamma_K}\sum_{\kappa q}{}_Q^K D_q^\kappa {}_a\rho_q^\kappa, \qquad (3.19)
\end{aligned}$$

where

$$
{}^K_Q D^\kappa_q = (-1)^\Delta \sqrt{\frac{2\kappa+1}{2K+1}} \sum_X \sqrt{2X+1} C^{X0}_{1\Delta1-\Delta} C^{\kappa0}_{X0K0} C^{KQ}_{XQ-q\kappa q} \Phi^X_{Q-q}.
$$

(3.20)

The kinetic equation corresponding to (3.16) is of the form

$$
{}_b\dot\rho^K_Q = -\Gamma_K {}_b\rho^K_Q + \Gamma_p \sum_{\kappa q} {}^K_Q D^\kappa_q {}_a\rho^\kappa_q.
$$

(3.21)

The most important feature of (3.19) follows from the properties of the Clebsch–Gordan coefficient $C^{\kappa0}_{X0K0}$. As can be readily appreciated from the triangle rule (Appendix C), since the rank X of the Dyakonov tensor Φ^X_{Q-q} does not exceed two (which, in turn, follows from Section 2.3 in the context of (2.22)), the connection between the ranks of the polarization moments of the ground and the excited states is described by the inequality $\kappa - 2 \le K \le \kappa + 2$. On the other hand, it follows equally from (2.24) that the intensity of radiation receives a contribution on the part of the polarization moments of the excited state ${}_b\rho^K_Q$ of rank $K \le 2$. It follows from here that only multipole moments of the lower level ${}_a\rho^\kappa_q$ of rank $\kappa \le 4$ can *directly manifest* themselves through fluorescence. For a given set of ground state polarization moments ${}_a\rho^\kappa_q$, $\kappa \le 4$ all coefficients necessary for calculations using (2.24) and (3.19) may be found in Table 2.1 and in Appendix C, Table C.1. In the case of linear or circular polarized excitation we will also use ${}_a\rho^\kappa_0$, $\kappa \le 4$, from Table 3.1, in order to obtain *precise* expressions for fluorescence intensity in the $J \to \infty$ limit, accounting for ground state optical pumping.

Let us consider a concrete example. Let Q-excitation in the geometry of Fig. 2.5 take place by linear polarized light, and let the alignment produced in the ground state be described by a corresponding set of multipoles; see Table 3.1.

Substituting the multiple moments ${}_a\rho^0_0$, ${}_a\rho^2_0$, ${}_a\rho^4_0$ of the lower state into (3.19) (the others, as just shown, entering into the equation for ${}_b\rho^K_Q$ with zero coefficients), we obtain the values of ${}_b\rho^0_0$ and ${}_b\rho^2_0$. The latter, in their turn, make it possible, with the aid of (2.24), to pass over to the expressions for the intensity of $Q \uparrow Q \downarrow$ laser-induced fluorescence, which is registered at right angles to the exciting beam (see Fig. 2.5) and is polarized parallel $({}_Q I_\parallel)$, or at right angles $({}_Q I_\perp)$, to the vector $\hat{\mathbf{E}}$ of the exciting light beam. The analytical expressions obtained for ${}_Q I_\parallel$ and ${}_Q I_\perp$ as functions of the pumping parameter χ, as well as the corresponding expressions for the degree of polarization ${}_Q\mathcal{P}$, are presented in Table 3.5.

An analysis of these formulae shows that the dimensionless normalized magnitude – the degree of polarization $\mathcal{P}$ – depends only on one, also dimensionless, parameter χ. Hence, the ${}_Q\mathcal{P}(\chi)$ dependence (see Fig. 3.8(a))

Table 3.5. The dependence of the intensity, anisotropy of polarization $\mathcal{R}$, degree of polarization $\mathcal{P}$ and degree of circularity $\mathcal{C}$ on the pumping parameter χ at the classical limit $J \to \infty$

Transition	$I_{\|,\perp}$
$Q \uparrow Q \downarrow$	$I_{\|} \sim \Gamma_p \left(\frac{1}{3} - \chi^{-1} + \chi^{-3/2} \arctan \chi^{1/2} \right)$ $I_{\perp} \sim \Gamma_p \frac{1}{2} \left[\frac{2}{3} + \chi^{-1} - \left(\chi^{-3/2} + \chi^{-1/2} \right) \arctan \chi^{1/2} \right]$
$(P,R) \uparrow (P,R) \downarrow$	$I_{\|} \sim \Gamma_p \left[5 - 3a^2 + 3 \left(a^{-1} - 2a + a^3 \right) \text{Arctha} \right]^*$ $I_{\perp} \sim \Gamma_p \frac{1}{2} \left[1 + 3a^2 + 3 \left(a^{-1} - a^3 \right) \text{Arctha} \right]$
$Q \uparrow (P,R) \downarrow$	$I_{\|} \sim \Gamma_p \left[6\chi^{-1} + 4 - 6 \left(\chi^{-3/2} + \chi^{-1/2} \right) \arctan \chi^{1/2} \right]$ $I_{\perp} \sim \Gamma_p \left[-3\chi^{-1} + 4 + 3 \left(\chi^{-3/2} - \chi^{-1/2} \right) \arctan \chi^{1/2} \right]$
$(P,R) \uparrow Q \downarrow$	$I_{\|} \sim \Gamma_p \left[-2 + 3a^2 + 3 \left(a - a^3 \right) \text{Arctha} \right]$ $I_{\perp} \sim \Gamma_p \frac{1}{2} \left[5 - 3a^2 + 3 \left(a^{-1} - 2a + a^3 \right) \text{Arctha} \right]$

Transition	$I_{s,o}$
$Q \uparrow Q \downarrow$	$I_{s,o} \sim \Gamma_p \left[\frac{5}{3} - a^2 + \left(a^{-1} - 2a + a^3 \right) \text{Arctha} \right]$
$P \uparrow P \downarrow$ $R \uparrow R \downarrow$	$I_s \sim \Gamma_p \left(1 - 3\chi^{-1} + 3\chi^{-3/2} \arctan \chi^{1/2} \right)$ $I_o \sim \Gamma_p \left[1 - 3\chi^{-1} - 3 \left(\chi^{-1/2} - \chi^{-3/2} \right) \arctan \chi^{1/2} + 3\chi^{-1} \ln(1+\chi) \right]$
$P \uparrow R \downarrow$ $R \uparrow P \downarrow$	$I_s \sim \Gamma_p \left[1 - 3\chi^{-1} - 3 \left(\chi^{-1/2} - \chi^{-3/2} \right) \arctan \chi^{1/2} + 3\chi^{-1} \ln(1+\chi) \right]$ $I_o \sim \Gamma_p \left(1 - 3\chi^{-1} + 3\chi^{-3/2} \arctan \chi^{1/2} \right)$
$Q \uparrow (P,R) \downarrow$	$I_{s,o} \sim \Gamma_p \left[1 + 3a^2 + 3 \left(a^{-1} - a^3 \right) \text{Arctha} \right]$
$(P,R) \uparrow Q \downarrow$	$I_{s,o} \sim \Gamma_p \left[1 + 6\chi^{-1} - 6\chi^{-3/2} \arctan \chi^{1/2} - 3\chi^{-1} \ln(1+\chi) \right]$

Transition	$\mathcal{R}$
$Q \uparrow Q \downarrow$	$\mathcal{R} = \dfrac{-3\chi^{-1} + \left(3\chi^{-3/2} + \chi^{-1/2} \right) \arctan \chi^{1/2}}{2 - 2\chi^{-1/2} \arctan \chi^{1/2}}$
$(P,R) \uparrow (P,R) \downarrow$	$\mathcal{R} = \dfrac{3 - 3a^2 + \left(3a^3 - 4a + a^{-1} \right) \text{Arctha}}{4 + 4(a^{-1} - a) \text{Arctha}}$
$Q \uparrow (P,R) \downarrow$	$\mathcal{R} = \dfrac{3 - \left(3\chi^{-1/2} + \chi^{1/2} \right) \arctan \chi^{1/2}}{4\chi - 4\chi^{1/2} \arctan \chi^{1/2}}$
$(P,R) \uparrow Q \downarrow$	$\mathcal{R} = \dfrac{-3 + 3a^2 - \left(a^{-1} - 4a + 3a^3 \right) \text{Arctha}}{2 + 2(a^{-1} - a) \text{Arctha}}$

Transition	$\mathcal{P}$
$Q \uparrow Q \downarrow$	$\mathcal{P} = \dfrac{3\chi^{-1/2} \left(3\chi^{-1} + 1 \right) \arctan \chi^{1/2} - 9\chi^{-1}}{3\chi^{-1/2} (\chi^{-1} - 1) \arctan \chi^{1/2} - 3\chi^{-1} + 4}$
$(P,R) \uparrow (P,R) \downarrow$	$\mathcal{P} = \dfrac{9 - 9a^2 + \left(9a^3 - 12a + 3a^{-1} \right) \text{Arctha}}{11 - 3a^2 + (3a^3 - 12a + 9a^{-1}) \text{Arctha}}$
$Q \uparrow (P,R) \downarrow$	$\mathcal{P} = \dfrac{9 - \left(9\chi^{-1/2} + 3\chi^{1/2} \right) \arctan \chi^{1/2}}{3 + 8\chi - \left(3\chi^{-1/2} + 9\chi^{1/2} \right) \arctan \chi^{1/2}}$
$(P,R) \uparrow Q \downarrow$	$\mathcal{P} = \dfrac{-9 + 9a^2 - 3 \left(a^{-1} - 4a + 3a^3 \right) \text{Arctha}}{1 + 3a^2 + 3(a^{-1} - a^3) \text{Arctha}}$

Transition	$\mathcal{C}$
$Q \uparrow Q \downarrow$	0
$P \uparrow P \downarrow$ and $R \uparrow R \downarrow$	$\mathcal{C} = \dfrac{\chi^{-1/2} \arctan \chi^{1/2} - \chi^{-1} \ln(1+\chi)}{\frac{2}{3} - 2\chi^{-1} - \left(\chi^{-1/2} - 2\chi^{-3/2} \right) \arctan \chi^{1/2} + \chi^{-1} \ln(1+\chi)}$
$P \uparrow R \downarrow$ and $R \uparrow P \downarrow$	$\mathcal{C} = \dfrac{\chi^{-1} \ln(1+\chi) - \chi^{-1/2} \arctan \chi^{1/2}}{\frac{2}{3} - 2\chi^{-1} - \left(\chi^{-1/2} - 2\chi^{-3/2} \right) \arctan \chi^{1/2} + \chi^{-1} \ln(1+\chi)}$
$Q \uparrow (P,R) \downarrow$	0

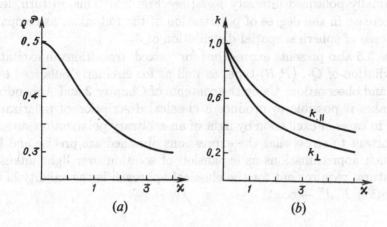

Fig. 3.8. Dependence of the degree of polarization $_Q\mathcal{P}$ (a) and of the non-linearity coefficient of fluorescence $k_{\|,\perp} = {}_QI_{\|,\perp}(\chi)/{}_QI_{\|,\perp}(0)$ (b) on the pumping parameter χ.

may serve as a calibration curve permitting us to pass from the experimentally measured quantity $\mathcal{P}$ to the quantity $\chi = \Gamma_p/\gamma_\Sigma$. Diminution of $_Q\mathcal{P}$ with increase in χ is due to the fact that the intensity $_QI_\|$ undergoes larger non-linear 'saturation' than does $_QI_\perp$; see Fig. 3.8(b). The reason for 'depolarization' due to alignment of the lower state can be convincingly visualized by the shape of the anisotropic spatial distribution $\mathbf{J}_a(\theta,\varphi)$; see Fig. 3.2(a) (bottom). Indeed, the deficit in the number of particles with $\mathbf{J}_a \parallel \hat{\mathbf{E}}$ ('negative' alignment) in the ground state leads to a deficit in particles with $\mathbf{J}_b \parallel \hat{\mathbf{E}}$ in the excited state. But these are just the molecules which produce the main contribution to $_QI_\|$. As a result the angular momenta $\mathbf{J}_b(\theta,\varphi)$ distribution of the upper level in Fig. 3.2(a) (top) is less 'extended' along the z-axis than at weak excitation; see Fig. 2.2(a).

Expressions obtained in a similar way for (P,R) transitions in excitation and radiation are also presented in Table 3.5. The reason for 'depolarization' with increase in χ for P-, R-type transitions may also be clarified by means of the classical model by the following considerations. For the (P,R)-transitions the absorbing dipoles $\hat{\mathbf{d}}$ are orthogonal to $\mathbf{J}$ and rotate around $\mathbf{J}$; see Fig. 1.5(b). Hence, under conditions of positive alignment of the moment $\mathbf{J}_a(\theta,\varphi)$ in Fig. 3.2(b) (bottom) there are fewer dipoles $\hat{\mathbf{d}}$ rotating in the xz and the yz planes and thus the $\hat{\mathbf{d}}$ are chiefly rotating in the xy plane, i.e. at right angles to the $\hat{\mathbf{E}}$-vector of the exciting light. These are the oscillators which give the larger contribution to the intensity $_{PR}I_\perp$ which is polarized in the same xy plane, as compared to the orthogonally polarized intensity $_P RI_\parallel$; see Fig. 2.5. This, in turn, leads to a decrease in the degree of polarization of the radiation, as compared to the case of spherical spatial distribution of $\mathbf{J}_a$.

Table 3.5 also presents expressions for 'mixed' transitions in excitation and radiation of Q-, (P, R)-type, as well as for circularly polarized excitation and observation. Using the contents of Chapter 2 and Appendix A also makes it possible to produce a classical description of polarization signals in cases of excitation by light of an arbitrary polarization state. It is important to stress that the expressions obtained are precise and free from such approximations as expansion of solution over light intensity. The natural requirement for the classical approach is the validity of the assumption $J', J'' \to \infty$.

2 Quantum mechanical description

Expressions which would be applicable for arbitrary angular momentum values can be obtained by using the quantum mechanical density matrix.

For the geometries considered in this chapter (see Figs. 2.5 and 2.6), which are peculiar in the sense that no external fields are applied and the quantization axis z is chosen along the symmetry axis of excitation, only diagonal elements of the excited state f_{MM} and ground state $\varphi_{\mu\mu}$ density matrix appear. The connection between f_{MM} and $\varphi_{\mu\mu}$, in analogy with the classical expression (3.16), may be written in the form

$$\dot{f}_{MM} = -\Gamma f_{MM} + \sum_\mu \langle M|\mathbf{E}^*\mathbf{d}|\mu\rangle\langle M|\mathbf{E}^*\mathbf{d}|\mu\rangle^* \varphi_{\mu\mu}, \tag{3.22}$$

where $\langle M|\mathbf{E}^*\mathbf{d}|\mu\rangle$ is the matrix element of the dipole transition. The density matrix elements f_{MM} may be obtained in explicit form, applying the Wigner–Eckart theorem to (3.22); see Appendix D. The stationary solution is in the form

$$f_{MM} = \frac{\Gamma_p}{\Gamma} \sum_{\mu q} |E^q|^2 \left(C^{J'M}_{J''\mu 1q}\right)^2 \varphi_{\mu\mu}. \tag{3.23}$$

The fluorescence intensity in the transition $b \to c$ with quantum numbers $J' \to J''_1$ and at given polarization $\hat{\mathbf{E}}'$ may be written as follows:

$$I = I_0 \sum_{\mu M} f_{MM} \langle \mu|\mathbf{E}'^*\mathbf{d}|M\rangle\langle \mu|\mathbf{E}'^*\mathbf{d}|M\rangle^*$$

$$= I_0 \sum_{\mu M q} f_{MM} |(E')^{-q}|^2 \left(C^{J'M}_{J''_1\mu 1q}\right)^2. \tag{3.24}$$

It might be useful to follow up once more the analogy between (3.24) and the classical expression (2.23) in which the role of summation over

μ, M and q is played by integrating over angular coordinates with angular coefficients $G'(\theta, \varphi)$ instead of $(C_{J_1''\mu 1q}^{J'M})^2$ characterizing the angular probability of radiation.

Applying (3.11), (3.23) and (3.24), as well as the explicit form of the Clebsch–Gordan coefficient which can be found in Table C.2, Appendix C, the expressions presented in Table 3.6 have been obtained, giving the degree of anisotropy $\mathcal{R}$ (2.26) and of polarization $\mathcal{P}$ (2.25) in the geometry according to Fig. 2.5, as well as the degree of circularity $\mathcal{C}$ (2.36) in the geometry according to Fig. 2.6. Table 3.6 represents the quantum mechanical analogue of Table 3.5 and contains expressions for arbitrary quantum numbers of angular momenta, as dependent on the pumping parameter χ. In the weak excitation limit $\chi = 0$, they coincide with the expressions presented in [148, 402].

Fig. 3.9 shows the $\mathcal{P}(J)$ and $\mathcal{C}(J)$ dependences for the types of transitions discussed. As can be seen for $Q \uparrow Q \downarrow$ transitions the quantum mechanical solutions already at $J \geq 5$ differ little from the $J \to \infty$ limit. At the same time, in the case of $P \uparrow P \downarrow$ or $P \uparrow R \downarrow$ type transitions, even at such high values as $J \cong 50$ the difference from the classical limit still constitutes a value of the order of 1% of that of the degree of polarization or of anisotropy. This may lead to considerable error if one uses 'classical' expressions from Table 3.5 for $\mathcal{P}(\chi)$ dependences of the type presented in Fig. 3.8 for the determination of parameter χ. Table 3.6 also presents the values of $\mathcal{R}(\chi)$, $\mathcal{P}(\chi)$ and $\mathcal{C}(\chi)$ in the limit of very strong optical depopulation for $\chi \to \infty$. It is noteworthy that in this case the corresponding values for P-excitation are identically equal to zero, independently of J, whilst in the other cases they behave either as $1/J$ or as $1/J^2$; see Fig. 3.9.

3.3 Examples of experimental studies

1 Excitation and relaxation parameters

We will now present some examples of experimental results concerning the viability of the model, which has been adopted as a basis of the description given in Sections 3.1 and 3.2.

For concrete estimates of the parameters of a reaction (3.1) let us turn to diatomic molecules, such as Na_2, K_2, Te_2, which have been studied most in experiments on optical pumping of molecules via depopulation. A number of data characterizing the states and transitions in these objects under conditions typical for such experiments are given in Table 3.7. These parameters are, to a certain extent, characteristic of diatomic molecules in thermal vapors of the first, sixth and seventh group of the periodic system of elements, such as alkali diatomics, S_2, Se_2, I_2, etc. These molecules may

Table 3.6. Expressions for anisotropy of polarization $\mathcal{R}(J)$, degree of polarization $\mathcal{P}(J)$ and degree of circularity $\mathcal{C}(J)$ for all types of dipole transitions, J being the quantum number of the initial level. Summation is over M from $-J$ to J.

	J''	J'	J''_1	$\mathcal{R}(J)$ $x=0$	$\mathcal{R}(\infty)$	$\dfrac{x\neq0}{\mathcal{R}(J)}$	$\dfrac{x\to\infty}{\mathcal{R}(J)}$
$Q{\uparrow}Q{\downarrow}$	J	J	J	$\dfrac{(2J-1)(2J+3)}{10J(J+1)}$	$\dfrac{2}{5}$	$\dfrac{\sum_M M^2(3M^2-J^2-J)}{\sum_M \left[1+x\frac{M^2}{J(J+1)}\right]\frac{M^2(J^2+J+1)}{J^2(J+1)^2}}$	$\dfrac{4}{7J}$; (0^*)
$R{\uparrow}R{\downarrow}$	J	$J+1$	J	$\dfrac{(J+2)(2J+5)}{10(J+1)(2J+1)}$	$\dfrac{1}{10}$	$\dfrac{\sum_M \left[1+x\frac{(J+1)^2-M^2}{(J+1)(2J+1)}\right]\frac{(J+1)^2+(J+1)^2-3M^2}{(J+1)(2J+1)}}{\sum_M 1+x\frac{(J+1)^2-M^2}{2(J+1)(2J+1)}}$	$\dfrac{1}{2J+1}$
$P{\uparrow}P{\downarrow}$	J	$J-1$	J	$\dfrac{(J-1)(2J-3)}{10J(2J+1)}$	$\dfrac{1}{10}$	$\dfrac{\sum_M 1+x\frac{(J^2-M^2)(J^2-J-3M^2)}{J^2-M^2}}{\sum_M 1+x\frac{(J^2-M^2)}{(2J+1)2J}}$	0
$R{\uparrow}P{\downarrow}$	J	$J+1$	$J+2$	$\dfrac{1}{10}$	$\dfrac{1}{10}$	$\dfrac{\sum_M 1+x\frac{(J+1)^2-M^2}{(J+1)(2J+1)}\,\frac{(J+1)^2-(J+2)^2-3M^2}{2(J+2)(2J+1)}}{\sum_M 1+x\frac{(J+1)^2-M^2}{2(J+1)(2J+1)}}$	$\dfrac{J+1}{2J^2+9J+10}$
$P{\uparrow}R{\downarrow}$	J	$J-1$	$J-2$	$\dfrac{1}{10}$	$\dfrac{1}{10}$	$\dfrac{\sum_M 1+x\frac{(J^2-M^2)(4J^2-10J+6)}{(J^2-M^2)(J-2J-3M^2)}}{\sum_M 1+x\frac{(J^2-M^2)}{(2J+1)2J}}$	0
$Q{\uparrow}R{\downarrow}$	J	J	$J-1$	$-\dfrac{2J+3}{10J}$	$-\dfrac{1}{5}$	$\dfrac{\sum_M 1+x\frac{M^2(J^2+J-3M^2)}{M^2(2J-1)}}{\sum_M 1+x\frac{M^2}{J(J+1)}}$	$-\dfrac{J+1}{4J(2J-1)}$; (0^*)
$Q{\uparrow}P{\downarrow}$	J	J	$J+1$	$-\dfrac{2J-1}{10(J+1)}$	$-\dfrac{1}{5}$	$\dfrac{\sum_M 1+x\frac{M^2(J^2+J-3M^2)}{M^2(2J^2+5J+3)}}{\sum_M 1+x\frac{M^2}{J(J+1)}}$	$-\dfrac{1}{4(2J+3)}$; (0^*)
$R{\uparrow}Q{\downarrow}$	J	$J+1$	$J+1$	$-\dfrac{2J+5}{10(J+1)}$	$-\dfrac{1}{5}$	$\dfrac{\sum_M 1+x\frac{(J+1)^2-M^2}{[(J+1)^2-M^2]3M^2-(J+1)^2-(J+1)}/[(J+1)(2J+1)]}{\sum_M 1+x\frac{(J+1)^2-M^2}{2(J+1)(2J+1)}}$	$-\dfrac{1}{J+2}$
$P{\uparrow}Q{\downarrow}$	J	$J-1$	$J-1$	$-\dfrac{2J-3}{10J}$	$-\dfrac{1}{5}$	$\dfrac{\sum_M 1+x\frac{(J^2-M^2)(J-1)J}{(J^2-M^2)(3M^2-J^2+J)}}{\sum_M 1+x\frac{(J^2-M^2)}{(2J^2+J)}}$	0

Table 3.6. continued

	J''	J'	J''_1	$\mathcal{P}(J)$ $x=0$	$\mathcal{P}(\infty)$	$\dfrac{x\neq 0}{\mathcal{P}(J)}$	$\dfrac{x\to\infty}{\mathcal{P}(J)}$
$Q\!\uparrow Q\!\downarrow$	J	J	J	$\dfrac{(2J-1)(2J+3)}{8J^2+8J-1}$	$\tfrac{1}{2}$	$\dfrac{\sum_M \dfrac{M^2(3M^2-J^2-J)}{1+\chi M^2/(J(J+1))}}{\sum_M \dfrac{M^2(M^2+J^2+J)}{1+\chi M^2/(J(J+1))}}$	$\dfrac{3}{8J+1}$; $(0^\ast)$
$R\!\uparrow R\!\downarrow$	J	$J+1$	J	$\dfrac{(J+2)(2J+5)}{14J^2+23J+10}$	$\tfrac{1}{7}$	$\dfrac{\sum_M \dfrac{[(J+1)^2-M^2][(J+1)^2+5J-M^2](2J+1)+2)]-3M^2}{1+\chi[(J+1)^2-M^2]/[(J+1)(2J+1)]}}{\sum_M \dfrac{(J^2-M^2)(3J^2-J-3M^2)}{1+\chi[(J+1)^2-M^2]/[(J+1)(2J+1)]}}$	$\dfrac{3}{4J+3}$
$P\!\uparrow P\!\downarrow$	J	$J-1$	J	$\dfrac{(J-1)(2J-3)}{14J^2+5J+1}$	$\tfrac{1}{7}$	0	0
$R\!\uparrow P\!\downarrow$	J	$J+1$	$J+2$	$\tfrac{1}{7}$	$\tfrac{1}{7}$	$\dfrac{\sum_M \dfrac{[(J+1)^2-M^2][(3J^2+13J-M^2)+14]}{1+\chi[(J+1)^2-M^2]/[(J+1)(2J+1)]}}{\sum_M \dfrac{[(J+1)^2-M^2][(J+2)^2-(J+1)^2-3M^2]}{1+\chi[(J+1)^2-M^2]/[(J+1)(2J+1)]}}$	$\dfrac{3(J+1)}{4J^2+19J+21}$
$P\!\uparrow R\!\downarrow$	J	$J-1$	$J-2$	$\tfrac{1}{7}$	$\tfrac{1}{7}$	$\dfrac{\sum_M \dfrac{(J^2-M^2)[(3J-4)(J-1)-M^2]}{1+\chi(J^2-M^2)/(2J^2+J)}}{\sum_M \dfrac{(J^2-M^2)(J^2+J-3M^2)}{1+\chi(J^2-M^2)/(2J^2+J)}}$	0
$Q\!\uparrow R\!\downarrow$	J	J	$J-1$	$-\dfrac{2J+3}{6J-1}$	$-\tfrac{1}{3}$	$\dfrac{\sum_M \dfrac{M^2(3J^2+J-3M^2)}{1+\chi M^2/(J(J+1))}}{\sum_M \dfrac{M^2(3J^2+J-3M^2)}{1+\chi M^2/(J(J+1))}}$	$-\dfrac{3(J+1)}{16J^2-9J-1}$; $(0^\ast)$
$Q\!\uparrow P\!\downarrow$	J	J	$J+1$	$-\dfrac{2J-1}{6J+7}$	$-\tfrac{1}{3}$	$\dfrac{\sum_M \dfrac{M^2(3J^2+7J-M^2+4)}{1+\chi M^2/(J(J+1))}}{\sum_M \dfrac{M^2[3M^2-(J+1)^2]}{1+\chi M^2/(J(J+1))}}$	$-\dfrac{3(J+1)}{16J^2+39J+23}$; $(0^\ast)$
$R\!\uparrow Q\!\downarrow$	J	$J+1$	$J+1$	$-\dfrac{2J+5}{6J+5}$	$-\tfrac{1}{3}$	$\dfrac{\sum_M \dfrac{[(J+1)^2-M^2][3M^2-(J+1)^2-(J+1)]}{1+\chi[(J+1)^2-M^2]/[(J+1)(2J+1)]}}{\sum_M \dfrac{(J^2-M^2)(J^2-J+M^2)}{1+\chi(J^2-M^2)/(2J^2+J)}}$	$-\dfrac{3}{2J+3}$
$P\!\uparrow Q\!\downarrow$	J	$J-1$	$J-1$	$-\dfrac{2J-3}{6J+1}$	$-\tfrac{1}{3}$	$\dfrac{\sum_M \dfrac{(J^2-M^2)(3M^2-J^2-J)}{1+\chi(J^2-M^2)/(2J^2+J)}}{\sum_M \dfrac{(J^2-M^2)(J^2-J+M^2)}{1+\chi(J^2-M^2)/(2J^2+J)}}$	0

Table 3.6. *continued*

	J''	J'	J''_1	$\chi = 0$ $C(J)$	$\chi = 0$ $C(\infty)$	$\chi \neq 0$ $C(J)$	$\chi \to \infty$ $C(J)$
$Q\!\uparrow\! Q\!\downarrow$	J	J	J	$\dfrac{5}{8J^2+8J-1}$	0	$\displaystyle\sum_M \frac{(J^2+J-M^2-M)(M+1)}{1+x(J^2+J-M^2-M)(J^2+J-M^2-2M-1)/(2J(J+1))}$	$-\dfrac{3}{4J^2+6J-1}$
$R\!\uparrow\! R\!\downarrow$	J	$J+1$	J	$\dfrac{5(J+2)(2J+1)}{14J^2+23J+10}$	$\dfrac{5}{7}$	$\displaystyle\sum_M \frac{(J+M+1)(J+M+2)(2J+1)}{1+x(J+M+1)(M+2)[(J+M+1)(J+M+2)/(2(J+1)(2J+1))]/(J+M+1)}$	$\dfrac{3(2J+1)}{4J^2+4J+3}$
$P\!\uparrow\! P\!\downarrow$	J	$J-1$	J	$\dfrac{5(J-1)(2J+1)}{14J^2+5J+1}$	$\dfrac{5}{7}$	$\displaystyle\sum_M \frac{(J-M-1)(J-M)(2J+1)}{1+x(J-M-1)(J-M)(J^2+J+M^2+2M+1)/[2J(2J+1)]}$	0
$R\!\uparrow\! P\!\downarrow$	J	$J+1$	$J+2$	$-\dfrac{5}{7}$	$-\dfrac{5}{7}$	$\displaystyle\sum_M \frac{(J+M+1)(J+M+2)[2(J+1)]}{1+x(J+M+1)(J+M+2)(2J^2-5J+M^2-2M+7)/(2(J+1)(2J+1))}$	$-\dfrac{3(2J+5)}{4J^2+16J+21}$
$P\!\uparrow\! R\!\downarrow$	J	$J-1$	$J-2$	$-\dfrac{5}{7}$	$-\dfrac{5}{7}$	$\displaystyle\sum_M \frac{(J-M-1)(J-M)(2J-3M)}{1+x(J-M-1)(J-M)(J^2+J+M^2-3M)/[2J(2J+1)]}$	0
$Q\!\uparrow\! R\!\downarrow$	J	J	$J-1$	$\dfrac{5}{6J-1}$	0	$\displaystyle\sum_M \frac{(J-M)(J-M+1)(J^2-3J+M^2+2M+3)}{1+x(J-M)(J-M+1)(J-M)/(2J(2J+1))}$	$\dfrac{3}{4J-1}$
$Q\!\uparrow\! P\!\downarrow$	J	J	$J+1$	$-\dfrac{5}{6J+7}$	0	$\displaystyle\sum_M \frac{(J-M)(J+M+1)(J^2-J+M^2+2M+1)}{1+x(J-M)(J+M+1)(M+1)/(2(J+1)(2J+1))}$	$-\dfrac{3(2J+3)}{8J^2+18J+13}$
$R\!\uparrow\! Q\!\downarrow$	J	$J+1$	$J+1$	$\dfrac{5}{6J+5}$	0	$\displaystyle\sum_M \frac{(J+M+1)(J+M+2)(J^2+3J+M^2-2M+3)}{1+x(J+M+1)(J+M+2)/[2(J+1)(2J+1)]}$	$\dfrac{3}{2J^2+8J+3}$
$P\!\uparrow\! Q\!\downarrow$	J	$J-1$	$J-1$	$-\dfrac{5}{6J+1}$	0	$\displaystyle\sum_M \frac{(J-M-1)(J-M)(J^2-J-M^2-2M-1)}{1+x(J-M-1)(J-M)/[2J(2J+1)]}$	0

* In some cases, where, in the column $\chi \to \infty$ the values of $\mathcal{R}(J)$ and $\mathcal{P}(J)$ are different for integer and semiinteger J, the results for semiinteger J are given in parentheses.

Table 3.7. Relaxation characteristics of rotational–vibrational levels of molecules Na_2, K_2, Te_2, as obtained in laser fluorescence experiments with optical polarization of the ground state angular momenta. Concentration N refers to the Na, K atoms and Te_2 molecules as partners of collisions producing cross-section σ in the saturated vapors of the respective elements

Parameters	$^{23}Na_2$		$^{39}K_2$		$^{130}Te_2$	
λ_{exc} nm	488.0		632.8		514.5	
α'', v''_a, J''_a	$X^1\Sigma_g^+, 3, 43$		$X^1\Sigma_g^+, 1, 73$		$X0_g^+, 6, 52$	
α', v'_b, J'_b	$B^1\Pi_u, 6, 43$		$B^1\Pi_u, 8, 73$		$A0_u^+, 11, 53$	
T, K	592		441		630	
N, cm^{-3}	$7 \cdot 10^{14}$	[298]	$0.3 \cdot 10^{14}$	[298]	$0.8 \cdot 10^{14}$	[81]
γ_Σ, 10^6 s^{-1}	0.8	[151]	$0.13 - 0.23$	[11, 178]	0.4	[370]
σ, 10^{-14} cm^2	2	[151]	3.3	[11]	4	[370]
Γ, 10^6 s^{-1}	140	[290]	86.2	[152]	1.49	[154]
Γ_p, 10^6 s^{-1}	$0.11 - 0.25$	[151]	$0.22 - 0.35$	[11, 178]	$0.16 - 0.22$	[370]
$\Gamma_{J'J''}$, 10^6 s^{-1}	11.2	[115]	3.78	[257]	0.037	[189]

be regarded, to a certain extent, as test objects in the laser spectroscopy of dimers. The data given in Table 3.7 are only for general orientation, since every single situation requires a concrete analysis. For quantitative estimates, which would permit us to forecast the ground state optical polarization in absorption, help can be found in monographs, books and reviews, in particular in [116, 202, 236, 256, 258] and sources cited therein.

The total relaxation process is usually characterized, as in (2.42), in approximation of an averaged effective cross-section σ:

$$\gamma_{col} = \sigma \langle v_{rel} \rangle N, \qquad (3.25)$$

where N is the concentration of collision partners X in the reaction (3.1). Typical values of σ for most of the diatomic molecules discussed in collisions with atoms or molecules are of the order of $(10^{-15} - 10^{-13})$ cm^2. Hence, at typical particle concentration in the bulk of the vapor, a fully viable estimate of γ_{col} would lie within the $(\gamma_{col} = 10^4 - 10^7)$ s^{-1} range, as may be seen from the examples given in Tables 3.7 and 3.8.

The details of the relaxation channels contributing towards the total effective cross-section of relaxation of the ground state σ requires measurements with fixed vibrational and rotational quantum numbers v''_i, J''_i and v''_a, J''_a of the reaction (3.1). Data on such measurements, e.g. in Na_2/Na beams in collisions with noble gases can be found in the monograph [116], and those on Li_2-containing vapour can be found in [306].

A description of reaction (3.1) rates with the aid of one constant γ_{col} (3.25) which does not depend on the orientation of the molecular angular momenta means that we neglect elastic collisions causing only a turn

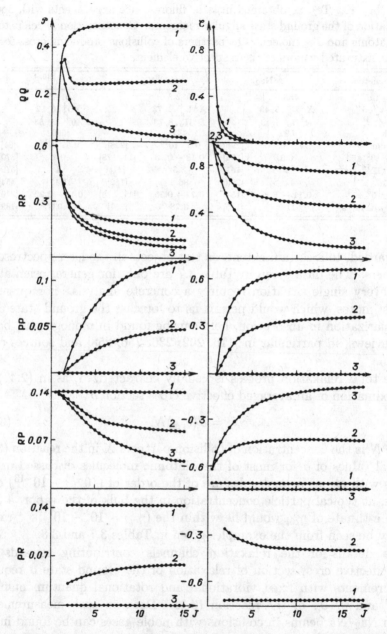

Fig. 3.9. Degree of polarization $\mathcal{P}$ or of circularity $\mathcal{C}$ as dependent on the angular momentum quantum number J for pumping parameter $\chi = 0$ (curve 1), 10 (curve 2), and ∞ (curve 3). The calculation was performed according to formulae from Table 3.6.

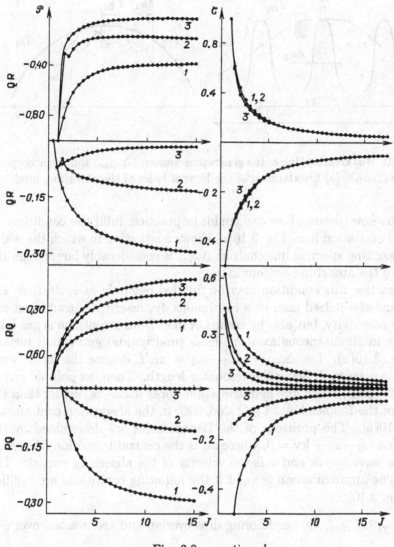

Fig. 3.9. *continued*

of the vector $\mathbf{J}_a(\theta, \varphi)$, without changing its value, or, more precisely, its vibrational and rotational state v_a'', J_a''. Such a process is called *reorientation* or 'pure' depolarization. The treatment, similar to those discussed in Section 2.5, on the excited states permits us to consider the 'pure' depolarization in elastic collisions as being less probable, compared to processes that change the rotational state, at least in the case of the molecules treated in this chapter.

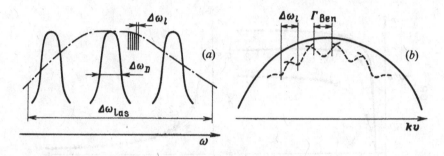

Fig. 3.10. (a) Overlapping of the generation contour ($\Delta\omega_{las}$) and the absorption contour ($\Delta\omega_D$). (b) Overlapping of the Bennet holes of the absorbing level.

Let us now discuss, how one could, in practice, fulfil the conditions of a broad excitation line. Fig. 3.10(a) shows a situation in which the width of the exciting spectral line contour $\Delta\omega_{las}$ is considerably larger than the width of the absorbing contour $\Delta\omega_D$.

In practice this condition may be fulfilled not only in excitation, e.g. by means of a pulsed laser or a continuous dye laser with insufficient frequency selectivity, but also by means of lines from a continuous gas laser working in simultaneous axial mode ω_l (multimode) generation regime; see Fig. 3.10(a). Let $\Delta\omega_l = \omega_{l+1} - \omega_l = \pi c/L$ denote the mode separation in a laser, L being the resonator length. Then, as pointed out in [110, 127, 231], *broad line approximation* works if $\Delta\omega_l$ is smaller than the width of the Bennet holes Γ_{Ben} [268, 320] in the absorption contour; see Fig. 3.10(b). The positions of the Bennet holes are determined by the condition $\omega_0 - \omega_l + \mathbf{k}\mathbf{v} = 0$, where ω_0 is the central transition frequency, $\mathbf{k}$ is the wave vector and $\mathbf{v}$ is the velocity of the absorbing particle. The broad line approximation is valid if the following conditions are fulfilled (see Fig. 3.10):

i $\Delta\omega_l < \Gamma_{Ben}$, i.e. neighboring dips overlap and are 'washed over';

ii many axial modes fall within the Doppler contour, i.e. $\Delta\omega_D \gg \Delta\omega_l$;

iii $\Delta\omega_{las} \gg \Delta\omega_D$, $\Gamma + \gamma_\Sigma$, and the intensity I_l of the modes changes little in $\Delta\omega_D$.

The data from Table 3.7 make it possible to make a few estimates. Let us consider a concrete example. If the length of the resonator is $L = 2$ m, the Doppler contour, say of a K_2 molecule in the $B^1\Pi_u$ state, contains about 11 axial modes, the distance between them being $\Delta\omega_l = \pi c/L \approx 4 \cdot 10^8$ s^{-1}. If we assume that the width of the Bennet dip is equal to the homogeneous line width, $\Gamma_{Ben} \approx \Gamma = 0.86 \cdot 10^8$s^{-1} (see

Table 3.7), i.e., at first glance the broad line approximation works unsatisfactorily. One must, however, keep in mind the instability of the modes due to the phase change under conditions of 'freely running' modes (without synchronization). Broadening of the Bennet dips due to saturation of optical transition also takes place [268]. This makes the broad line approximation more realistic considering that the width of the amplification contour of the laser is $\approx 6 \cdot 10^9$ s^{-1}.

Moreover, in recent years broad band lasers have appeared which lack any frequency modal structure, at the same time retaining such common properties of lasers as directivity and spatial coherence of the light beam at sufficiently high spectral power density. The advantages of such a laser consist of fairly well defined statistical properties and a low noise level. In particular, the authors of [245] report on a tunable modeless direct current laser with a generation contour width of ≈ 12 GHz, and with a spectral power density of ≈ 50 μW/MHz. The constructive interference which produces mode structure in a Fabry–Perot-type resonator is eliminated by phase shift, introduced by an acoustic modulator inserted into the resonator.

Thus, broad line approximation, being sufficiently realistic in a number of cases, permits us to introduce one dynamic constant Γ_p to characterize the absorption process. It is this fact which makes separation of dynamic and angular variables following (2.1) possible. This approximation will always be assumed in the course of our further discussion.

We wish to add that there exists a wide variety of literature that considers the opposite case of monochromatic excitation by an infinitely narrow line causing velocity selection, such as [261, 268, 269, 320, 362] and the sources quoted therein. This description has been developed basically in connection with laser theory; it refers most often to stabilized single-mode excitation. The intermediate case between monochromatic and broad line excitation is the most complex one, requiring integration over the modal structure of the laser inside the bounds of the absorption contour [28, 231, 243].

These examples of experimental studies demonstrate the possibility of estimating the validity of the simple excitation and relaxation model (Section 3.1) for a concrete molecule in definite experimental conditions.

2 Detection of ground state momenta polarization

The possibility of optical alignment by absorption of angular momenta in the ground state of a molecule was first demonstrated by Drullinger and Zare [124] in 1969. The authors used the Q-transition $(X^1\Sigma_g^+, v_a'' = 3, J_a'' = 43) \rightarrow (B^1\Pi_u, v_b' = 6, J_b' = 43)$ in a Na$_2$ molecule in sodium vapor bulk (see Table 3.7) excited by the 488.0 line from an Ar$^+$ laser. Non-

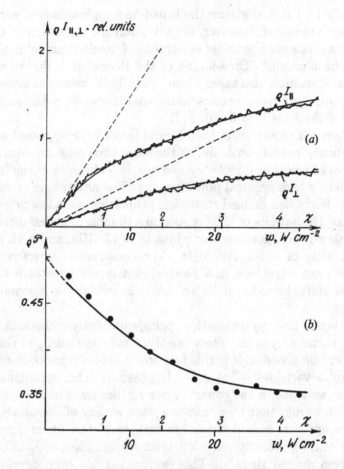

Fig. 3.11. (a), (b) Manifestation of non-linear ground state optical alignment of Na$_2$ in fluorescence intensity $_QI_\parallel$, $_QI_\perp$ (a) and in degree of polarization $_Q\mathcal{P}$ (b); w is the power density of the exciting laser. The dotted lines refer to the linear dependence expected at $\chi \to 0$.

linear dependence of fluorescence intensity was observed together with a decrease in the degree of linear polarization $\mathcal{P}$ (in the traditional geometry of the experiment, as shown in Fig. 2.5, on the power density w of the exciting radiation). The same authors pursued these studies in 1973 [125] in order to determine relaxation parameters. The work was also taken up by the authors of [368]. Fig. 3.11 shows the dependences of intensities $_QI_\parallel$ and $_QI_\perp$, as well as the degree of polarization $_Q\mathcal{P}$ on w, obtained by them under conditions close to those given in Table 3.7. Fig. 3.11(a) shows that the deviation from linearity of the radiation intensity dependence is stronger for $_QI_\parallel$ than for $_QI_\perp$, thus leading to simul-

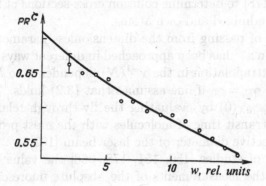

Fig. 3.12. Measured [369] values of degree of circularity C as dependent on the excitation power density w.

taneous decrease in the degree of polarization $_QP(\chi)$, in accordance with the calculation formulae from Table 3.5; see Fig. 3.11(b), solid curve. Actually, the experimental dependence shown in Fig. 3.11 is in accordance with the predictions presented in Fig. 3.8. This is the manifestation of optical polarization of angular momenta (in this case — *alignment*) of the level $X^1\Sigma_g^+, v_a'' = 3, J_a'' = 43$ in Na$_2$.

For the case of a (P, R)-transition a similar 'depolarization' of radiation with increase in χ was first registered by the authors of [386] on iodine molecules. In a later paper [369] optical *orientation* in ground state Na$_2$ molecules in transition $(X^1\Sigma_g^+, v_a'' = 2, J_a'' = 45 \rightarrow A^1\Sigma_u^+, v_b' = 14, J_b' = 44$, at 632.8 nm excitation) was reported, from decrease in circularity $_{PR}C(\chi)$ at circularly polarized excitation, by the geometry of Fig. 2.6 (P-transition). The $C(w)$ dependence obtained is presented in Fig. 3.12, where the solid curve is calculated using the corresponding formulae from Table 3.6.

Results obtained with these and other molecules, in particular $K_2(X^1\Sigma_g^+)$ [178], confirmed that the model as employed under the above conditions may be used for a description of ground state optical polarization. Hence, the parameter $\chi = \Gamma_p/\gamma_\Sigma$ thus obtained may be used to determine the relaxation constant γ_Σ. The first such application of ground state optical alignment consisted of determining relaxation rates γ_Σ and total effective relaxation cross-sections σ using (3.2) and (3.25) in the experiment by Drullinger and Zare [125]. They determined the dependence of $\chi^{-1} \propto \gamma_\Sigma$ on a concentration N of noble gas atoms introduced in the cell, as well as on that of H_2, N_2, CH_4. Similar measurements were carried out by Clark and McCaffery [102] with $I_2(X^1\Sigma_g^+, v_a'' = 0, J_a'' =13$ and 15) molecules. The linearity region of $\chi^{-1}(N)$ was used by the authors of

[102, 125, 151, 178] to determine collision cross-sections of Na_2, K_2 and I_2 molecules with admixed and own atoms.

The problem of passing from the dimensionless parameter $\chi = \Gamma_p/\gamma_\Sigma$ to values of γ_Σ in s^{-1} has been approached in different ways. The simplest idea is to use extrapolation in the $\chi^{-1}(N)$ dependence to $N = 0$, leading to $\gamma_{col} = 0$ and $\gamma_\Sigma = \gamma_0$ if one assumes that (3.2) holds. This allows us to obtain $\Gamma_p = \gamma_0\chi(0)$ by evaluating the fly-through relaxation rate γ_0 as a reciprocal transit time of molecules with the most probable velocity through the effective diameter of the laser beam [102].

In a number of studies [150, 151, 178, 368] the value of Γ_p was determined from the measurements of the absolute fluorescence intensity. Indeed, in the linear limit of weak excitation $\chi \to 0$, we can obtain from (2.11), (2.23) and (3.18) the intensity for the $Q \uparrow Q \downarrow$ transition $J_a'' \to J_b' \to J_c''$:

$$Q I_\parallel^0 = \frac{h\nu n_a^0 \Gamma_p \Gamma_{J_b' J_c''}}{4\pi\Gamma} \int_0^{2\pi} \int_0^\pi \cos^4\theta \sin\theta \, d\theta \, d\varphi. \tag{3.26}$$

Hence, knowledge of the concentration of molecules n_a^0 at level a, the rate $\Gamma_{J_b' J_c''}$ of the spontaneous transition $b \to c$, and the upper state b relaxation rate Γ allows us to determine Γ_p from the absolute intensity measurements.

All three methods have their drawbacks. Extrapolation to fly-through conditions in (3.2) approximation requires a homogeneous profile of the exciting beam, as will be discussed in Section 3.6. The method of absolute intensity measurements introduces considerable error, partly connected with the spatial profile of the beam. Registration of kinetics is more feasible in the case of direct measurements of γ_Σ, as will be discussed further in greater detail (Section 3.6). Nevertheless, the values obtained by the authors of [102, 125, 151, 178] for the effective relaxation cross-sections, being of the order of magnitude of $(0.3–5)\cdot10^{-14}$ cm^2 permit us to estimate the total effect of the relaxation of ground state angular momenta polarization in collisionally induced ground state vibrational–rotational transitions.

The previously discussed papers [102, 124, 125, 151, 178, 368, 369, 386, 401] form a series of, as one might put it, early, 'demonstration' works performed in the 1970s by similar methods aimed at the detection of the optical polarization effects of the angular momenta in the ground state of diatomic molecules in a bulk via absorption of light. The method consisted of observing the changes in intensity and polarization produced by the same pumping light, as follows from Tables 3.5 and 3.6 and Figs. 3.8, 3.9, 3.11 and 3.12. Another aim consisted of the determination of relaxation parameters. The further development of the method took several directions, consisting, firstly, of direct studies of the relaxation kinetics of polarization in the ground state, as will be considered later (Section 3.6). Secondly, it deals with studies of the precession of light-produced distribution of momenta $\mathbf{J}_a$ in an external magnetic field, i.e. with the non-linear Hanle effect in ground state and related interference phenomena, as will be discussed in Sections 4.3 and 4.4. Finally, it implies the use of other methods of creation and registration of angular momenta optical polarization in gas phase molecules (Sections 3.4, 3.5).

To detail the relaxation channels, a 'two-laser' method has been applied which includes a pumping beam and a probe beam with a different wavelength. Ottinger and Schröder [306] used an intensive beam from an Ar^+ laser (476.5 nm) for optical depopulation of $Li_2(X^1\Sigma_g^+, v_a'' = 1, J_a'' = 24)$, registering at the same time the fluorescence excited by a dye laser 598–611 nm from the levels $J_i'' = J_a'' + \Delta J''$, $\Delta J'' = 2, 4, \ldots, 14$; see Fig. 3.13. The relative intensity changes $(I^{(0)} - I)/I^{(0)}$ of this fluorescence were determined; I, or $I^{(0)}$ correspond either to the presence or absence of the pumping beam which is being interrupted at 1 kHz frequency. By numerical solution of balance equations the constants γ_i and the cross-sections σ_i of transfer processes between rotational levels with fixed J_a'' and different $\Delta J''$ in collisions with He, Ne, Ar, Kr and Xe atoms were found, equal to $(5–8) \cdot 10^{-16}$ cm^2 for $\Delta J'' = 2$. Optical depopulation was used as a 'label' of a certain rovibronic level of the electronic ground state. This approach is of a rather general nature, underlying such methods as signals of *laser-induced fluorescence* with *modulated population* [282, 335], as well as *laser polarization spectroscopy* [373, 396]. It might be noted that in [306] there is only talk of the effects of change in population, although multipoles of non-zero rank are also generated.

Principally, new results for studying collisional transitions between rovibronic ground state levels of dimers are obtained by the application of two-laser modulation spectroscopy to *crossed atom–molecule beams*, with depopulation of the lower level. In the first place, this raises the efficiency of depopulation of the lower level considerably, since relaxation is basically determined only by fly-through time. Secondly, a combination of optical depopulation of the v_a'', J_a'' level with the method of crossed beams yields the most detailed characteristics of the elementary act of the collisional transition $v_a'', J_a'' \rightarrow v_i'', J_i''$ in the form of relative differential cross-sections $\sigma_{ai}(\theta)$ with fixed initial and final rovibronic numbers and scattering angle θ. This problem has been solved in the above-mentioned extensive series of papers by Bergmann and coworkers [61, 62, 63], as well as by Serri *et al.* [345], who measured differential cross-sections in crossed Na/Na$_2$ jet and noble gas beams. Peaks have been observed in the differential cross-section for definite θ values, the angles θ_{max} being different for different $\Delta J'' = J_i'' - J_a''$. This phenomenon has been named, in analogy with the optical one, the 'halo' effect [345], or the 'rotational rainbow' effect

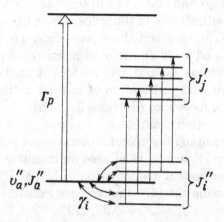

Fig. 3.13. Application of a pumping beam and a tunable probe beam for detailing relaxation channels in the electronic ground state.

in other papers [62, 339]. It is important to mention that the measured cross-sections allow the determination of interaction potential.

In 1986 Bergmann and coworkers [191], using a technique previously evolved by them [287], succeeded in achieving practically complete selection of molecules in a state with fixed projection $|M''| = J_a''$ in the Na_2 electronic ground state, using supersonic beams containing Na_2. This repeats, in a certain sense, the success achieved with atoms (see Fig. 1.1). However, in the case of molecules, owing to the large number of M-sublevels, and owing to less favourable transition rates, as well as to the extremely serious demands on power and spectral properties of the laser source, the situation presents a much more serious problem. The authors of [191] employed a technique of 'total P-depopulation', when, according to Table 3.6, the state with $|M''| = J_a''$ does not absorb light polarized along $\hat{\mathbf{E}}$. Hence, the molecules are aligned, in the $\chi \gg 1$ limit, in a state where $M'' = \pm J_a''$ with respect to the quantization axis, which is parallel to the $\hat{\mathbf{E}}$-vector of exciting light. The method of detection with the aid of an intensive depopulating second beam permitted demonstration of the effect on levels with rotational quantum number up to $J'' = 28$. An effect of hyperfine structure was observed in the case of small J'' values, as could have been expected. The above technique of $|M''|$-selection of Na_2 molecules with fixed v_a'', J_a'' has been applied in [191, 287] to study rotationally inelastic collisions in the Na_2–Ne system. A strong tendency towards following the $|\Delta M''| \ll J''$ rule in the collisional process was observed, which conforms with data for excited states [213, 215] (Section 2.5), and does not contradict the $\Delta M'' = 0$ rule.

3.4 Fluorescent population of vibronic levels

As may be seen from Fig. 1.1, optical polarization of the angular momenta of atoms may be created through population in spontaneous transitions from the upper state. Until now in the present chapter we have discussed how polarization of the angular momenta is created on the lower level at absorption of light, when reverse spontaneous transitions are negligible. However, *the spontaneous transitions* can produce polarization of angular momenta in molecules as well. This takes place predominantly in the case, where, owing to spontaneous transitions, molecules find themselves on 'empty' (thermally unpopulated) rovibronic levels. In such a situation the angular momenta of sufficiently high-positioned rovibronic levels of the electronic ground state may become polarized, e.g., in the cycle $a \rightarrow b \rightarrow c$ (see Fig. 1.2), as well as those of levels belonging to some other, intermediate, electronic state. Both cases are of interest. Thus, for instance, in [293] thermally non-populated rovibronic levels of H_2 and D_2 molecules were effectively populated through fluorescence with the aim of studying rotational relaxation. Special interest consists of populating levels near the dissociation limit in this manner. One may expect peculiarities in relaxation constants here as well as in magnetic properties.

Let us now consider the following experimental scheme (see Fig. 3.14). Let the level J_1'' be populated by fluorescence in the cycle $J'' \rightarrow J' \rightarrow J_1''$, that is, in a three-state cycle of so-called Λ-configurations, where state J' has greater energy than J'' and J_1''. To probe the polarization of the angular momenta created on level J_1'' laser-induced fluorescence in the second cycle $J_1'' \rightarrow J_1' \rightarrow J_2''$ may be used (see the broken lines in Fig. 3.14).

In this scheme the distribution of the angular momenta of the molecules on level J_1'' is characterized not only by zero rank polarization moment $_{J_1''}\rho_0^0$, i.e., by population, but also by polarization moments of higher rank. In other words, we may obtain orientation or alignment on level J_1''. We will start discussing this kind of populating via fluorescence by elucidating what kind of polarization moments may arise on level J_1'' at given polarization $\hat{\mathbf{E}}$ of the exciting light, and what their magnitudes are.

When molecules arrive at the state with rotation quantum number J_1'' from the state J' in the process of spontaneous radiation at rate $\Gamma_{J'J_1''}$ (see Fig. 3.14), a photon possessing unit spin is emitted in an arbitrary direction. Let us assume that the angular momentum carried away by the emitted photon is small, as compared with both $\mathbf{J}_1''$ and $\mathbf{J}'$. This means that the angular momentum vector of each separate molecule does not change its value and does not turn in space as a result of the spontaneous transition. Consequently, the angular momenta distribution $\rho_{J'}(\theta, \varphi)$ is

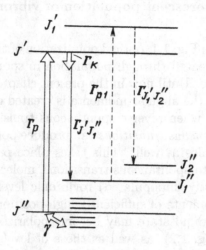

Fig. 3.14. Scheme for creating and registering optical polarization by 'population'.

transferred from the upper state J' to the lower one J_1'' without change in its shape. Thus polarization moments $_{J_1''}\rho_q^\kappa$ are created on level J_1'' through spontaneous transition. It is important that $_{J_1''}\rho_q^\kappa$ coincide with excited state multipole moments $_{J'}\rho_Q^K$ with an accuracy up to the factor that does not depend on the rank κ. The mechanism of the appearance of polarization moments $_{J'}\rho_Q^K$ in the process of light absorption and the methods of their calculation have already been discussed in Chapter 2. After their appearance at level J_1'' the polarization moments $_{J_1''}\rho_q^\kappa$ are, naturally, further subjected to relaxation processes taking place at the rate $_{J_1''}\gamma_\kappa$. We thus obtain the following form of the equation of motion:

$$_{J_1''}\dot\rho_q^\kappa = \Gamma_{J'J_1''}\,_{J'}\rho_Q^K\delta_{K\kappa}\delta_{Qq} - \,_{J_1''}\gamma_\kappa\,_{J_1''}\rho_q^\kappa. \qquad (3.27)$$

Here the first term in the righthand side determines the rate of creation of polarization moments in the spontaneous transition process, whilst the second term describes their relaxation. In writing Eq. (3.27) it is assumed that collisions do not lead to population of the state J_1'', since it lies sufficiently high and is surrounded by non-populated rovibronic levels within the range of thermal energy kT.

Let us find the magnitude of orientation and alignment of level J_1'' at its fluorescent population $J' \rightarrow J_1''$, when J' is excited by the action of weak light causing a transition $J'' \rightarrow J'$ (see Fig. 3.14). The magnitude of moments $_{J'}\rho_Q^K$ created on level J' may be found from (2.21), hence we know the rate of the generation of moments $_{J_1''}\rho_q^\kappa$ on level J_1''. The stationary solution of Eq. (3.27) has the following form in this case:

$$J_1'' \rho_q^\kappa = (-1)^{\Delta + \kappa} \left(\frac{\Gamma_p \Gamma_{J'J_1''}}{\sqrt{2\kappa + 1} \Gamma_{\kappa J_1''} \gamma_\kappa} \right) C_{1\Delta 1 - \Delta}^{\kappa 0} \Phi_q^\kappa(\hat{\mathbf{E}}). \qquad (3.28)$$

Eq. (3.28) makes it possible to obtain the value of all desired polarization moments created on level J_1'' by substituting numerical values of Clebsch–Gordan coefficients from Table C.1, Appendix C, and the values of the Dyakonov tensor for the chosen polarization of light from Table 2.1. In particular, if excitation is linearly polarized with the $\hat{\mathbf{E}}$-vector along the z-axis, the degree of alignment at Q-type absorption is $_{J1''}\rho_0^2 / _{J_1''}\rho_0^0 = 2/5$, and at (P, R)-type absorption it is $_{J1''}\rho_0^2 / _{J''}\rho_0^0 = -1/5$. The degree of orientation and alignment at excitation with circularly polarized light propagating along the z-axis is, respectively, $_{J_1''}\rho_0^1 / _{J_1''}\rho_0^0 = \pm 1/2$ and $_{J_1''}\rho_0^2 / _{J_1''}\rho_0^0 = 1/10$ at (P, R)-type absorption. Here the plus sign corresponds to absorption of righthand polarized light (r) in P-transition, or to lefthand polarized light (l) in R-transition, whilst a minus sign corresponds to combinations (l, P) or (r, R). In the case of absorption of circularly polarized light in Q-transitions no orientation is created, but the degree of alignment is $_{J''}\rho_0^2 / _{J_1''}\rho_0^0 = -1/5$. All the numerical values have been obtained under the assumption that $\Gamma_K = \Gamma$, $_{J_1''}\gamma_\kappa = _{J_1''}\gamma$, i.e. that the relaxation rate of the corresponding polarization moment does not depend on its rank. In the opposite case the respective values mentioned above have to be multiplied by the ratio $(\Gamma_0 {}_{J''}\gamma_0)/(\Gamma_\kappa {}_{J_1''}\rho_\kappa)$. We have thus shown that considerable polarization of the angular momenta of molecules emerges on the J_1'' level. It coincides in value with the polarization of the upper level J' excited by weak light.

If we compare these values with the corresponding ones, obtained in the optical polarization of angular momenta of the lower level by depopulation (see Figs. 3.3, 3.4 and 3.5), then we will see that similar values can be obtained only in the case of a large non-linearity parameter, $\chi \cong 10$ and more.

We will now consider the possibility of setting up an experiment to study the relaxation characteristics of level J_1'' by applying fluorescent population (see Fig. 3.14).

One must first touch upon the question of the attainable effectivity of populating level J_1'' through fluorescence. In order to obtain an estimate of the part of the equilibrium population $n_{J''}$ of level J'' which can be pumped over to level J_1'' using weak excitation approximation, we may apply Eq. (3.28), from which we obtain:

$$\frac{n_{J_1''}}{n_{J''}} = \frac{\Gamma_p \Gamma_{J'J_1''}}{3 {}_{J_1''}\gamma_0 \Gamma_0}. \qquad (3.29)$$

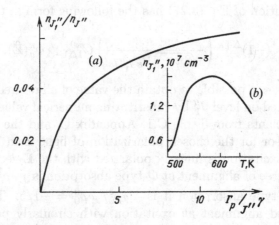

Fig. 3.15. (a) Dependence of the population of level J_1'' with respect to the population of level J'' on the parameter Γ_p/γ. (b) Calculation of $n_{J_1''}$ for $K_2(X^1\Sigma_g^+, 16, 73)$ in thermal potassium vapor.

In deriving (3.29) we make use of the fact that $n_i \propto {}_i\rho_0^0$. One of the important parameters affecting the magnitude of $n_{J_1''}$ is the rate of decay $\Gamma_{J'J_1''}$ which is proportional to the Franck–Condon factor for the radiative transition $J' \to J_1''$. As a matter of fact, this is where the term 'Franck–Condon pumping' comes from (the term is used occasionally for denoting fluorescent population processes).

In order to increase the population of level J_1'', it is necessary, according to (3.29), to increase the rate of light absorption Γ_p; in other words, one must intensify the exciting light beam. This leads, however, to saturation of absorption and to optical polarization of angular momenta on the initial level J''. This means, in turn, that Eq. (3.29) is no longer valid. An example of the numerical calculation of $n_{J_1''}/n_{J''}$ has been given in [30] using parameters characteristic of fluorescent population for a K_2 molecule through the cycle $X^1\Sigma_g^+(1,73) \to B^1\Pi_u(8,73) \to X^1\Sigma_g^+(16,73)$ and taking account of the optical depopulation of the initial level. The results are presented in Fig. 3.15(a), which show that an increase in Γ_p is effective only up to $\Gamma_p/_{J''}\gamma_0 \sim 10$. After that with further increase in Γ_p the value of $n_{J_1''}/n_{J''}$ grows only insignificantly. On the other hand, in order to raise the $n_{J_1''}$ value, it seems advantageous to increase the concentration of $n_{J''}$ on the initial level J'' by raising the temperature of the gas cell. At the same time, however, we also obtain an increase in the collisional relaxation rate of all levels involved in the cycle $J'' \to J' \to J_1''$ (see Fig. 3.14). In the case of alkali dimers, in vapor at thermal equilibrium these rates depend on the concentration of own atoms of the alkali metal which also grows rapidly with increase in temperature [298]. As a result the temperature dependence of $n_{J_1''}$ is of the shape shown in Fig. 3.15(b), in the case of the above transition for K_2. One may see that the $n_{J_1''}(T)$ curve shows a sufficiently well-discernable peak. It follows that it is better to use molecular beams where collisions are less effective for achieving high $n_{J''}$ values.

Apart from the above discussed method of fluorescent population, which has been shown to be effective in practice in the cases of Na_2 and Li_2 ([282] and references therein) it is also possible to realize other no less, but sometimes even more, effective ways of populating highly situated rovibronic molecular levels. For heteronuclear molecules possessing a non-zero permanent dipole moment and hence permitting optical transitions within one electronic state [194], such a population can be effected by direct optical excitation in the IR region. In particular, in this fashion rotational relaxation was studied in HF [196], CO [82], NO [367] and other molecules, applying double IR–IR and IR–UV resonance methods. In the case of dimers, population may be achieved, employing stimulated Raman scattering as was done on H_2 [293] and Na_2 [55, 163] molecular beams.

If $J'' \rightarrow J'$ excitation is accompanied or followed by deexcitation $J' \rightarrow J_1''$ in a 'stimulated emission process' (SEP), then the population efficiency of the J_1'' level can be increased considerably. It is now known [248, 347] that the process might be made more effective by applying the Λ-configuration scheme in which the first-step ($J'' \rightarrow J'$) excitation pulse is applied *after* the second-step ($J' \rightarrow J_1''$) pulse which, at first glance, seems surprising. This process is called 'stimulated Raman scattering by delayed pulses' (STIRAP). The population transfer here takes place coherently and includes coordination of the Rabi nutation phase in both transitions.

In the process of applying the STIRAP method [248], one also creates polarization of angular momenta, both on the J'' and on the J_1'' levels. The mechanism of its formation can easily be explained by analyzing the transition probability between different magnetic sublevels M_J [52]. Let us, for instance, consider a situation where both light waves are linearly polarized along the z-axis and both waves produce P-type transitions $J'' \rightarrow (J' = J'' - 1) \rightarrow (J_1'' = J'')$. Let the rotational levels J'' and J_1'' be different, e.g. belonging to different vibrational states. The probability of a transition between different magnetic sublevels produced by the light wave is proportional to $\left(C_{J''M10}^{J'M} \right)^2 \propto (J'' + M_{J''})(J'' - M_{J''})$ for the first transition, and to $\left(C_{J_1''M10}^{J'M} \right)^2 \propto (J_1'' + M_{J_1''})(J_1'' - M_{J_1''})$ for the second one; see Table C.2. As a result we obtain, for the example under consideration, the following picture presented in Fig. 3.16. As can

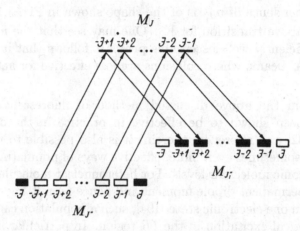

Fig. 3.16. Population of M_J-sublevels by the STIRAP method in the cycle $J'' \to (J' = J'' - 1) \to (J_1'' = J'')$.

be seen, in the interaction with the first laser the sublevels $M_{J''} = J''$ and $M_{J''} = -J''$ remain unaffected. If absorption is sufficiently intensive, as it is in the STIRAP method [248], all magnetic sublevels in state J'', except $M_{J''} = \pm J''$, are completely emptied. Longitudinal alignment $_{J''}\rho_0^2$ is therefore created, the magnitude of which can be calculated in quantum mechanical formalism in a similar way to (3.13), and equals

$$_{J''}\rho_0^2 = \sum_M C_{J''M20}^{J''M} n_{M_{J''}} = 2n_{J''}\sqrt{\frac{J''(2J''-1)}{(J''+1)(2J''+3)}}. \qquad (3.30)$$

In a similar way the whole population is transferred from the magnetic sublevels of state J' to state J_1'' according to the scheme in Fig. 3.16, under conditions of intensive stimulated radiational transitions. As a result, all magnetic sublevels $M_{J_1''}$, except $M_{J_1''} = \pm J_1''$, will be equally populated, whilst the sublevels $M_{J_1''} = \pm J_1''$ will be completely 'empty'. In this situation we have a picture which is directly opposite to that of the J'' level, where only the sublevels $M_{J''} = \pm J''$ are populated. The alignment created on level J_1'' may be calculated as $_{J_1''}\rho_0^2 = \sum_M C_{J_1''M20}^{J_1''M} n_{M_{J''}}$.

It ought to be pointed out that the methods of transfer of population and also of polarization between the states are being developed further. One might mention, in particular, the cw 'all-optical triple resonance' (AOTR) method evolved by Stwalley's group [282] and tried out on Na_2. The method consists of the use of three tunable single-frequency continuous wave lasers allowing flexible movement of both population and polarization from the thermally populated level to any other level of the ground state.

3.5 Laser-interrogated dichroic spectroscopy

So far we have been judging orientation and alignment of angular momenta both in the excited and in the ground state from laser-induced fluorescence. If we aim at ground state studies such an approach must be considered as a not quite direct one, because at first absorption of light transfers the molecule into an excited state, in which it may be subjected to all sorts of influences (collisions, various external field effects, etc.), and only after all this do we obtain emission from which we make judgements about the ground state of the molecule.

Clearly, if, instead of fluorescence, we register absorption of a probe beam of different polarization, we may obtain direct information on the distribution of the ground state angular momenta of the molecule. The dependence of the absorption coefficient on the angular momenta distribution of the absorbing level, and on the polarization of the probe beam, can be found from (3.19) in the following way. As can be easily understood, the population of the upper level $_b\rho_0^0$ of the absorbing transition is directly proportional to the coefficient $\alpha(\hat{\mathbf{E}})$ of absorption of light of polarization $\hat{\mathbf{E}}$. Now applying the properties of the Clebsch–Gordan coefficients, as given in Appendix C, from (3.19), we obtain

$$\alpha(\hat{\mathbf{E}}) = A \sum_{\kappa} (-1)^{\Delta+\kappa} \sqrt{2\kappa+1} C_{1\Delta1-\Delta}^{\kappa 0} \sum_{q} \Phi_q^{\kappa}(\hat{\mathbf{E}})_a \rho_{-q}^{\kappa}, \qquad (3.31)$$

where A is the proportionality coefficient.

For Na atoms the dependence of the absorption coefficient of the probe light on the multipole moments $_a\rho_q^{\kappa}$ of the 3^2S_F state produced by the pumping light was registered long ago [126]. Measurement of light absorption from a given molecular rovibronic level is a difficult task. It demands considerable concentrations of molecules and narrow laser lines. Hence, the method of absorption is much inferior to the fluorescence one, as regards sensitivity, at least in the visible part of the spectrum. The main difficulties here are connected to the necessity of registering small changes in the light flux against the background of a strong signal passing through. These difficulties can, however, be overcome if one uses a method called *laser-interrogated dichroic spectroscopy* [48]. The method permits measurement of the polarization of angular momenta of ground (initial) molecular states and is based on using a probe beam passing through crossed polarizers with the gas cell in between. Thus this method is associated with *laser polarization spectroscopy* which has gained wide popularity in investigations of atoms, as well as of molecules [116, 336]. The methods of polarization spectroscopy make determination of various rank relaxation rate constants possible for atomic levels [8, 346].

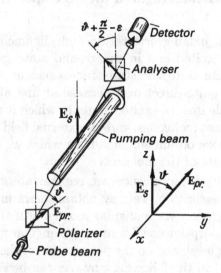

Fig. 3.17.   Scheme of experiment on laser-interrogated dichroism.

The authors of [297] have presented the most consistent classical approach for the description of the polarization of a molecular gas by a strong light wave. The basic and principal difference between the methods of polarization spectroscopy and laser probing of dichroism consists of the fact that polarization spectroscopy demands the application of monochromatic single-frequency radiation, whilst probing of dichroism is based on the employment of a broad spectral irradiation.

Let us discuss in greater detail the problem as to what information on the initial rovibronic level of a molecule can be obtained by the methods of laser-interrogated dichroic spectroscopy. Let us consider the following situation. Let there exist a strong laser field, the $\hat{\mathbf{E}}_\mathbf{S}$ vector of which is directed along the z-axis (see Fig. 3.17). This field creates an anisotropic distribution of angular momenta in the ground state, which is described by the polarization moments $_a\rho_0^\kappa$ entering into (3.31). The anisotropy thus created is tested by a weak test wave which is also polarized. The orientation of its polarization vector $\hat{\mathbf{E}}_\mathbf{pr}$ is shown in Fig. 3.17. The pump and probe beams propagate at a small angle with respect to each other in such a way as to provide only for the probe beam to hit the analyzer and then the detector. The cell with atoms or molecules is placed between the crossed (at angle $\varepsilon = 0$, see Fig. 3.17) polarizer and analyzer. Unlike polarization spectroscopy, in the case of probing of anisotropy both beams may propagate in opposite directions, as well as in the same direction. Thus, for instance, in studying Yb atoms by the laser-interrogated dichroism method in [260] both beams were propagating in the same direction.

At first, following the experimental scheme according to Fig. 3.17, we will give a qualitative assessment of the expected signal. The strong pump beam 'eats a hole' in the angular distribution of molecules in the ground state. At Q-transition the shape of the 'eaten out' distribution is shown in Fig. 3.2(a) (bottom). Probing the created distribution by means of the probe beam, we obtain different absorption probabilities for the $\hat{E}_{prz}$ and the $\hat{E}_{pry}$ components of the polarization vector $\hat{\mathbf{E}}_{\mathbf{pr}}$. As a result the detector receives a signal depending on the anisotropy produced by the pumping beam. If the pumping beam is switched off, the probe beam only becomes weakened, passing through the medium, but no turn of its $\hat{\mathbf{E}}_{\mathbf{pr}}$ vector takes place. Accordingly, the probe beam does not pass through the analyzer, and no signal is registered by the detector.

For quantitative analysis of the expected signals it is convenient to apply the methods of matrix optics [167]. Then, as is shown in [31], the intensity I of the signal, which has passed through the analyzer in the process of probing by light with $\hat{\mathbf{E}}_{\mathbf{pr}}$, may be found:

$$
I = I_0 \left[\frac{\varepsilon^2}{4} + \frac{l^2}{16}(\alpha_z - \alpha_y)^2 \sin^2 2\theta \right.
$$
$$
\left. + \frac{\pi^2 l^2}{\lambda^2}(n_z - n_y)^2 \sin^2 2\theta + \frac{\varepsilon l}{4}(\alpha_z - \alpha_y) \sin 2\theta \right], \qquad (3.32)
$$

where I_0 is the initial intensity of the probe beam, l is the path length in the optically pumped gas, λ is the wavelength of the probe light and ε is the small 'discrossing' angle between the $\hat{\mathbf{E}}_{\mathbf{pr}}$ vector (before interacting with the medium) and the analyzer. The quantities $\alpha_{z,y}$ and $n_{z,y}$ are the absorption and refraction coefficients for light which is linearly polarized along the z- and y-axes. For a rarefied molecular gas the values of $\alpha_i l$, and $[\pi l/(2\lambda)]n_i$ are of the same order of magnitude, namely $\approx 10^{-3}$. Thus, by choosing the value of discrossing angle ε adequately one may achieve such a situation, where the decisive role in the signal (3.32) would belong to the last term, which is proportional to $(\alpha_z - \alpha_y)$. Substituting corresponding α values in (3.31) one can easily connect the observed absorption signal with the polarization moments of the ground level $_a\rho_q^\kappa$. An example will be presented in the next section. A similar expression can be obtained for the light intensity I that has passed through the crossed analyzer and polarizer with gas between them when the polarization of the probe beam is arbitrary.

3.6 Transient process

1 Dynamic method

In Section 3.3 we mentioned the difficulties of determining the absolute values of rate constants and cross-sections of collisional relaxation of optically aligned molecules in the ground state by means of *stationary optical excitation*. It is therefore important to develop *dynamic methods* of relaxation time measurements in *non-stationary processes*. Methods of optical pumping offer various ways of determining the relaxation times of oriented systems by modulation of the pumping light intensity [65, 316, 356]. One can, for instance, suddenly switch on the pumping light, with subsequent detection of the kinetics from absorption or from fluorescence, as was achieved by Cagnac [93] with Hg atoms. The problem consists of complex time dependence under conditions of intensive pumping. This drawback was eliminated by the method proposed by Bouchiat and Grossetete [76], which consisted of the use of two beams, a pumping one, and a probe one, i.e., a sufficiently weak one, for detection after termination of the pumping action, so as not to create any orientation effect.

The first application of the kinetic method for studying relaxation in the molecular ground state was attempted with HF molecules for investigating rotational relaxation, using an IR source [195]. Employing optical depopulation of the electronic ground state of dimers the transient process method was applied in Schawlow's laboratory [143]. The 'repopulation' time (restitution of population) of the $Na_2(X^1\Sigma_g^+, v'', J'')$ level, emptied by a strong dye laser pulse, causing $X \to B$ absorption, was measured. In the experiment the absorption coefficient of a weak probing pulse from another dye laser, causing $X \to A$ transition from the same v'', J'' level was measured as being dependent on the time of delay of the probe pulse with respect to the depopulating pulse. The scanning of the delay time between probe and pumping pulses yielded the characteristic time of population.

The main drawback of the method proposed in [143] consists of registration from absorption, which lowers sensitivity. The laser fluorescent version, proposed simultaneously in [11, 239] and applied to K_2 and Na_2 molecules, proved to be considerably more sensitive. In its simplest manner [11] (see Fig. 3.18) an intensive laser beam is periodically 'suddenly' weakened, thus becoming a probe beam, with correspondent durations T_p and T_{pr} (see Fig. 3.18(a)). The time dependence of the $a \to b \to c$ laser-induced fluorescence (see Fig. 1.2) in this type of modulation is demonstrated in Fig. 3.18(b). Region 1 corresponds to the establishment of pumping to a steady state, i.e. to the level of region 2. The most convenient region for determining γ_Σ is region 3, since the increase in intensity

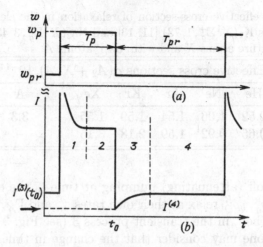

Fig. 3.18. Illustrating the transient process method of time dependence of exciting radiation (a) and of fluorescence intensity (b).

of fluorescence at $t > t_0$ directly reflects the *transient process of thermalization* of the depopulated lower level a to the equilibrium state (region 4). The transient process sets in after sudden attenuation of the pumping beam from w_p to 'probe' level w_{pr} at time t_0 (see Fig. 3.18(a)). The response $I^{(3)}$ of the system in region 3 is linear with respect to absorption, and if certain conditions discussed further are fulfilled, it can be considered as monoexponential, which leads to a simple dependence $I^{(3)}(t - t_0)$ in the form

$$I^{(3)}(t - t_0) = I^{(4)} - \left[I^{(4)} - I^{(3)}(t_0) \right] \exp[-\alpha(t - t_0)]. \qquad (3.33)$$

The above variant of the method is the most simple, although not the only one. Pulsed lasers can be applied in a two-beam variant – a pump and a probe one with scanning of the delay time between the pulses; this method was employed in [239]. The results of determining total effective cross-sections σ, in assumption (3.2) and (3.25) for Na_2 and K_2 in [11, 239] are presented in Table 3.8.

2 Relaxation kinetics of polarization moments

Let us now consider [16], how, in the transient process, angular momenta polarization in the ground state manifests itself, emerging alongside its emptying. Let us assume that anisotropic distributions of angular momenta of the upper and the ground states have been created as a result of absorption of light of certain polarization (see Fig. 3.18(b), region 2).

Table 3.8. Total effective cross-section of relaxation in the electronic ground state of the dimers $K_2(X^1\Sigma_g^+, 1, 73)$ [11, 13] and $Na_2(X^1\Sigma_g^+, 3, 43)$ [239], in collisions with admixture atoms X and with 'own' atoms A

| | Effective cross sections $\sigma(A_2 + X, A), 10^{-14}\mathrm{cm}^2$ | | | | | |
	He	Ne	Ar	Kr	Xe	A
K_2	0.52	1.06	1.54	1.59	1.55	3.3
Na_2	0.65	0.92	1.59	2.18	3.15	–

After switching off (attenuating) pumping at time t_0 the created multipole moments $_a\rho_q^\kappa$, $_b\rho_Q^K$ relax at their own rates γ_κ and Γ_K. If we assume that $\gamma_\Sigma \ll \Gamma_K$, then, in the transient process 3 (see Fig. 3.18(b)) on the γ_Σ^{-1} time scale, one may consider that the change in time of the multipole moments $_b\rho_Q^K$ of the excited state and, hence, the fluorescence (2.29), (2.30) caused by them, is governed following (3.19) by slower changes in the multipole moments of the ground state $_a\rho_q^\kappa(t-t_0)$. For $\kappa = 0$ we have

$$_a^{(3)}\rho_0^0(t-t_0) = -\left[_a^{(4)}\rho_0^0 - _a^{(2)}\rho_0^0\right]\exp\left[-\gamma_0(t-t_0)\right] + _a^{(4)}\rho_0^0, \qquad (3.34)$$

and for $\kappa \neq 0$

$$_a^{(3)}\rho_0^\kappa(t-t_0) = _a^{(2)}\rho_0^\kappa \exp\left[-\gamma_\kappa(t-t_0)\right]. \qquad (3.35)$$

In particular, in the case of linearly polarized excitation through $_b\rho_0^2$, $_b\rho_0^0$, and according to (3.19), *three ground state polarization moments* may emerge with $\kappa = 0, 2$ and 4. Indeed, substituting coefficients Φ_ξ^X from Table 2.1 and Clebsch–Gordan coefficients from Table C.1 into (3.19) we obtain

$$_b\rho_0^0 = \frac{\Gamma_p}{3\Gamma_0}\left(_a\rho_0^0 + 2_a\rho_0^2\right),$$

$$_b\rho_0^2 = \frac{\Gamma_p}{\Gamma_2}\frac{1}{105}\left(14_a\rho_0^0 + 55_a\rho_0^2 + 36_a\rho_0^4\right). \qquad (3.36)$$

If we insert (3.34) and (3.35) into (3.36) and then the result into (2.29) and (2.30) we obtain *three exponentials* in time dependence of the observable signals. However, in many cases, and in analogy with the data for excited states (Section 2.5), one ought not to expect considerable differences between γ_0, γ_2 and γ_4, which, as a rule, justify the *monoexponential approximation* (3.33).

However, the alignment relaxation constant γ_2 may be measured separately by means of the method of *laser-interrogated dichroism*, as discussed in the preceding Section (Section 3.5). Let us assume that in the scheme shown in Fig. 3.17 the strong field $\hat{\mathbf{E}}_{\mathbf{S}}$ is interrupted at time t_0, whilst the probe beam $\hat{\mathbf{E}}_{\mathbf{pr}}$ is acting continuously. In this case one can describe the behavior of the absorption coefficient $\alpha_i(\hat{\mathbf{E}}_{\mathbf{pr}})$ of the probe beam in the transient process at $t > t_0$ by substituting into (3.31) the polarization moments ${}_a^{(3)}\rho_0^\kappa$, as defined by (3.35). Substituting α_y, α_z thus obtained into (3.32) in which, according to the analysis performed in Section 3.5, only the first and the last term are left, we obtain for Q- and (P, R)-absorption respectively:

$$_QI^{(3)}(t - t_0) \;\propto\; \frac{\varepsilon^2}{4} + \frac{\varepsilon l}{4} \sin 2\theta\, {}_a\rho_0^2(t_0) \exp\left[-\gamma_2(t - t_0)\right],$$

$$_{P,R}I^{(3)}(t - t_0) \;\propto\; \frac{\varepsilon^2}{4} - \frac{\varepsilon l}{8} \sin 2\theta\, {}_a\rho_0^2(t_0) \exp\left[-\gamma_2(t - t_0)\right]. \quad (3.37)$$

Here ${}_a\rho_0^2(t_0)$ are the polarization moments created on the ground level a by the strong beam $\hat{\mathbf{E}}_{\mathbf{S}}$. The time dependence (3.37) shows that we have here a transient process after switching off dichroism, being described by one exponential, with alignment relaxation rate γ_2. This is unlike fluorescent registration in which the polarization moment of the ground state a appears only through the moments of the upper state b and the signal consists of several exponential terms (see (3.36)). Eq. (3.37) is also valid in the case where the strong field $\hat{\mathbf{E}}_{\mathbf{S}}$ has the form of a short pulse creating a polarization moment ${}_a\rho_0^2(t_0)$. The probe beam is assumed to be continuous.

It is simple to arrange the registration of the transient process in the case of fluorescent population considered in Section 3.4. For this purpose it is, for instance, sufficient that the first beam, which causes a transition $J'' \to J'$ (Fig. 3.14) should act in short pulses $\Gamma_p = G\delta(t)$. Then in the time dependence of the fluorescence produced by the continuous probe beam (cycle $J_1'' \to J_1' \to J_2''$; dotted line in Fig. 3.14) we observe the relaxation of the polarization moments of level J_1'' directly.

3 Fly-through relaxation: ground state

If the role of collisions diminishes (rarified gas), it becomes important to account for *fly-through relaxation* due to the thermal motion of the molecules optically polarized by the laser beam and possessing given power distribution $w_p(y, z)$ across its cross-section (excitation geometry corresponds to Fig. 2.5). Let us consider the kinetics of such a process, for which purpose we will turn again to the balance equation for classical density probability (3.4) and try to take molecular motion [13] into consideration. Let a group of molecules (see Fig. 3.19) with a given orientation of angular momentum $\mathbf{J}(\theta, \varphi)$ move along the y-axis with velocity v_y and cross the laser beam. The laser field action on this group is determined by its displacement along the y-axis and is different at different points

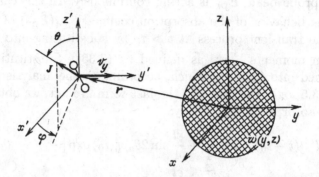

Fig. 3.19. Motion of a molecule through a laser beam with power distribution $w(y, z)$.

in time because the power distribution $w_p(y, z)$ is not uniform. Here we assume that the laser radiation is strong and that its action is stationary. Taking this into account, we can pass from the time dependence to the coordinate dependence in (3.4) by the substitution $dt = dy/v_y$. As a result the solution of Eq. (3.4) ought to be rewritten as follows:

$$\rho_a(y, z, \theta, \varphi) = \rho_a^0 \exp\left\{-\int_{-\infty}^{y}\left[G(\theta, \varphi)\frac{Bw(\xi, z)}{v_y} + \frac{\gamma_{col}}{v_y}\right]d\xi\right\}$$

$$\times\left\{1 + \frac{\gamma_{col}}{v_y}\left[\int_{-\infty}^{y}\exp\left(\int_{\zeta}^{y}G(\theta, \varphi)\frac{Bw(\eta, z)}{v_y} + \frac{\gamma_{col}}{v_y}\right)d\eta\right]d\zeta\right\}, \quad (3.38)$$

where ξ, ζ, η are the integration variables and $Bw(y, z)$ is the absorption rate. The initial condition is: $\rho_a = \rho_a^0$ at $y \to -\infty$. The stationary radiation signal with polarization $\hat{\mathbf{E}}'$ becomes considerably more complex, as compared to (2.23). Indeed, one must, firstly, average over the velocity $\mathbf{v}$, keeping in mind the fact that the projection of the Maxwellian distribution upon the yz plane is $F(u) = 2u\exp(-u^2)$, $u = v/v_p$, v_p being the most probable velocity. One must, secondly, integrate over the y, z coordinates of the space within which both the pumping beam $w_p(y, z)$ and the probing beam $w_{pr}(y, z)$ are acting, and average over all the orientations of the angular momenta. As a result we have:

$$I(\hat{\mathbf{E}}') = \frac{1}{4\pi\Gamma}\int_u\int_y\int_z\int_\theta\int_\varphi F(u)G(\theta, \varphi)G'(\theta, \varphi)Bw_{pr}(y, z)$$

$$\times\rho_a(y, z, \theta, \varphi)\sin\theta\, du\, dy\, dz\, d\theta\, d\varphi, \quad (3.39)$$

where $\rho_a(y, z, \theta, \varphi)$ is found from (3.38).

In the transient process 3 (see Fig. 3.18) the probability density $\rho_a^{(3)}$ relaxes from $\rho_a^{(2)}$ to $\rho_a^{(0)} = \rho_a^{(4)}$ according to region 3, Fig. 3.18(b):

$$\rho_a^{(3)}(y, z, \theta, \varphi) = \rho_a^{(4)} - \left\{\rho_a^{(4)} - \rho_a^{(2)}\left[y - v_y(t - t_0), z, \theta, \varphi\right]\right\}$$
$$\times \exp\left[-\gamma_{col}(t - t_0)\right], \qquad (3.40)$$

It is assumed here that the molecule in its ground (initial) state a either crosses the light beam without collisions, or passes, as the result of collisions with rate γ_{col}, into another state with different rovibronic numbers. Elastic collisions, including disorienting collisions, are neglected in this case if they do not remove molecules from state a. The corresponding fluorescence intensity has to be calculated from (3.39), where ρ_a is replaced by $\rho_a^{(3)}$ according to (3.40).

Let us consider the limiting case, namely the *absence of collisions*, assuming that $\gamma_{col} = 0$ in (3.40), in other words that the mean free path considerably exceeds the characteristic diameter of the beam.

For the case of a *rectangular beam profile*, where the power density $w_p(y, z) = w_{pr}(y, z) = 0$ at $y^2 + z^2 > r_0^2$ (see Fig. 3.19) and possesses constant value w_p^0 inside the circle $y^2 + z^2 \leq r_0^2$, it can be found, within the limit $Bw_p^0 r_0 / v_p \rightarrow \infty$, that the analytical solution has the following form (with accuracy up to a constant):

$$I^{(3)}(\tau) \propto \left[I_0(2/\tau^2) + I_1(2/\tau^2)\right] \exp(-2/\tau^2), \qquad (3.41)$$

where I_0, I_1 are modified Bessel functions and $\tau = tv_p/r_0$ is a dimensionless time parameter. For the realistic situation of arbitrary pumping efficiency, numerical integration of (3.39) is presented in Fig. 3.20(a) (solid lines). In all cases the dependence is practically indistinguishable from the exponential form:

$$I^{(3)}(t) \propto C_1 - C_2 \exp(-At), \qquad (3.42)$$

where the exponent power A does not depend on the pumping parameter and is easily determined through v_p, r_0:

$$A = T_0^{-1} \cong 0.854 v_p/r_0. \qquad (3.43)$$

In other words, one may speak here of a certain 'fly-through lifetime' T_0 of the optically depopulated lower level.

A completely different situation arises when the profile of the beam differs considerably from a rectangular one. Fig. 3.20(b) shows numerically calculated curves $_QI_{\parallel}(\tau)$ for a Gaussian distribution $w_{p,pr}(y, z) = w_{p,pr}^0 \exp\left[-(y^2 + z^2)/r_0^2\right]$, which is cut off assuming that $w_p(y, z) = w_{pr}(y, z) = 0$ at $y^2 + z^2 > 2r_0^2$. Fig. 3.20(c) shows the intensity $_QI_{\parallel}(\tau)$

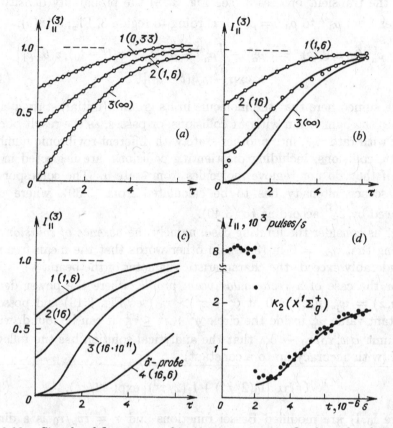

Fig. 3.20. Signals of fluorescence kinetics representing fly-through relaxation of an optically depopulated initial level: (*a*) rectangular profile of the beam; (*b*) limited Gaussian profile; (*c*) unlimited Gaussian profile; (*d*) experimentally registered signal. Values of the non-linearity parameter $Bw_p^0 v_p / r_0$ are shown in brackets.

at unlimited distributions $w_{p,pr}(y, z)$. As can be seen, at large $Bw_p^0 v_p / r_0$ values the dependence becomes visibly non-exponential with pronounced 'delay' at the first stage of the development of the transient process. The difference in shape of the curves in Fig. 3.20(*b, c*) is due to the effect of the 'wings' of the Gaussian distribution. If we have a very narrow probe beam in the form of a δ-function, i.e. differing from zero only in the centre of the depopulating beam, then relaxation does not take place for a certain 'lapse of time', see curve 4 in Fig. 3.20(*c*). This is due to the fact that there exists a region around the probe beam which is practically free of molecules in the labelled state, and a certain time must pass until such molecules will have arrived from the ambient space.

At small values of $Bw_p^0 v_p/r_0$ the dependence approaches an exponential one (see Fig. 3.20(b, c), curve 1). However, the exponent of the approximating exponential curve depends on $Bw_p^0 v_p/r_0$, which deprives the parameter A^{-1} in (3.42) of the conventional meaning of 'lifetime'.

The kinetics signal registered in [13] for $K_2(X^1\Sigma_g^+, 1, 73)$ under condition of prevalence of fly-through relaxation is shown in Fig. 3.20(d) and demonstrates the expected 'delay'. The solid line represents calculated data in accordance with (3.38), (3.39) and (3.40), applying experimentally measured distributions $w_{p,pr}(y, z)$ by scanning the diaphragm (0.1 mm diameter). The distributions obtained were close to the Gaussian with $2r_0 \cong 3$ mm.

Generally speaking, the influence of fly-through effects on the optical pumping of the lower level is essential for various non-linear phenomena [268]. These problems, on application to atomic systems, are treated, for instance, in [10].

A practically important case arises when the time of *collisional relaxation* is comparable to that of *fly-through*, i.e. $\gamma_{col} \approx v_p/r_0$. The shape of $\rho_a^{(2)}$ as determined by (3.38) then becomes more complex. In this case the curve representing the kinetics of the transient process is not very indicative. We will therefore follow up the validity of our assumption regarding the additivity (see (3.2)) by applying the following procedure. We will calculate, according to (3.38) and (3.39), the degree of fluorescent polarization $_Q\mathcal{P}^{theor}$ excited by a stationary pumping beam $w_p(y, z)$ possessing a Gaussian profile. We then use the $_Q\mathcal{P}^{theor}$ values obtained from Table 3.5 for determining χ. The χ^{-1} values thus calculated are presented in Fig. 3.21, as being dependent on γ_{col}. The non-linearity of the dependence in Fig. 3.21(a) indicates non-validity of (3.2) under these conditions, within the $\gamma_{col} \leq \gamma'_{col}$ region, i.e. in the transition region from a collisional to a 'fly-through' mechanism.

Note that such a 'break' may clearly be predicted in the concentration dependence of χ^{-1} also, because $\gamma_{cal} \sim N$. This has, most likely, been observed in earlier works [102, 125, 401] on optical polarization of the ground state $(X^1\Sigma_g^+)$ of Na_2 and I_2 molecules. Indeed, at large $N \gg N'$, when we have $\gamma_{col} \gg v_p/r_0$, collisions are dominating, and linear $\chi^{-1}(N)$ dependence is observed with a slope coefficient $\langle \sigma v_{rel} \rangle$. With diminishing N down to $N \leq N'$, i.e. in the break region, we observe a passover to $\gamma_{col} \propto v_p/r_0$. Hence, at sufficiently large Bw_p^0, due to 'delay' (see Fig. 3.20) χ^{-1} decreases faster than in the 'collisional' region. If we lower the absorption rate Bw_p^0 by an order of magnitude (see Fig. 3.21(b)), the dependence approaches a linear dependence. However, the exponent $\alpha = \alpha_0 + \gamma_{col}$ in the transient process (3.33) does not by any means coincide with γ_Σ from (3.2). One may, nevertheless, assume equality of

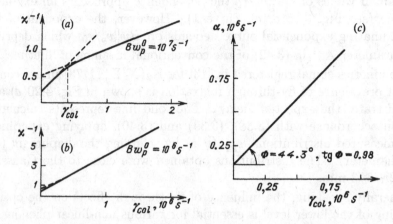

Fig. 3.21. Connection between the pumping parameter χ^{-1} and the rate of collisional relaxation γ_{col} in the 'fly-through' region $\gamma_{col} < v_p/r_0$ (a), and the 'collisional' region $\gamma_{col} > v_p/r_0$ (b). $\alpha(\gamma_{col})$ shows the calculated dependence of the exponent (3.33) on the collisional relaxation rate (c). The beam diameter is assumed to be $2r_0 = 3$ mm.

the slopes of the straight lines $\Delta\alpha/\Delta N = \Delta\gamma_\Sigma/\Delta N$, which follows from the result given in Fig. 3.21(c) obtained from numerical modeling where the exponent α is obtained by exponential approximation of the 'real' kinetics curves assuming that $Bw_p^0 = 10^6 \text{s}^{-1}$. As can be seen from the diagram, the equality $\Delta\alpha = \Delta\gamma_{col}$ holds with an accuracy of not less than $\approx 2\%$, which may be taken as the criterion of the applicability of the use of a transient process for the determination of *total effective cross-sections of thermalization* after [11, 13, 239]. One must be careful in interpreting the values α and α_0, since they themselves may depend on the pumping rate Γ_p, i.e. on excitation power.

4 Fly-through relaxation: excited state.

So far we have been discussing the fly-through effects of molecules in a non-excited state. An extremely large amount of work has been performed on measuring the *spontaneous decay* time of *excited states* of molecules in the gaseous phase, applying the procedure of detection of laser-induced fluorescence kinetics after pulsed excitation. This has stimulated us to mention here some ideas connected with the preceding considerations but dealing with weak excitation. It is essential to the investigator to know when he might expect the exponential law of decay to be invalid. We will leave aside the effects of narrow line (monochromatic) laser excitation effects, assuming, as is the tradition in the present work, the condition of broad line excitation (3.3.1). We will, further, single out the same two sources of non-exponentiality as the ones discussed above for the case of the ground state.

Firstly, there may be a difference in the decay rates of the polarization moments $_b\rho_Q^K$ of various ranks K, which is frequently the case with atoms, as was demonstrated for instance in [6, 45, 96, 135]. In order to avoid this effect, one uses well-known methods of separating the moment $_b\rho_0^0$, as described in Sections 2.4 and 2.5, namely using the 'magic' angle $\theta_0 \cong 54.7°$, or convenient linear combinations $I_\parallel + 2I_\perp$ which follow from (2.29), (2.30) and the corresponding discussion. In this way we can obtain signals depending only on $_b\rho_0^0$.

Secondly, it is necessary to analyze possible distortion of the result due to the 'exit' of molecules from the zone of observation in the course of thermal motion in a time lapse shorter than the characteristic lifetime, as already pointed out in [131]. One must keep in mind that here we speak of the volume from which radiation is being collected, since, if the particle does not leave this region during its lifetime, the role of fly-through will not manifest itself. The practically important case of a Gaussian spatial profile of the energy distribution in the beam with account taken of diffusion has been discussed in [78] for the case of NO ($v = 5$). Here, as before, we will discuss, mainly following [32], the case of 'fly-through' of molecules across the beam directed along the x-axis with the Gaussian spatial energy distribution function, the characteristic diameter of the beam $2r_0$ being much smaller than the mean free path. Applying (3.40) to the excited state, we obtain:

$$_b\rho_Q^K(t - t_0) = \frac{1}{\pi r_0^2} {}_b\rho_Q^K(t_0)$$

$$\times \exp\left[\frac{[y - uv_p(t - t_0)]^2 + z^2}{r_0^2} - \Gamma_K(t - t_0)\right],$$

$$(3.44)$$

where polarization moments $_b\rho_Q^K(t_0)$ at time point t_0 can be calculated using (2.21).

Let the radiation be registered along the y-axis, and light pass through the entrance slit of a monochromator from a region which is limited in width by z values $-l/2$ to $l/2$ (assuming that the slit is positioned parallel to the light beam direction (x-axis) and the magnification of the illuminating lens is equal to unity, we denote the width of the entrance slit by l). In order to find the time dependence of the fluorescence, we average (3.44) over the bulk from which the fluorescence is being observed, as well as over the projections of the velocity v_z, assuming that the depth of definition along y is usually quite large, and only motion along z is in

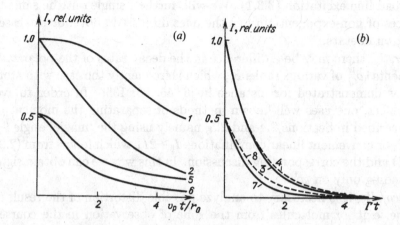

Fig. 3.22. $(a),(b)$ The kinetics of fluorescence calculated, following (3.46), for $l/r_0 = 5$ (curves 1–4) and for $l/r_0 = 1$ (curves 5–8) with parameter ratios $\Gamma = 0$ (1 and 5), $\Gamma = 0.25 v_p/r_0$ (2 and 6), $\Gamma = v_p/r_0$ (3 and 7). The dotted lines (4 and 8) refer to exponential decay at a rate Γ, when $v_p t/r_0 = 0$.

a state to 'lead away' the radiating particle from the zone of observation. The required averaged value of the multipole moment is

$$
{}_b\rho_Q^K(t - t_0) = \int_{-\infty}^{\infty} du \int_{-l/2}^{l/2} dx \int_{-\infty}^{\infty} dy \, {}_b\rho_Q^K(t - t_0)\sqrt{\pi} \exp(-u^2)
$$

$$
= {}_b^{(0)}\rho_Q^K \exp[-\Gamma_K(t - t_0)]\mathrm{erf}\left(\frac{l}{2\sqrt{r_0^2 + v^2(t - t_0)^2}}\right).
$$

$$(3.45)$$

Here Maxwellian distribution of particle velocities is assumed. It can be seen that widening of the entrance slit l diminishes the role of fly-through relaxation. Substituting the expressions obtained into (2.24), it becomes possible to calculate the intensity $I(t - t_0)$. If we assume that $\Gamma_0 = \Gamma_1 = \Gamma_2 = \Gamma$ the expression becomes independent of the light polarization and transition type of the observation and excitation, and is quite simple:

$$
I(t) \propto e^{-\Gamma t}\mathrm{erf}\left(\frac{l}{2\sqrt{r_0^2 + v_p^2 t^2}}\right),
$$

$$(3.46)$$

putting $t_0 = 0$.

The time dependence of such a signal is presented in Fig. 3.22. For curves 1–4 the width of the entrance slit l is five times larger than the Gaussian radius of the beam r_0; for curves 5–8 it is equal to it, and lower intensity is, naturally, registered in the second case. Cases 1 and 5, for which it is assumed that $\Gamma = 0$, describe only a fly-through relaxation process. Another limiting case, where we have $v_p t/r_0 = 0$ is demonstrated by the exponential curves 4 and 8 (dotted lines), with decay constant Γ. It may be seen from the diagram that, other conditions being equal (equal values of l/r_0), the role of the fly-through relaxation is determined only by the relation between Γ and v_p/r_0. Thus, for instance, the dependence presented in Fig. 3.22(a) in I, $v_p t/r_0$ coordinates refers to a situation where fly-through dominates (v_p/r_0 larger than Γ). The dependence in Fig. 3.22(b), presented in I, Γt coordinates makes comparison possible between the case where $v_p/r_0 = \Gamma$ and that of a 'pure' exponential $I \propto \exp(-\Gamma t)$, where we have $v_p/r_0 = 0$. The signals for $v_p/r_0 < \Gamma$ are positioned between the curves 3 and 4, or between 7 and 8 for the corresponding ratio l/r_0, i.e. approaching an exponential rather closely. For instance, for $v_p/r_0 = 0.1\Gamma$ the deviation from an exponential does not exceed 0.5% at $\Gamma t = 1$. Accounting for the effect of slit width, the role of fly-through can be estimated by means of a criterion which presumes a comparison with the parameter

$$A_0 = \frac{\frac{l\Gamma}{v_p} \exp\left(-\frac{l\Gamma^2 v_p^{-2}}{4+4r_0^2\Gamma^2 v_p^{-2}}\right)}{\left(1+r_0^2\Gamma^2 v_p^{-2}\right)^{3/2}}, \tag{3.47}$$

the smallness of which ensures that exponential decay is well satisfied.

One may say, in conclusion to this chapter, that the material presented permits us to follow up the conditions for which optical transitions may lead to polarization of the angular momenta of a molecular ensemble in the ground state. Such phenomena are not unusual and may take place in rarefied gases, particularly in molecular beams at rather modest power densities of the exciting light. Their manifestations can be studied through fluorescence (intensity, polarization, kinetics) or absorption, and may be described by the analytical expressions applying appropriate approximations. The most self-consistent description is based on polarization moments posessing definite relaxation rates.

4

Effect of external magnetic field on angular momenta distribution

4.1 Basic concepts

The preceding chapter closed with a discussion on kinetic methods which presume investigations of non-stationary time dependent relaxation processes of optical polarization of the angular momenta in a molecular ensemble. Another possibility, which also permits us to introduce a time scale, consists of the application of an *external magnetic field*. Indeed, since an angular momentum $\mathbf{J}$ produces a corresponding proportional (collinear) magnetic moment μ_J:

$$\mu_J = g_J \mu_B \frac{\mathbf{J}}{\hbar}, \qquad \mu_J = g_J \mu_B \sqrt{J(J+1)}, \qquad (4.1)$$

the external field $\mathbf{B}$ creates precession of the angular momentum $\mathbf{J}(\theta, \varphi)$ around $\mathbf{B} \parallel \mathbf{z}$ (see Fig. 4.1(a)) with angular velocity

$$\omega_J = -\frac{g_J \mu_B}{\hbar} \mathbf{B}. \qquad (4.2)$$

We assume here that the Bohr magneton μ_B is a positive quantity. The numerical value of the Bohr magneton and some other fundamental constants, as recommended by CODATA, the Committee on Data for Science and Technology of the international Council of Scientific Unions, are presented in Table 4.1 [103]

Hence, the Landé factor g_J, say for an atomic state $^1P_1$, which is determined only by the orbital momentum of the electron, is negative, $g_J = -1$, and the magnetic moment is directed in opposition to the mechanical one. The magnetic field $\mathbf{B}$ interacts directly with the magnetic moment μ_J, and as a result a moment of forces $\mathbf{M} = \mu_J \times \mathbf{B}$ emerges, leading to a change in angular momentum:

$$\frac{d\mathbf{J}}{dt} = \mathbf{M} = \frac{g_J \mu_B}{\hbar}(\mathbf{J} \times \mathbf{B}) = -(\mathbf{J} \times \omega_J). \qquad (4.3)$$

Table 4.1. Recommended values of some fundamental physical constants

Quantity	Symbol	Value	Units	Relative uncertainty ppm
Speed of light in vacuum	c	299 792 458	$m \cdot s^{-1}$	exact
Planck constant	h	6.626 075 5(40)	$10^{-34} J \cdot s$	0.60
in electron volts		4.135 669 2(12)	$10^{-15} eV \cdot s$	0.30
$h/2\pi$	$\hbar$	1.054 572 66(63)	$10^{-34} J \cdot s$	0.60
in electron volts		6.582 122 0(20)	$10^{-16} eV \cdot s$	0.30
Elementary charge	e	1.602 177 33(49)	$10^{-19} C$	0.30
Electron mass	m_e	9.109 389 7(54)	$10^{-31} kg$	0.59
in atomic units		5.485 799 03(13)	$10^{-4} u$	0.023
in electron volts		0.510 999 06(15)	MeV	0.30
Electron g-factor	g_s	2.002 319 304 386(20)		1×10^{-5}
Bohr magneton	μ_B	9.274 015 4(31)	$10^{-24} J \cdot T^{-1}$	0.34
in electron volts		5.788 382 63(52)	$10^{-5} eV \cdot T^{-1}$	0.089
in hertz		1.399 624 18(42)	$10^{10} Hz \cdot T^{-1}$	0.30
in wavenumbers		46.686 437(14)	$m^{-1} \cdot T^{-1}$	0.30
in kelvins		0.671 709 9(57)	$K \cdot T^{-1}$	8.5

Hence, at a *positive* g-factor, precession of **J** takes place in a *clockwise* direction if viewed from the tip of the **B**-vector, and in a *counterclockwise* direction at *negative* g_J (see Fig. 4.1(a)).

The problem of the Landé factors in molecules is complex, and its signs may differ. Some information on this point will be presented in Section 4.5. We wish to remind the reader that the direction of **B** naturally determines the quantization axis z. The frequency ω_J fixes the time scale, and this permits us, as we will see presently, to obtain direct (in one sense) information on *relaxation processes* and/or to study *molecular magnetism*.

We will first try to understand the basic outlines of the phenomena on the basis of the framework of *polarization moments*, as treated in the preceding chapters. In order to avoid overloading the text with excessive formalism and in order to achieve easier understanding, we will consider a simplified model in the present chapter which gives an idea of the essence of the phenomena. The possibility of a more comprehensive quantum mechanical description will be offered by the equations presented in the following chapter.

Let the distribution $\rho_b(\theta, \varphi)$ of angular momenta in the excited state (see Fig. 4.1(b)) arise under the action of light in the form of a short pulse (δ-pulse). The magnetic field **B** creates precession around **B** not only of the angular momentum of a separate molecule, but also of the distribution over the ensemble. Since, however, all momenta **J** are in precession with one and the same angular velocity ω_J (4.2), their mutual positions with respect to each other remain the same. Hence, the whole 'rose of vectors' **J** is in precession as a single entity, which means that the

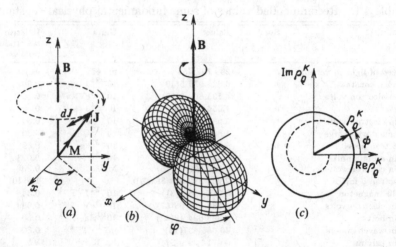

Fig. 4.1. Precession of angular momenta in a magnetic field: (*a*) rotation of **J** for $g_J > 0$; (*b*) change in probability density $\rho_b(\theta, \varphi)$; (*c*) behaviour of the polarization moment ρ_Q^K in a complex plane (broken curve – in the presence of relaxation).

distribution $\rho_b(\theta, \varphi)$ rotates around **B**, maintaining its shape and size. This holds if no relaxation processes take place (i.e., relaxation does not manifest itself within the ω_J^{-1} time scale), and only the coordinate φ of the angular momentum changes.

It ought to be clear now how to include the effect of the magnetic field in the simplest balance equations of type (3.4) and (3.16). To this end one must take into consideration the fact that the change in time caused by precession around **B** of the probability density $\rho(\theta, \varphi)$ of molecules is

$$\left.\frac{\partial \rho(\theta, \varphi)}{\partial t}\right|_{magn} = \frac{\partial \rho(\theta, \varphi)}{\partial \varphi}\frac{\partial \varphi}{\partial t} = \omega_J \frac{\partial \rho(\theta, \varphi)}{\partial \varphi}. \tag{4.4}$$

Adding this term to the righthand side of the kinetic equations (3.4) and (3.16), we obtain for the lower state

$$\dot{\rho}_a(\theta, \varphi) = -\Gamma_p G(\theta, \varphi)\rho_a(\theta, \varphi) + \gamma_\Sigma \left[\rho_a^0 - \rho_a(\theta, \varphi)\right] + \omega_{J''}\frac{\partial \rho_a(\theta, \varphi)}{\partial \varphi}, \tag{4.5}$$

and for the upper (excited) state

$$\dot{\rho}_b(\theta, \varphi) = \Gamma_p G(\theta, \varphi)\rho_a(\theta, \varphi) - \Gamma\rho_b(\theta, \varphi) + \omega_{J'}\frac{\partial \rho_b(\theta, \varphi)}{\partial \varphi}. \tag{4.6}$$

One may say that the *magnetic precession* ('the magnetic term') is added to the relaxation at a rate Γ_p, γ_Σ or Γ. One must keep in mind the fact that, since all approximations adopted in Section 3.1 are in force, the additional energy

$$E_{magn} = -\mu\mathbf{B} \tag{4.7}$$

of the interaction of μ with the magnetic field $\mathbf{B}$ does not affect the absorption or emission probability, including the angular coefficients $G(\theta, \varphi)$. In other words, the broad excitation line approximation holds. In quantum mechanical terms this means that the Zeeman splitting caused by the magnetic field does not affect the absorption and emission probability.

It is again very fruitful here to expand the probability densities $\rho_a(\theta, \varphi)$ and $\rho_b(\theta, \varphi)$ over spherical functions (2.14) and to pass to equations for classical polarization moments $_a\rho_q^\kappa$, $_b\rho_Q^K$, as determined by (2.16). The decisive simplification of equations consists of the fact that the derivative of the spherical function $Y_{KQ}(\theta, \varphi)$ over the angle φ (see Appendix B) is of the form

$$\frac{\partial Y_{KQ}(\theta, \varphi)}{\partial \varphi} = iQ Y_{KQ}(\theta, \varphi), \tag{4.8}$$

since the dependence of Y_{KQ} on φ is determined only by the exponential factor $e^{iQ\varphi}$. It follows that the additive 'magnetic terms' in the expansion of the kinetic equations (4.5) and (4.6) over $_a\rho_q^\kappa$, $_b\rho_Q^K$ are of the form $iq\omega_{J''}{}_a\rho_q^\kappa$ and $iQ\omega_{J'}{}_b\rho_Q^K$ respectively. For the excited state, taking account of the effect of the constant magnetic field in Eq. (3.21) leads to

$$_b\dot\rho_Q^K = \Gamma_p \sum_{\kappa q} {}_Q^K D_q^\kappa {}_a\rho_q^\kappa - \Gamma_K {}_b\rho_Q^K + iQ\omega_{J'}{}_b\rho_Q^K, \tag{4.9}$$

where $_Q^K D_q^\kappa$ is defined according to (3.20). The analogue of Eq. (4.9) for the ground state may be written as an expansion of (4.5) over spherical functions:

$$_a\dot\rho_q^\kappa = -\Gamma_p \sum_{\kappa' q'} {}_q^\kappa D_{q'}^{\kappa'} {}_a\rho_{q'}^{\kappa'} - \gamma_{\kappa a}\rho_q^\kappa + \lambda_q^\kappa \delta_{\kappa 0}\delta_{q0} + iq\omega_{J''}{}_a\rho_q^\kappa. \tag{4.10}$$

Here $\lambda_q^\kappa \delta_{\kappa 0}\delta_{q0}$ is the rate of 'supply' of population from the thermostat on to level a.

The application of equations to the description of optico-magnetic effects will be discussed in Sections 4.2, 4.3 and 4.4. At the same time it is possible, without resorting to the solutions of the equations, to offer certain general considerations on the behavior of a molecular ensemble in a magnetic field.

It follows from (4.9) and (4.10) that the magnetic field does not affect the *longitudinal component* of the polarization moments $_b\rho_0^K$, $_a\rho_0^\kappa$ (they are real quantities) where $Q = q = 0$, since the magnetic terms become equal to zero. This can easily be understood from the following. In accordance with Section 2.3, an ensemble of angular momenta possessing only longitudinal components of alignment or orientation has axial (cylindrical) symmetry with respect to $\mathbf{B} \parallel \mathbf{z}$, i.e. it is non-coherent. Such are, for instance, the shapes of the angular momenta distribution in Figs. 2.2 and 3.2, the precession of which around the z-axis in a magnetic field does not change the distribution of the momenta in any way.

Let us now imagine that the function $\rho(\theta, \varphi)$ does not possess axial symmetry with respect to the z-axis, as, e.g., in Fig. 4.1(b), and that it is created by pulsed excitation at $t = 0$. In this case its expansion over multipoles includes *transversal polarization moments* ρ_Q^K (these are complex quantities) where $Q \neq 0$. If the precession frequency $\omega_{J'}$ is much larger than the relaxation rate Γ_K, then the components ρ_Q^K will depend on time, in the time scale $\omega_{J'}^{-1}$, following (4.9), according to the simple harmonic law:

$$\rho_Q^K(t) = \left[\text{Mod}\rho_Q^K(0)\right] e^{i(Q\omega_{J'}t + \psi_0)}, \qquad (4.11)$$

$$Q\omega_{J'}t + \psi_0 = \psi. \qquad (4.12)$$

Here $\text{Mod}\rho_Q^K$ is the module or magnitude of the polarization moment, but ψ_0 is its initial phase at the moment of excitation. As can be seen, the magnetic field itself does not alter the *absolute value* of the multipole moment, changing only its *phase* ψ.

The definition of the phase ψ and of $\text{Mod}\rho_Q^K$ are usual for those complex quantities (see Fig. 4.1(c)):

$$\tan \psi = \frac{\text{Im}\,\rho_Q^K}{\text{Re}\,\rho_Q^K}, $$

$$\text{Mod}\,\rho_Q^K = \sqrt{(\text{Re}\,\rho_Q^K)^2 + (\text{Im}\,\rho_Q^K)^2}. \qquad (4.13)$$

We thus see that the 'purely magnetic' evolution (4.11) of the polarization moment is, in essence, a linear change in time of its phase ψ according to (4.12), with conservation of the module $\text{Mod}\rho_Q^K$ (circle in Fig. 4.1(c)). The factor $e^{i\psi}$ means that the dependence $\rho_Q^K(t)$ is *periodic* with a period $T_Q = 2\pi/Q\omega_{J'}$, i.e. that each transversal component of a polarization moment passes into itself with its own frequency $Q\omega_{J'}$. This is in full agreement with what has been said before in Section 2.3 on the connection between the *coherence* and *symmetry* of $\rho(\theta, \varphi)$. The model presented affords the conservation of the shape of the angular momenta distribution $\rho(\theta, \varphi)$ in the course of precession (see Fig. 4.1(b)). Incidentally, it may not seem quite appropriate in this context to maintain the statement that the magnetic field itself 'destroys coherency', as described by the transversal components ρ_Q^K, $Q \neq 0$. Indeed, it follows from (4.11) that at excitation by a δ-pulse the polarization moments ρ_Q^K are not *destroyed* by an external magnetic field, but *oscillate* with frequency $Q\omega_{J'}$.

Of course, along with magnetic precession, *relaxation processes* take place in molecules after pulsed excitation. If the relaxation of a K-rank polarization moment is characterized by a rate constant Γ_K, then the absolute value, i.e., $\mathrm{Mod}\,\rho_Q^K$ of ρ_Q^K, diminishes over time ('decays') exponentially:

$$\mathrm{Mod}\,\rho_Q^K(t) = \mathrm{Mod}\,\rho_Q^K(0)e^{-\Gamma_K t}. \tag{4.14}$$

This means that $\mathrm{Mod}\,\rho_Q^K$ in Eq. (4.11) now changes over time according to (4.14). It is of considerable importance here that the precession (4.11) and relaxation (4.14) are independent, in the sense that the decay does not have any effect on the change of phase ψ. Simultaneous actions of turn and relaxation are shown by the broken line in Fig. 4.1(c). This corresponds to rotation of the figure $\rho(\theta, \varphi)$, as presented in Fig. 4.1(b), with a simultaneous decrease in its size.

Let us now consider continuous wave (cw) light creating a distribution of angular momenta in the excited state coherent with respect to the z-axis. The stationary distribution of angular momenta $\rho_b(\theta, \varphi)$ may be determined by solving Eqs. (4.9) and (4.10). For $_b\dot{\rho}_Q^K = 0$ we obtain from (4.9)

$$_b\rho_Q^K = \frac{\Gamma_p}{\Gamma_K - iQ\omega_{J'}} \sum_{\kappa q} {}^K_Q D^\kappa_q {}_a\rho_q^\kappa, \tag{4.15}$$

where $_a\rho_q^\kappa$ is the stationary solution of (4.10). Substituting (4.15) into (2.24) yields a description of the effect of a constant magnetic field on the observed signal in those approximations for which we used (4.9) and (4.10).

Thus, the method of description presented in this section permits us to preserve the clarity of the picture and, at the same time, to elucidate thoroughly the essence of the phenomena connected with the effect of a magnetic field on the intensity and polarization of the radiation of a system with a large angular momentum. This is going to be the subject of the following two sections.

4.2 The Hanle effect in molecules: excited state

1 Calculation of observed signals

Let the cw exciting light be weak in the sense that one may neglect all the multipole moments in the ground state, except for the equilibrium population $_a\rho_0^0$. We then obtain from (4.15), allowing for (3.20):

$$_b\rho_Q^K = \frac{\Gamma_p}{\Gamma_K - iQ\omega_{J'}}(-1)^\Delta C_{1\Delta1-\Delta}^{K0} C_{K0K0}^{00} C_{KQ00}^{KQ} \Phi_Q^K(\hat{\mathbf{E}})_a\rho_0^0. \qquad (4.16)$$

The expression becomes simpler if we consider $C_{KQ00}^{KQ} = 1$, $C_{K0K0}^{00} = (-1)^K/\sqrt{2K+1}$ (see Appendix C (C.5)). When in addition the normalization $_a\rho_0^0 = 1$ is taken into account, see (2.17), we obtain

$$_b\rho_Q^K = \frac{\Gamma_p}{\Gamma_K - iQ\omega_{J'}} \frac{(-1)^{\Delta+K}}{\sqrt{2K+1}} C_{1\Delta1-\Delta}^{K0} \Phi_Q^K(\hat{\mathbf{E}}). \qquad (4.17)$$

Substituting (4.17) into (2.24), we are able to calculate the intensity of radiation for any geometry and polarization of excitation, as determined by $\Phi_Q^K(\hat{\mathbf{E}})$, and also the observations, as determined by $\Phi_Q^K(\hat{\mathbf{E}}')$, for all types of transitions (taken into consideration by respective Clebsch–Gordan coefficients with Δ, Δ'); the necessary numerical values of $C_{1\Delta1-\Delta}^{K0}$ and $C_{1-\Delta'1\Delta'}^{K0}$ entering into (4.17) and (2.24) may be found in Table C.1 (Appendix C).

Let us take a look at the most typical conditions of registration. Imagine the exciting light beam to be orthogonal to the magnetic field $\mathbf{B}$ (z-axis) and linearly polarized, the vector $\hat{\mathbf{E}}$ being along the y-axis (see Fig. 4.2). Fluorescence is detected in the direction of the z-axis. Most frequently one measures the changes produced by the magnetic field in the *intensity difference* $I_\| - I_\perp$, or the *degree of polarization* $\mathcal{P}$ (2.25). In this geometry, Eq. (2.24) for the fluorescence intensity $I(\hat{\mathbf{E}}')$ includes the *transversal components* of alignment $_b\rho_{\pm2}^2$, leading to

$$I_\| = A(-1)^{\Delta'} \left[C_{1-\Delta'1\Delta'}^{00} \left(-\frac{1}{\sqrt{3}}\right)_b\rho_0^0 \right.$$
$$\left. +\sqrt{5}C_{1-\Delta'1\Delta'}^{20} \left(-\frac{1}{\sqrt{30}}{}_b\rho_0^2 - \frac{1}{2\sqrt{5}}{}_b\rho_2^2 - \frac{1}{2\sqrt{5}}{}_b\rho_{-2}^2\right) \right], \quad (4.18)$$

$$I_\perp = A(-1)^{\Delta'} \left[C_{1-\Delta'1\Delta'}^{00} \left(-\frac{1}{\sqrt{3}}\right)_b\rho_0^0 \right.$$
$$\left. +\sqrt{5}C_{1-\Delta'1\Delta'}^{20} \left(-\frac{1}{\sqrt{30}}{}_b\rho_0^2 + \frac{1}{2\sqrt{5}}{}_b\rho_2^2 + \frac{1}{2\sqrt{5}}{}_b\rho_{-2}^2\right) \right]. \quad (4.19)$$

It is essential that here the difference between $I_\|$ and $I_\perp$ be determined only by the difference in sign with which $_b\rho_{\pm2}^2$ enter the equations. This difference is determined by the sign of the components $\Phi_{\pm2}^2$ for $\varphi' = \pi/2$ and $\varphi' = 0$, respectively; see Table 2.1 (here $\theta = \pi/2$). On this basis, and applying (2.15), one can see that the dependence of $I_\| - I_\perp$ on the magnetic field is determined only by the action of B on the $\mathrm{Re}\,_b\rho_2^2$,

$$I_\| - I_\perp = \text{const} \times \mathrm{Re}\,_b\rho_2^2, \qquad (4.20)$$

Applying (4.17) we obtain

$$\mathrm{Re}\,_b\rho_2^2 \propto \frac{\Gamma_p\Gamma_2}{\Gamma_2^2 + 4\omega_{J'}^2}. \tag{4.21}$$

At the same time, as follows from (4.18) and from (4.19), the sum $I_\parallel + I_\perp$ does not contain any transversal components where $Q \neq 0$, i.e. it does not depend on the magnetic field. The full expression for the degree of polarization as obtained from (4.18) and (4.19) is of the form

$$\mathcal{P}(B) = \frac{\mathcal{P}(0)}{1 + 4\omega_{J'}^2/\Gamma_2^2}, \tag{4.22}$$

where $\mathcal{P}(0)$ is the degree of polarization in the absence of a magnetic field ($B = 0$). The values of $\mathcal{P}(0)$ can be found in Table 3.6, case $\chi = 0$. This statement may appear to conflict with the fact that the formulae in this table were obtained for a different geometry (see Fig. 2.5) but, of course, in the absence of a magnetic field both geometries are equivalent.

We can see from (4.22) that the effect of a magnetic field (i.e. increase of $\omega_{J'}/\Gamma_2$) leads to a diminution of $\mathcal{P}$ down to complete depolarization at $\omega_{J'}/\Gamma_2 \to \infty$, the dependence $\mathcal{P}(B)$ showing a *Lorentz-type* shape. Eq. (4.17) and its corollary (4.21) show that at stationary excitation, when, unlike (4.11), the moment of excitation is different in time for different molecules of the ensemble, precession in the magnetic field destroys the transversal components $_b\rho_{\pm Q}^K$, namely $_b\rho_{\pm 2}^2$, if the precession frequency $\omega_{J'}$ exceeds the relaxation rate Γ_2 considerably. Alongside depolarization, the coherence which is linked with $_b\rho_{\pm 2}^2$ also disappears. The aforesaid is clearly illustrated by Fig. 4.2, which shows how, in the dependence on the ratio $\omega_{J'}/\Gamma_2$ of the precession frequency and relaxation rate of alignment, the distribution of the angular momenta gradually *turns*, becoming more and more *isotropic*, up to full axial symmetry with respect to the z-axis, i.e. in the xy plane. The turning angle α (see Fig. 4.2) of the distribution in the xy plane is

$$\alpha = \frac{1}{2}\arctan\left(\frac{\mathrm{Im}\,_b\rho_2^2}{\mathrm{Re}\,_b\rho_2^2}\right) = 0.5\arctan(2\omega_{J'}/\Gamma_2). \tag{4.23}$$

The origin of this formula can be understood from the form of expansion (2.14) if one remembers that only 'transversal' spherical functions $Y_{KQ\neq 0}$ are φ-dependent; see (B.1), Appendix B. In other words, the shape of the figures in Fig. 4.2 on increase of the magnetic field is not conserved, and the symmetry of the distribution is connected to the transition from the situation when *coherence* in the J-distribution is present, Fig. 4.2(a) down to the fully incoherent case in Fig. 4.2(e). Hence in this case one can speak about the destruction of coherence by a magnetic field, contrary to the case in Fig. 4.1(b).

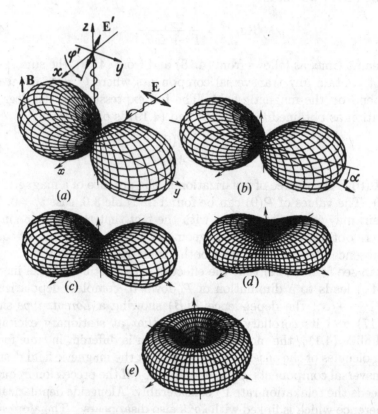

Fig. 4.2. Effect of the magnetic field on the distribution $\rho_b(\theta, \varphi)$ of the angular momenta of the excited state (linearly polarized Q-excitation, weak light). Values of $\omega_{J'}/\Gamma$ are: (a) 0; (b) 0.5; (c) 1.0; (d) 1.5; (e) ∞.

A similar illustration for the case of an ensemble *oriented* by circularly polarized light is presented in Fig. 4.3. In this case the magnetic field **B** is orthogonal with respect to the direction of the exciting beam.

It ought to be mentioned that in all cases the '*volume*' of the figures depicted in Figs. 4.2 and 4.3 is not changed by the magnetic field. This holds for weak excitation due to the independence of the light absorption probability on B, when the ground state angular momenta distribution remains isotropic. The total fluorescence intensity is therefore independent of the magnetic field, the latter being able to cause only angular redistribution of emitted light and changes in the polarization of fluorescence. This circumstance is essential for the *linear* Hanle effect.

Let us now consider a more general case, when we register in the z-direction the intensity components of the radiation with mutually perpendicular linear polarization $I_{\varphi'}$ and $I_{\varphi'+\pi/2}$, the index denoting the az-

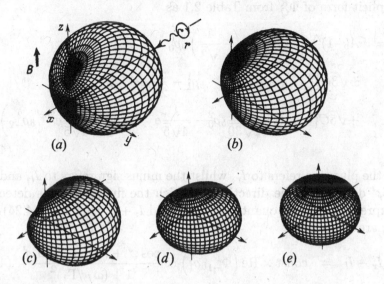

Fig. 4.3. Effect of magnetic field on the distribution $\rho_b(\theta, \varphi)$ of the angular momenta of the excited state (righthanded circularly polarized R-excitation, weak light). Values of $\omega_{J'}/\Gamma_1$ are: (a) 0; (b) 0.5; (c) 1.5; (d) 5.0; (e) ∞.

imuthal angle of the $\hat{\mathbf{E}}'$-vector of the detected radiation (see Fig. 4.2(a)). We have, in this case,

$$I_{\varphi'+\pi/2} - I_{\varphi'} = \text{const} \times \left(\text{Re}\,\Phi^2_{-2b}\rho^2_2 \right), \qquad (4.24)$$

$$I_{\varphi'+\pi/2} + I_{\varphi'} \neq f(B). \qquad (4.25)$$

This leads to

$$\mathcal{P}_{\dot{\varphi}'}(B) = \frac{I_{\varphi'+\pi/2} - I_{\varphi'}}{I_{\varphi'+\pi/2} + I_{\varphi'}} = \mathcal{P}(0)\frac{\cos 2\varphi' + 2\,(\omega_{J'}/\Gamma_2)\sin 2\varphi'}{1 + 4\,(\omega_{J'}/\Gamma_2)^2}. \qquad (4.26)$$

For $\varphi' = 0$ we obtain from (4.26) a *Lorentzian-type* shape (4.22), and for $\varphi' = \pm\pi/4$ we obtain a *dispersion* shape of the signal:

$$\mathcal{P}_{\pm\pi/4}(B) = \mathcal{P}(0)\frac{\pm 2\omega_{J'}/\Gamma_2}{1 + 4\,(\omega_{J'}/\Gamma_2)^2}, \qquad (4.27)$$

which allows us to determine the sign of $\omega_{J'}$ and, along with it, the sign of $g_{J'}$; see Eq. (4.2).

A similar analysis of expected signals can be performed in the case of circularly polarized excitation. If we observe the circularly polarized fluorescence in the xy plane, the righthanded I_r and lefthanded I_l polarized fluorescence components can be written in accordance with (2.24), and in the explicit form of Φ_Q^K from Table 2.1 as

$$
\begin{aligned}
I_{r,l} = A(-1)^{\Delta'} \Bigg[& C_{1-\Delta'1\Delta'}^{00} \left(-\frac{1}{\sqrt{3}}\right) {}_b\rho_0^0 \\
& \pm \sqrt{3} C_{1-\Delta'1\Delta'}^{10} \left(\frac{1}{2\sqrt{3}} e^{-i\varphi'} {}_b\rho_1^1 - \frac{1}{2\sqrt{3}} e^{i\varphi'} {}_b\rho_{-1}^1 \right) \\
& + \sqrt{5} C_{1-\Delta'1\Delta'}^{20} \left(\frac{1}{2\sqrt{30}} {}_b\rho_0^2 - \frac{1}{4\sqrt{5}} e^{-2i\varphi'} {}_b\rho_2^2 - \frac{1}{4\sqrt{5}} e^{2i\varphi'} {}_b\rho_{-2}^2 \right) \Bigg],
\end{aligned}
$$

$$(4.28)$$

where the plus sign refers to I_r, whilst the minus sign refers to I_l, and the angle φ' determines the direction in which the fluorescence is detected. The expressions for the quantities $I_r - I_l$ and $I_r + I_l$, similar to (4.24) and (4.25), are of the form:

$$I_r - I_l = \text{const} \times \text{Re}\left(\Phi_{-1b}^1 \rho_1^1\right) \propto \frac{\cos\varphi' + (\omega_{J'}/\Gamma_1)\sin\varphi'}{1 + (\omega_{J'}/\Gamma_1)^2}, \quad (4.29)$$

$$I_r + I_l = f\left({}_b\rho_{\pm 2}^2\right) = f(B). \quad (4.30)$$

We can see some important peculiarities in Eqs. (4.29) and (4.30). The intensity difference $I_r - I_l$ is characterized by the parameter $\omega_{J'}/\Gamma_1$, which contains the relaxation rate of orientation (in the case of linearly polarized excitation the difference $I_\| - I_\perp$ contains the relaxation rate of alignment; see (4.21)). At the same time the sum of intensities $I_r + I_l$ in this case still contains transversal components of alignment ${}_b\rho_{\pm 2}^2$ and is now magnetic field dependent.

The dependence $\mathcal{C}(B)$ thus becomes more complicated in comparison to $\mathcal{P}(B)$, as defined by (4.26). Namely, if we have the righthanded circularly polarized exciting light directed along the x-axes, as shown in Fig. 4.3, we obtain

$$
\mathcal{C} = \frac{60 C_{1\Delta 1-\Delta}^{10} C_{1\Delta'1-\Delta'}^{10} \frac{\Gamma_0}{\Gamma_1} \frac{\cos\varphi' + (\omega_{J'}/\Gamma_1)\sin\varphi'}{1+(\omega_{J'}/\Gamma_1)^2}}{40(-1)^{\Delta-\Delta'} + C_{1\Delta 1-\Delta}^{20} C_{1\Delta'1\Delta'}^{20} \frac{\Gamma_0}{\Gamma_2} \left(3 + 9\frac{\cos 2\varphi' + 2(\omega_{J'}/\Gamma_2)\sin 2\varphi'}{1+4(\omega_{J'}/\Gamma_2)^2}\right)},
$$

$$(4.31)$$

being dependent on two parameters $\omega_{J'}/\Gamma_1$ and $\omega_{J'}/\Gamma_2$. As far as $C_{1010}^{10} \equiv 0$ (see Table C.1, Appendix C), $\mathcal{C}(B) \equiv 0$ if we have a Q-type transition in excitation or absorption. Hence circularly polarized excitation can be applied only if we use (P, R)-type transitions in molecules.

The dependence of $\mathcal{C}(B)$ on two parameters $\omega_{J'}/\Gamma_1$ and $\omega_{J'}/\Gamma_2$ is of course connected to the fact that, as can be seen from (2.37) and (2.39), alignment is always created along with orientation under circularly polarized excitation. Hence, for the purpose of determining the orientation relaxation rate Γ_1, one has to register the intensity difference $I_r - I_l$

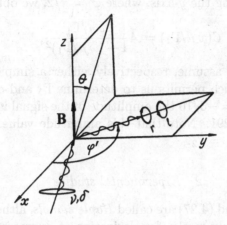

Fig. 4.4. The geometry adopted in the calculation of the degree of circularity $\mathcal{C} = (I_r - I_l)/(I_r + I_l)$ under elliptically polarized excitation.

(4.29), and not the degree of circularity $\mathcal{C}$. If, however, one prefers to keep to the normalized parameter $\mathcal{C}$ (see, e.g., [332]), the situation when the level crossing signal is determined only by $g_{J'}/\Gamma_1$ can be created [35] by applying elliptically polarized light for excitation. As can be seen from Table 2.1, one may choose such a relation between the ellipticity parameters v, δ and the exciting light beam direction characterized by θ, φ (see Fig. 4.4) that $\Phi_{\pm 2}^2 = 0$ and no $_b\rho_{\pm 2}^2$ emerges, but at the same time the transversal orientation $_b\rho_{\pm 1}^1$ is created. Indeed, if we put

$$\delta = \pi/2, \quad \varphi = 0, \quad \cos^2 v = \frac{1}{1 + \cos^2 \theta}, \tag{4.32}$$

then the magnetic field dependence of the degree of circularity $\mathcal{C}(B)$ of fluorescence, as registered at right angles to the magnetic field, will be determined solely by the destruction of the transversal orientation $_b\rho_{\pm 1}^1$ by the magnetic field. One must remember that only the $_b\rho_{\pm 2}^2$ components of alignment are absent in this geometry, whilst $_b\rho_{\pm 1}^2$, $_b\rho_0^2$ are still present. Nevertheless, $_b\rho_{\pm 1}^2$ components do not manifest themselves in the case where circularly polarized fluorescence is observed in the xy plane; see (4.28). The longitudinal alignment $_b\rho_0^2$ does not depend on the magnetic field B. Thus, observing along the x-axis, where $\varphi' = 0$ (see Fig. 4.4), we have

$$\mathcal{C}(\omega_{J'}/\Gamma_1) = \frac{\mathcal{A}}{1 + (\omega_{J'}/\Gamma_1)^2}, \tag{4.33}$$

whilst observing along the y-axis, where $\varphi' = \pi/2$, we obtain

$$C(\omega_{J'}/\Gamma_1) = \mathcal{A}\frac{\omega_{J'}/\Gamma_1}{1 + (\omega_{J'}/\Gamma_1)^2}, \tag{4.34}$$

i.e. the dependences assume, respectively, either a simple Lorentzian or dispersion shape which permits us to determine Γ_1 and σ_1. At the optimum angle $\cos 2\theta_m = -3/10$ the amplitude of the signal is large, reaching $|\mathcal{A}| = 5/\sqrt{91} \cong 0.5291$. Note that this amplitude value is obtained assuming that $\Gamma_0 = \Gamma_1 = \Gamma_2$.

2 Experimental studies

Eqs. (4.22), (4.26) and (4.27) are called *Hanle signals*, although in the past this term referred only to the depolarization of fluorescence in the form of (4.22). Depolarization of atomic radiation in a weak magnetic field was first observed by Wood and Ellet in 1923 [398]. Around the same time Hanle worked on the same problem in Heidelberg, at the James Franck Institute [184, 185]. He studied the effect of a magnetic field of intensity $\approx 10^{-4}$ T on polarized resonance fluorescence $6^3P_1 - 6^1S_0$ of Hg atoms. Hanle had better optics, a field of high homogeneity at his disposal and could compensate for background magnetic fields, but the main point is that he was able to give a correct *classical explanation* of the effect (quantum mechanics had not been developed at that time). The phenomenon was later named the *Hanle effect* and was applied to the determination of the magnitude of the ratio $g_{J'}/\Gamma_2$. The *quantum description* of the Hanle effect was given by Breit in 1933 [83].

A new approach to the phenomenon appeared in 1959, when Colegrove *et al.* [107] discovered and explained, in terms of crossing of Zeeman sublevels M' and M, the signals which were similar to the Hanle effect but which appeared at non-zero magnetic field; see Fig. 4.5. Its essence is mostly represented by a quantum mechanical *density matrix* $f_{MM'}$ (see [73, 139, 140, 304]). Its non-diagonal elements denote the existence of coherence between sublevels M and M'. Such coherence can be created, e.g., at linearly polarized excitation $\hat{\mathbf{E}} \perp \mathbf{B}$, when a light $\hat{\mathbf{E}}$-vector can be resolved into two components E^{+1} and E^{-1} (see Section 2.1); one may speak of a superpositional transition upon $M = \mu - 1$, $M' = \mu + 1$, i.e., from one lower state $|\mu\rangle$ upon the superposition of states $|M\rangle$ and $|M'\rangle$, $|M - M'| = 2$. As already mentioned in Section 2.3, and as will be shown in Chapter 5, the non-zero matrix elements $f_{MM\pm2}$ correspond to the existence of transversal alignment $f_{\pm2}^2$ where $Q = \pm2$. As may be seen from Fig. 4.5, the changes in $\mathcal{P}(B)$ are observed in the vicinity of $B = B_0$ when the sublevels $|M\rangle$ and $|M'\rangle$ approach each other within a Γ_2 width. Splitting to a distance corresponding to $\omega_{MM'} \gg \Gamma_2$ in the

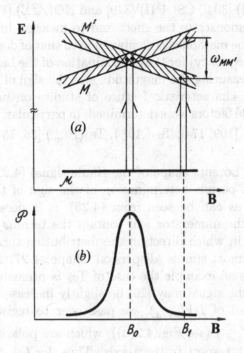

Fig. 4.5. (a) Crossing of magnetic sublevels M' and M. (b) Signal of level crossing in degree of polarization $\mathcal{P}$.

frequency scale $\omega_{MM'} = \omega_M - \omega_{M'}$, $\omega_M = E_M/\hbar$, $\omega_{M'} = E_{M'}/\hbar$ destroys coherence, reducing $f_{MM'}$ (or f_Q^K, $Q = M - M'$) to zero and thus leads to depolarization of radiation. It would therefore be logical to characterize the phenomenon by the term *magnetic level crossing* [107]. The Hanle effect now becomes a special case (at $B_0 = 0$) which is frequently named *zero field level crossing*. This latter term is, generally speaking, used in a wider sense, since the term 'Hanle effect' is not always applied to cases of circular or elliptic polarization of light.

Level crossing has been used rather widely in the case of atoms and ions in order to determine lifetime, Landé factors, fine and hyperfine structure constants and relaxation cross-sections of coherence (σ_1, σ_2). For deeper acquaintance with these questions we recommend monographs and reviews [6, 96, 228, 296, 300, 301, 314] and the literature cited therein.

Let us now turn to molecules. In 1966 Zare [399] was the first to treat in detail the problem of applying magnetic level crossing to molecular objects. He used both the classical and quantum mechanical approachs. Three years later he and his collaborators obtained the first experimental results on the Hanle effect with $Na_2(B^1\Pi_u)$ [290], $OH(A^2\Sigma^+)$ [165]. Among the works performed through 1969–1970 one may mention the studies on $CO(A^1\Pi)$ [394], $CS(A^1\Pi)$ [349] and $NO(A^2\Sigma)$ [171]. The basic aim here was to demonstrate the effect and to measure lifetimes. Occasionally [166, 349] the method was combined with that of double resonance (optical and radiofrequency) for the determination of the Landé factor $g_{J'}$. The necessity of measuring the magnitude and the sign of $g_{J'}$ separately is in many cases a characteristic feature of studies on molecules (Section 4.5). The Landé factors were determined, in particular, for $I_2(B^3\Pi_{0_u^+})$ [88], for $Se_2(B^3\Sigma_u^-)$ [109, 174], $Te_2(A0_u^+)$, $Te_2(B0_u^+)$ [26, 154, 234] and for $Na_2(A^1\Sigma_u^+)$ [365].

The 'traditional' Lorentz shape of the Hanle signal (4.22), which contains $\omega_{J'}^2$, does not permit determination of the sign of the Landé factor. To this end, as can be seen from (4.26), it is necessary to have $\varphi' \neq 0$, and then the numerator will contain the term $\omega_{J'}$. The registration then 'feels' in which direction the distribution turns in Fig. 4.2. The signal has its most simple 'dispersion' shape (4.27) at $\varphi' = \pm\pi/4$ (see Fig. 4.7(b)); as an example the case of Te_2 is presented in Fig. 4.6. The amplitude of the signal may [26] be slightly increased as compared to (4.27), if, instead of $I(\hat{\mathbf{E}}'_{\pm\pi/4})$, we pass over to registration of the components of $I(\hat{\mathbf{E}}'_{\pm\varphi'_m})$ (see Fig. 4.2(a)), which are polarized at the optimal angle φ'_m with respect to the z-axis. Thus, for $Q\uparrow$, $Q\downarrow$ transitions if we measure $\mathcal{P} = (I_{-\varphi'_m} - I_{\varphi'_m})/(I_{-\varphi'_m} + I_{\varphi'_m})$ at the optimum angle $\varphi'_m \cong 52.8°$ the amplitude of the signal increases by 3.5% in comparison with $\mathcal{P}_{-\pi/4} = (I_{-\pi/4} - I_{\pi/4})/(I_{-\pi/4} + I_{\pi/4})$.

A specific property of the Hanle effect in molecules (unlike atoms) is the large values of magnetic fields required for $\omega_{J'}/\Gamma_2 \sim 1$. These are of the order of $B \sim 0.01$–10 T, and sometimes even higher. The experimenter does not as a rule have fields exceeding 1 T. In these cases one may use the dispersion shape of the signal (4.27). Then, for $B = 0$, the derivative over $\xi_2 = \omega_{J'}/\Gamma_2$ equals $d\mathcal{P}/d\xi_2 = 2\mathcal{P}(0)$. Hence, for magnetic fields at which $\xi_2 \ll 1$, one may find from $\mathcal{P}_{\pi/4}(B) \cong 2\mathcal{P}(0)\omega_{J'}/\Gamma_2$ the ratio ξ_2 (in the case of a Lorentz contour (4.22) the signal will be negligibly small at $\xi_2 \ll 1$, owing to $d\mathcal{P}(B)/d\xi_2 = 0$ at the point $B = 0$; see Fig. 4.7(a)). Thus, in the case of $Na_2(A^1\Sigma_u^+)$ the halfwidth of the Hanle signal may exceed 10 T. Nevertheless the application of the dispersion signal made it possible [365] to determine the extremely small value of the product $g_{J'}/\Gamma_1 \cong 3 \cdot 10^{-17}$ s performing measurements in the $B \leq 1$ T region. In the case of (P, R)-type transitions leading to $\mathcal{P}(0) = 1/7$, the amplitude of the signal is larger if one measures the dispersion-type dependence in the degree of circularity $\mathcal{C}(B)$, observation being conducted at right angles to the exciting beam and the magnetic field $\mathbf{B}$ [22]; see Fig. 4.7(c, d). In this case the signal in the vicinity of $B = 0$ turns out to be considerably larger, since $d\mathcal{C}(B)/d\xi_1 = 10/13$ at the point where $B = 0$ (here $\xi_1 = \omega_{J'}/\Gamma_1$).

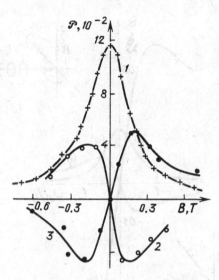

Fig. 4.6. Hanle signal on the molecule $^{130}\text{Te}_2(A0_u^+)$ of Lorentz shape at $v' = 11, J' = 52$ (1) and dispersion shape at $v' = 11, J' = 53$ (2) and $v' = 6, J' = 87$ (3).

This geometry allows us to register quite large $\mathcal{C}(B)$ changes at $B < 1$ T for $\text{Na}_2(A^1\Sigma_u^+)$ in the case where $\omega_{J'}/\Gamma_1 \leq 0.05$; see Fig. 4.7(e).

The determination of the concentration dependence of the width of the Hanle signal is widely used for measuring the relaxation cross-section of alignment σ_2, applying (2.42), where $K = 2$. As an example we may quote data from [312] given in Fig. 4.8, where a very large value of $\sigma_2 \cong 1000$ Å^2 was obtained for collisions $\text{K}_2(B^1\Pi_u) + \text{K}$.

In the case of molecules, just as in the case of atoms one may observe level crossing signals at *non-zero fields*, when $B_0 \neq 0$ (see Fig. 4.5). At first sight, there should not be any principal difference between this and the Hanle effect. Nevertheless, a serious problem arises in the case of molecules, due to weakening of the signal by at least a factor of $2J' + 1$ for a level with angular momentum J', since the signal appears at a crossing of only two sublevels, unlike the Hanle effect, when at the point $B = 0$ we have crossing of all Zeeman sublevels. This leads to an increase of the 'background', which does not contain a useful signal. Nevertheless, signals in non-zero fields have been registered and studied using diatomic molecules as early as 1970–1973 on CS ($A^1\Pi$) [349] (in a mixed Stark–Zeeman crossing variant), on CN between the states $B^2\Sigma^+$ and $A^2\Pi$ [270], as well as on the radical OD $A^2\Sigma^+$ [393]. For instance, Weinstock and Zare [393] determined the position of three level crossings with $|\Delta M| = 2$,

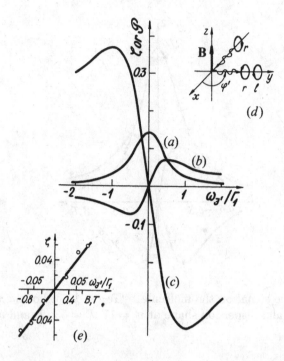

Fig. 4.7. Calculated Hanle signal curves for P, R transitions: (a) linear polarized excitation (Fig. 4.2, $\varphi' = \pi/2$); (b) the same conditions, except that $\varphi' = \pi/4$; (c) circularly polarized excitation at geometry (d); (e) experimentally measured [365] $C(B)$ dependence for $Na_2(A^1\Sigma_u^+, v' = 16, J' = 17)$, $R\uparrow, P\downarrow$-transition.

belonging to hyperfine states with $F' = 1/2, 3/2$ and $5/2$, originating from $N = 1, J = 3/2$. Measuring the positions of the crossings made it possible to find the hyperfine structure constants. Frequently, as, e.g., in [270, 393], we actually have *anticrossing of levels*, where precise crossing is disturbed by an internal or external *perturbation*. This phenomenon has been described in detail for the case of atoms, for instance in [6, 70, 96, 314]. The authors of [393], for example, determined the dipole moment of the OD $(A^2\Sigma^+)$ molecule from Stark shifts of the anticrossing positions.

4.3   Hanle effect: ground state

1   Depopulation effect

Let us now consider the situation, discussed in Chapter 3, where the pumping parameter $\chi = \Gamma_p/\gamma_\kappa$ is sufficiently large to create an anisotropic distribution of angular momenta in the ground state, as described by the

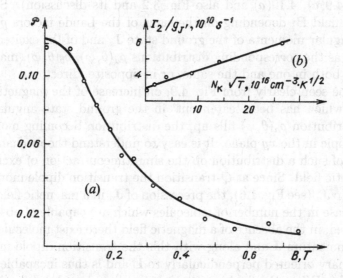

Fig. 4.8. Hanle signal from a $K_2(B^1\Pi_u)$ molecule (weak excitation) (a) and dependence of relaxation rate Γ_2 on potassium atom concentration (b).

set of coefficients $_a\rho_q^\kappa$. The exciting light connects them with the excited state polarization moments $_b\rho_Q^K$ determining the intensity of fluorescence. In the presence of a magnetic field this connection is expressed by Eq. (4.15), where in the presence of a magnetic field $_a\rho_q^\kappa$ are the stationary solutions of (4.10) and are of the form

$$_a\rho_q^\kappa = \frac{\Gamma_p \sum_{\kappa'q'} {}^\kappa_q D^{\kappa'}_{q'} {}_a\rho^{\kappa'}_{q'} + \lambda^\kappa_q \delta_{\kappa 0}\delta_{q0}}{\gamma_\kappa - iq\omega_{J''}}. \tag{4.35}$$

After substituting $_b\rho_Q^K$ thus obtained into the intensity $I(\hat{\mathbf{E}}')$ expression (2.24), it becomes obvious that through $_b\rho_Q^K$ the magnetic field dependence of the transversal components of the multipole moments of the ground (initial) state $_a\rho_q^\kappa$, where $q \neq 0$, may manifest itself in fluorescence, this dependence being determined by the ratio $q\omega_{J''}/\gamma_\kappa$. One may therefore speak of the manifestation of the *level crossing* signal in fluorescence from the *electronic ground state* containing information on $g_{J''}/\gamma_\kappa$.

A clear picture of this may be obtained in the case of linearly polarized excitation from the diagrams in Figs. 4.9 and 4.10, where the angular momenta distributions of the upper state $(\rho_b(\theta, \varphi))$ and the lower state $(\rho_a(\theta, \varphi))$ are shown for Q-type and (P, R)-type transitions under conditions where $g_{J''}/\gamma_\kappa \gg g_{J'}/\Gamma_K$. In the absence of a magnetic field, we have $\omega_{J'}/\Gamma_K = \omega_{J''}/\gamma_\kappa = 0$. The exciting light creates an anisotropy in the momenta $\mathbf{J}$ distribution, both in the excited and in the ground state (see Figs. 4.9(a), 4.10(a) and also Fig. 3.2 and its discussion). Switching on the field $\mathbf{B}$, depending on the signs of the Landé factors $g_{J'}$ and $g_{J''}$, the angular momenta of the ground state $\mathbf{J}_a$ and of the excited state $\mathbf{J}_b$, as well as the corresponding distributions $\rho_a(\theta, \varphi)$, $\rho_b(\theta, \varphi)$, may turn around $\mathbf{B}$, both in one and the same, or in opposite, directions.

As can be seen clearly from Fig. 4.9, on increase of the magnetic field the 'hole' which has been 'eaten out' in the ground state angular momenta distribution $\rho_a(\theta, \varphi)$ fills up, the distribution becoming more and more isotropic in the xy plane. It is easy to understand the mechanism of formation of such a distribution on the simultaneous action of excitation and magnetic field. Since at Q-transition the transition dipole moment $\mathbf{d}$ is parallel to $\mathbf{J}_a$ (see Fig. 1.6), the precession of $\mathbf{J}_a$ in a magnetic field leads to an increase in the number of molecules which are capable of absorbing light. Indeed, in the absence of a magnetic field there exist molecules with angular momentum $\mathbf{J}$ and along with this the transition dipole moment $\mathbf{d}$ is stationary oriented perpendicularly to $\hat{\mathbf{E}}$ and is thus incapable of absorbing light. When the field is switched on, no such molecules exist and the total absorption rate increases. The volume of the ground state angular momentum distribution $\rho_a(\theta, \varphi)$ has therefore been decreased, but the excited state distribution $\rho_b(\theta, \varphi)$ has increased in volume (see Fig. 4.9(b, c)), leading to an increase in the total fluorescence intensity. In this sense one may speak of the *non-linear* magnetic field effect, as opposed to the linear effect at weak excitation, considered in Section 4.2, where the volume of the figures shown in Fig. 4.2 is always conserved, and the total fluorescence intensity is also conserved. With a sufficiently strong magnetic field where $\omega_{J''}/\gamma_\kappa \gg 1$ (see Fig. 4.9(d, e)), during the absorption time γ_κ^{-1} the ground state molecule has managed to rotate many times around the magnetic field $\mathbf{B}$ and as a result the ground state angular momenta distribution $\rho_a(\theta, \varphi)$ in the bottom pictures possesses the axial symmetry with respect to the z-axes. The angular momenta distribution of the excited state $\rho_b(\theta, \varphi)$ is formed in much the same way. With a very strong field $\mathbf{B}$ the excited molecules also manage to rotate many times around $\mathbf{B}$ during the lifetime Γ_K^{-1} and thus the axial symmetric distribution $\rho_b(\theta, \varphi)$ is formed.

A similar situation arises at (P, R)-type linearly polarized excitation (see Fig. 4.10) if one remembers that the transition dipole moment is in this case orthogonal to $\mathbf{J}_a$ (see Fig. 1.6) and $\rho_a(\theta, \varphi)$, $\rho_b(\theta, \varphi)$ at $B = 0$ are now of the form presented in Fig. 3.2(b).

Due to the fact that the effect of a magnetic field on the ground state angular momenta distribution $\rho_a(\theta, \varphi)$ causes changes in the excited state distribution $\rho_b(\theta, \varphi)$ (see Figs. 4.9 and 4.10), one may expect to observe the *ground state Hanle effect* in fluorescence intensity difference $I_\parallel - I_\perp$ or in the degree of polarization $\mathcal{P}(B)$. Indeed, since we have $g_{J''}/\gamma_\kappa \gg$

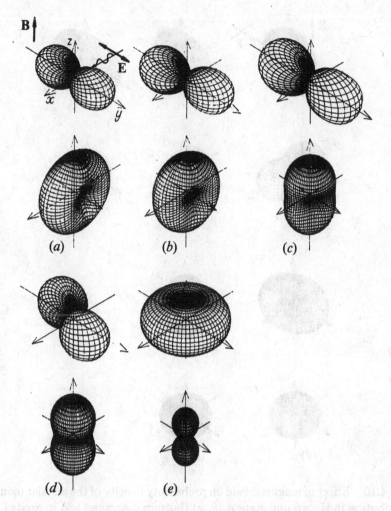

Fig. 4.9. Effect of magnetic field on probability density of the angular momenta distribution in the ground state $\rho_a(\theta, \varphi)$ (bottom diagrams) and in the excited state $\rho_b(\theta, \varphi)$ (top diagrams) in the case of linearly polarized Q-excitation (strong pumping, $\chi = 10/3$). The $\omega_{J''}/\gamma$ and $\omega_{J'}/\Gamma$ values are, respectively: (a) 0; (b) 0.6 and 0.03; (c) 2.0 and 0.1; (d) 12.0 and 0.6; (e) ∞.

$g_{J'}/\Gamma_K$, the ground state distribution $\rho_a(\theta, \varphi)$ (Figs. 4.9 and 4.10(b, c), bottom pictures) is turning around **B**, whilst $\rho_b(\theta, \varphi)$ does not yet turn due to $\omega_{J'}/\Gamma_K \ll 1$. This leads to 'elongation' of $\rho_b(\theta, \varphi)$ along the $\hat{\mathbf{E}}$-direction and to a subsequent increase in $I_\parallel - I_\perp$. If we continue judging the anisotropy of angular momenta distribution from the degree of linear polarization $\mathcal{P}(B) = (I_\parallel - I_\perp)/(I_\parallel + I_\perp)$, observing fluorescence along **B** (see Fig. 4.11), then the conditions in Fig. 4.10 correspond to the

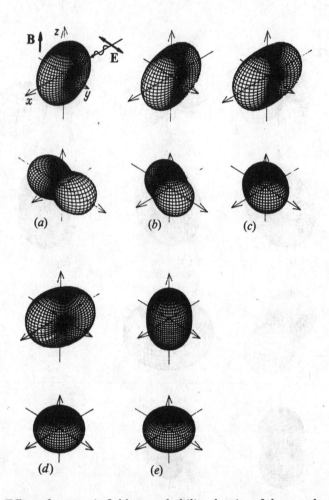

Fig. 4.10.   Effect of magnetic field on probability density of the angular momenta distribution in the ground state $\rho_a(\theta, \varphi)$ (bottom diagrams) and in excited state $\rho_b(\theta, \varphi)$ (top diagrams) in the case of linearly polarized (P, R)-excitation (strong pumping, $\chi = 10/3$). The $\omega_{J''}/\gamma$ and $\omega_{J'}/\Gamma$ values are, respectively: (*a*) 0; (*b*) 1.0 and 0.05; (*c*) 2.4 and 0.12; (*d*) 12 and 0.6; (*e*) ∞.

dependence $\mathcal{P}(B)$ shown by curve 1 in Fig. 4.11, which can be calculated from (4.15), (4.18), (4.19) and (4.35). The value $\mathcal{P}(0) \cong 0.098$ for $\chi = 10/3$ can be obtained from Table 3.5. On switching on the magnetic field, $\mathcal{P}(B)$ increases with increasing $\omega_{J''}/\gamma_\kappa$; see curve 1 in Fig. 4.11 where points b, c correspond to respective figures in Fig. 4.10. Further along, with increase in $\omega_{J'}/\Gamma_K$ we arrive at diminution of $\mathcal{P}(B)$ due to turn and isotropization of distributions shown in Figs. 4.10 and 4.11. One may conclude that the observed dependence $\mathcal{P}(B)$ shown by curve 1 in

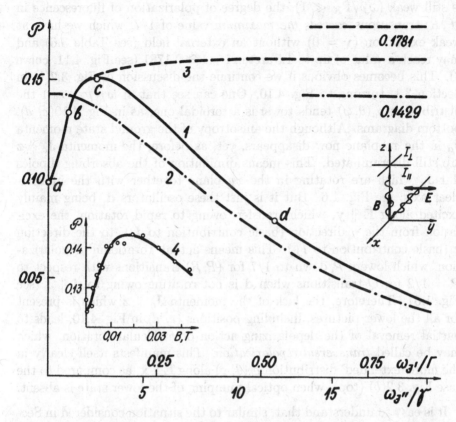

Fig. 4.11. Hanle effect on the degree of linear polarization $\mathcal{P} = (I_\parallel - I_\perp)/(I_\parallel + I_\perp)$ at (P,R)-excitation: 1 – superpositional signal calculated at the same conditions as Fig. 4.10, dots refer to the positions a, b, c, d as in Fig. 4.10; 2 – 'pure' excited state signal at $\chi = 0$; 3 – 'pure' ground state signal at $g_{J'} = 0$; 4 – experimentally measured dependence for Te_2 under conditions as given in Fig. 4.6, curve 1, but in the region of weaker magnetic field and at strong pumping ($\chi \approx 3$).

Fig. 4.11 reflects a *superpositional* Hanle effect from both the ground (initial) and excited states. To demonstrate this in Fig. 4.11 we depict the 'pure' ground state effect (supposing $g_{J'} = 0$) (see curve 3), as well as the 'pure' excited state effect (supposing $\chi = 0$) (see curve 2). In this favorable situation both effects are well distinguished in the observable superpositional signal.

The superpositional Hanle effect may lead to some, at first glance, unexpected peculiarities. Firstly we wish to draw attention to one interesting fact [17]: under conditions where the effect has already developed from the ground state ($\omega_{J''}/\gamma_\kappa \gg 1$), but that from the excited state is still weak ($\omega_{J'}/\Gamma_K \ll 1$), the degree of polarization of fluorescence in (P, R)-transitions *exceeds the maximum* value of $1/7$, which we have at weak excitation ($\chi = 0$) without an external field (see Table 3.6) and may reach a value of up to $1.5 \arctan 1 - 1 \cong 0.1781$ (see Fig. 4.11, curve 3). This becomes obvious if we continue the discussion of Fig. 3.2(b) in Section 3.1, turning to Fig. 4.10. One can see that at $\omega_{J''}/\gamma_\kappa \gg 1$ the distribution $\rho_a(\theta, \varphi)$ tends towards a toroidal one, as in Fig. 4.10(c, d), bottom diagrams. Although the anisotropy of the ground state momenta $\mathbf{J}_a$ in the xy plane now disappears, yet, as before, the momenta $\mathbf{J}_a \parallel \mathbf{z}$ are still discriminated. This means diminution of the absorbing dipoles $\mathbf{d} \perp \mathbf{J}_a$ which are rotating in the xy plane together with the internuclear axis; see Fig. 1.6. But it is just these oscillators $\mathbf{d}$, being mainly excited along $\hat{\mathbf{E}} \parallel \mathbf{y}$, which transfer, owing to rapid rotation, the excitation from the y direction (basic contribution to $I_\parallel$), to the direction x (basic contribution to $I_\perp$). This means actual 'rotational' depolarization, which lowers $\mathcal{P}$ down to $1/7$ for (P, R)-transitions with respect to $\mathcal{P} = 1/2$ for Q-transitions when $\mathbf{d}$ is not rotating owing to $\mathbf{d} \parallel \mathbf{J}$ (see Fig. 1.6). Therefore, the lack of the momenta $\mathbf{J}_a \parallel \mathbf{z}$ which is present for all the lower pictures, including positions (c, d) in Fig. 4.10, leads to partial removal of the depolarizing action of molecular rotation, which may be called *transversal repolarization*. This manifests itself clearly in the more 'extended' distribution $\rho_b(\theta, \varphi)$ along $\hat{\mathbf{E}} \parallel \mathbf{x}$, as compared to the case Fig. 3.2(b) (top), when optical pumping of the lower state is absent.

It is easy to understand that, similar to the situation considered in Section 3.1, the polarization moments $_b\rho_0^0$, $_b\rho_Q^2$ which determine the fluorescence intensities $I_\parallel$, $I_\perp$ (see (4.18) and (4.19)) are now directly connected by linearly polarized optical excitation with ground state polarization moments $_a\rho_q^4$. This means that the *fourth rank* (hexadecapole) moments $_a\rho_q^4$ can be directly manifested in the Hanle signal yielding, in some favorable conditions, an additional narrow structure in the form of a peak in the vicinity of $B = 0$, as shown in Fig. 4.12 (solid line). One must, however, take into account the fact that the polarization moment $_a\rho_q^4$ is, according to (4.35), connected to the sixth rank moments $_a\rho_q^6$ and through them also to higher rank moments. As shown in [12], the peak in $\mathcal{P}(B)$ (see Fig. 4.12) disappears if we neglect $_a\rho_q^6$, and the additional narrow structure may therefore be interpreted as a manifestation in fluorescence of the ground state polarization moments $_a\rho_q^\kappa$ with $\kappa \geq 4$.

Note that the superpositional Hanle signal, reflecting overlapping of effects from both levels, coupled with optical excitation, is sensitive to the *signs* of the Landé factors $g_{J'}$ and $g_{J''}$ even with 'Lorentzian' geometry, where $\varphi' = \pi/2$ in Fig. 4.2(a), contrary to the linear Hanle effect. This is easily understood, since there is a large difference between the cases

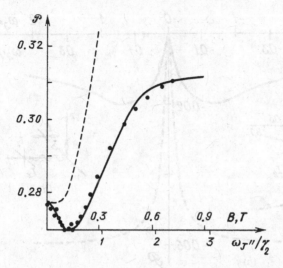

Fig. 4.12. Experimental signal (dots) showing additional structure of the non-linear ground state Hanle effect for K_2. The solid line is the result of approximation at $\gamma = 0.35 \cdot 10^6$ s^{-1}, $\Gamma_p = 2.4 \cdot 10^6$ s^{-1}, the other data are taken from Tables 3.7 and 4.2. The broken line is calculated for the same parameters, but with $g_{J''}$ and $g_{J'}$ of equal sign.

where the distribution $\rho_a(\theta, \varphi)$ and $\rho_b(\theta, \varphi)$ in the magnetic field **B** turn in *opposite* or in the *same* directions. Just how much the signal in these two cases differ one may judge from the curves in Fig. 4.12, which differ only in the signs of $g_{J'}$ and $g_{J''}$.

We will now consider the case where the effect of the magnetic field is stronger in the excited state, i.e. $g_{J'}/\Gamma_K \gg g_{J''}/\gamma_\kappa$. Owing to this, the rotation around the magnetic field **B** $\parallel$ **z** averages the angular momenta distribution in the xy plane much faster in the excited state than in the ground state. Therefore, one cannot observe any manifestation of the ground state Hanle effect if fluorescence is viewed in the direction of the magnetic field. If, however, one registers the intensity difference $I_z - I_x$ or the degree of polarization $\mathcal{P} = (I_z - I_x)/(I_z + I_x)$ of fluorescence emitted, say in the direction of the $\hat{\mathbf{E}}$-vector of the exciting light, i.e., along the y-axis (see Fig. 4.13(a)), the ground state signal will be pronounced in the Hanle contour (solid line) [27, 29, 129, 149, 152]. The calculated signal is presented in Fig. 4.13(b).

Let us turn to experimental investigations. The ground state Hanle signal was registered on Na$_2(X^1\Sigma_g^+)$, K$_2(X^1\Sigma_g^+)$ [149, 152] for linearly polarized Q-excitation under conditions close to those given in Fig. 4.9. In particular, the detailed studies [12, 18] clearly demonstrated a narrow

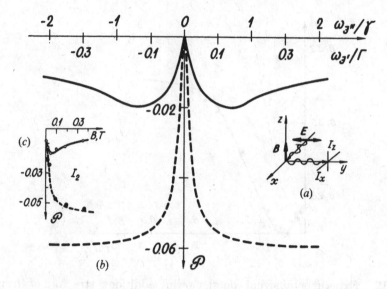

Fig. 4.13. Magnetic field dependence of the degree of polarization $\mathcal{P} = (I_z - I_x)/(I_z + I_x)$ for $\omega_{J''}/\gamma_\kappa < \omega_{J'}/\Gamma_K$, view from the 'tip' of the vector $\hat{\mathbf{E}}$: (a) experimental scheme; (b) calculation, $\omega_{J''}/\gamma = 5\omega_{J'}/\Gamma$; (c) experiment with I_2, excitation $(X^1\Sigma_g^+, v'' = 0, J'' = 13$ and $15) \to (B^3\Pi_{0_u^+}, v' = 43, J' = 12$ and $15)$; broken line refers to weak excitation.

peak at $B = 0$; see dots in Fig. 4.12. The results of experiments on Te$_2$ [370] are shown in Fig. 4.11, curve 4, for linearly polarized (P, R)-excitation, where the conditions correspond to the distributions $\rho_b(\theta, \varphi)$, $\rho_a(\theta, \varphi)$ shown in Fig. 4.10. The case $g_{J'}/\Gamma_K > g_{J''}/\gamma_\kappa$ is valid, in particular, for the I_2 molecule, and the results of experiments [149, 150] demonstrating Hanle signals of $I_2(X^1\Sigma_g^+)$ are presented in Fig. 4.13(c).

The *value* of non-linear superpositional Hanle signals consists of the fact that they contain information on the ground state, which is not easy accessible: namely, on relaxation constants γ_κ and on Landé factors $g_{J''}$, the latter being of particular interest in our opinion; see Section 4.5. The main problem consists of the fact that the description of the signals of the types described in Figs. 4.11, 4.12 4.13 and 4.14 include, following (4.35), the ratio $g_{J''}/\gamma$ as one of the non-linear parameters. Hence, it is necessary for the determination of $g_{J''}$ to measure the relaxation rate independently, which in itself is far from simple (Sections 3.3 and 3.6). One certain possibility of making the task easier consists of the variant [18] suggesting the use of 'fly-through' relaxation (3.43), where $\gamma_0 = T_0^{-1}$. The sign of the Landé factor is easily found [26] from the dispersion shape contours, as in Fig. 4.14, presenting signals registered in the $\mathcal{P}_{\pi/4}(B)$ value, as determined according to (4.26). As may be seen, the shape of

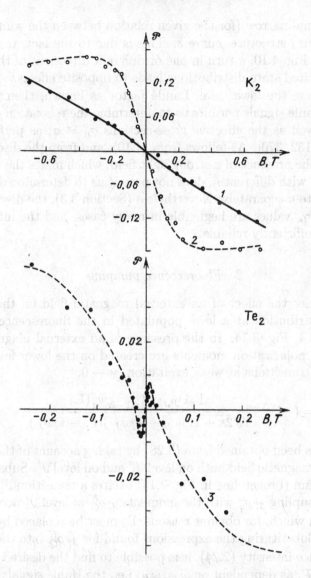

Fig. 4.14. Registered 'dispersion' Hanle signals $\mathcal{P}_{\pi/4}(B)$: 1 and 2 – K$_2$, Q-excitation $(X^1\Sigma_g^+, 1, 73) \rightarrow (B^1\Pi_u, 6, 73)$, $\chi \ll 1$ and $\chi = 3$; 3 – ^{130}Te$_2$, P-excitation $(X0_g^+, 1, 132) \rightarrow (A0_u^+, 11, 131)$, $\chi = 3$.

the superpositional signals changes considerably with dependence on the relation of the signs of $g_{J'}$ and $g_{J''}$. For K$_2$ in the $^1\Sigma_g^+ \leftrightarrow {}^1\Pi_u$ transition where we have $g_{J'}g_{J''} > 0$ (Section 4.5), the ground state contribution leads to a larger slope of curve 2, as compared to that of curve 1 in the case of weak excitation. For ^{130}Te$_2$, when we have $g_{J'}g_{J''} > 0$, the non-linear ground (initial) level contribution manifests itself in the appearance of an additional narrow (for the given relation between the widths of the Hanle contours) structure, curve 3. This is due to the fact, that, as can be seen from Fig. 4.10, a turn in one or different directions of the ground state and excited state distributions leads to opposite effects.

If we assume the lower level Landé factor as known, then the fit of non-linear Hanle signals permits us to determine the relaxation constants $\gamma_\kappa \cong \gamma$, as well as the effective cross-sections $\sigma_\kappa \cong \sigma$, as performed in [18, 27, 149, 152, 370]. As follows from (4.10), and from the discussion in Section 3.2, the presence of a strong light field, which mixes the multipole moments ${}_a\rho_q^\kappa$ with different κ, does not permit us to determine the γ_κ values for definite κ separately. Nevertheless (Section 3.3), the discrepancies between the γ_κ values are negligible in many cases, and the information obtained is sufficiently reliable.

2   Fluorescence pumping

Let us consider the effect of an external magnetic field on the angular momenta distribution at a level populated in the fluorescence process; see Section 3.4, Fig. 3.14. In the presence of an external magnetic field the following polarization moments are created on the lower level J_1'' via spontaneous transitions at weak excitation, $\chi \to 0$:

$$J_1'' \rho_q^\kappa = (-1)^{\Delta+\kappa} \frac{\Gamma_p \Gamma_{J'J_1''} C_{1\Delta 1-\Delta}^{\kappa 0} \Phi_q^\kappa(\hat{\mathbf{E}})}{\sqrt{2\kappa+1}(J_1''\gamma_\kappa - iq\omega_{J_1''})(\Gamma_\kappa - iq\omega_{J'})}. \qquad (4.36)$$

Eq. (4.36) has been obtained from (3.28) by taking account of the effect of the external magnetic field both on level J_1'' and on level J'. Subsequently, the probe beam (broken line in Fig. 3.14) excites a transition $J_1'' \to J_1'$ at a rate Γ_{p_1}, coupling $J_1''\rho_q^\kappa$ with the moments $J_1'\rho_Q^K$ of level J_1' according to Eq. (4.15), in which, for obvious reasons, Γ_p must be replaced by Γ_{p_1} and $\omega_{J'}$ by $\omega_{J_1'}$. Substituting the expressions found for $J_1''\rho_Q^K$ into the formula for fluorescence intensity (2.24), it is possible to find the desired degree of polarization $\mathcal{P}$, as dependent on $\omega_{J''}, \omega_{J'}$, i.e., the Hanle signals observed in the cycle $J_1'' \to J_1' \to J_2''$; see Fig. 3.14. In particular, if both the populating beam (with light vector $\hat{\mathbf{E}}$) and the probe beam (with light vector $\hat{\mathbf{E}}_1$) produce Q-type molecular transitions, then we have, according to [30],

$$Q\mathcal{P} = \frac{\beta_3 \pm \beta_2}{2\tau_1 \pm \beta_1}, \qquad (4.37)$$

$$\tau_1 = \frac{\Gamma_{p_1}\Gamma_p\Gamma_{J'J_1''}}{J_1'\Gamma_{J_1''}\gamma\Gamma}, \quad \beta_1 = \frac{\Gamma_{p_1}\Gamma_p\Gamma_{J'J_1''}(J_1''\gamma\Gamma - 4\omega_{J_1''}\omega_{J'})}{J_1'\Gamma(J_1''\gamma^2 + 4\omega_{J_1''}^2)(\Gamma^2 + 4\omega_{J'}^2)},$$

$$\beta_2 = \frac{\Gamma_{p_1}\Gamma_p\Gamma_{J'J_1''J_1'}\Gamma}{\Gamma_{J_1''}\gamma(_{J_1'}\Gamma^2 + 4\omega_{J_1'}^2)},$$

$$\beta_3 = \frac{\Gamma_{p_1}\Gamma_p\Gamma_{J'J_1'}(_{J_1'}\Gamma_{J_1''}\gamma\Gamma - 4\Gamma\omega_{J_1'}\omega_{J_1''} - 4_{J_1'}\Gamma\omega_{J'}\omega_{J_1''} - 4_{J_1''}\gamma\omega_{J'}\omega_{J_1'})}{(_{J_1''}\gamma^2 + 4\omega_{J_1''}^2)(\Gamma^2 + 4\omega_{J'}^2)(_{J_1'}\Gamma^2 + 4\omega_{J_1'}^2)}.$$

We assume here, for the sake of brevity, that $_{J_1'}\Gamma_K = {}_{J_1'}\Gamma$, $\Gamma_\kappa = \Gamma$ and $_{J_1''}\gamma_\kappa = {}_{J_1''}\gamma$. The expressions have been obtained for the geometry of Fig. 4.2(a) ($\varphi' = \pi/2$). The plus signs in (4.37) correspond to the probe beam having $\hat{\mathbf{E}}_1 \parallel \hat{\mathbf{E}}$, and the minus signs correspond to $\hat{\mathbf{E}}_1 \perp \hat{\mathbf{E}}$.

What kind of information do these Hanle signals contain? Fig. 4.15 might help to make this clear. There we see the expected shape of the Hanle signals at $\omega_{J'} = \omega_{J_1'}$ and at $_{J'}\Gamma = {}_{J_1'}\Gamma$. The curves 1 ($\hat{\mathbf{E}}_1 \parallel \hat{\mathbf{E}}$) and 4 ($\hat{\mathbf{E}}_1 \perp \hat{\mathbf{E}}$) demonstrate the signal from the lower level J_1'' in $\omega_{J_1''}/_{J_1''}\gamma$ scale on the assumption that $\omega_{J_1''}/_{J_1''}\gamma \gg \omega_{J_1'}/_{J_1'}\Gamma$, meaning that the upper level magnetism is not essential. In this case the signal is determined by the parameter $g_{J_1''}/_{J_1''}\gamma$ only and is described by the following two simple expressions:

$$_Q\mathcal{P}(\hat{\mathbf{E}}_1 \parallel \hat{\mathbf{E}}) = \frac{2 + 4\left(\omega_{J_1''}/_{J_1''}\gamma\right)^2}{3 + 8\left(\omega_{J_1''}/_{J_1''}\gamma\right)^2}, \tag{4.38}$$

$$_Q\mathcal{P}(\hat{\mathbf{E}}_1 \perp \hat{\mathbf{E}}) = \frac{-4\left(\omega_{J_1''}/_{J_1''}\gamma\right)^2}{1 + 8\left(\omega_{J_1''}/_{J_1''}\gamma\right)^2}. \tag{4.39}$$

Curves 2 ($\hat{\mathbf{E}}_1 \parallel \hat{\mathbf{E}}$) and 5 ($\hat{\mathbf{E}}_1 \perp \hat{\mathbf{E}}$) in Fig. 4.15 refer to the $\omega_{J_1'}/_{J_1'}\Gamma \sim 1$ scale, in which the *excited state* Hanle effect manifests itself (the ground state Hanle effect is already fully developed and does not manifest itself in this scale). The signal is of Lorentz shape:

$$_Q\mathcal{P} = \frac{\pm 0.5}{1 + 4\left(\omega_{J_1'}/_{J_1'}\Gamma\right)^2},$$

where the plus sign refers to $\hat{\mathbf{E}}_1 \parallel \hat{\mathbf{E}}$, and the minus sign refers to $\hat{\mathbf{E}}_1 \perp \hat{\mathbf{E}}$. In this way one can obtain information separately on parameters $g_{J_1''}/_{J_1''}\gamma$ and on $g_{J_1'}/_{J_1'}\Gamma$ at a definite ratio between the signal widths of levels J_1'' and J_1'.

The *superpositional* signal of Hanle levels J_1', J' and J_1'' is demonstrated by curves 3 ($\hat{\mathbf{E}}_1 \parallel \hat{\mathbf{E}}$) and 6 ($\hat{\mathbf{E}}_1 \perp \hat{\mathbf{E}}$), for which we assume that $\omega_{J_1''}/_{J_1''}\gamma = 15\omega_{J_1'}/_{J_1'}\Gamma$. Such a relation can be met at $B \leftrightarrow X$ transitions in Na$_2$ and K$_2$ dimers (see [114, 152] and Tables 3.7 and 4.2). At

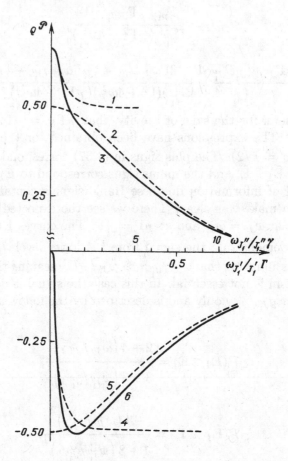

Fig. 4.15.   Hanle signals produced by the probe beam at fluorescence pumping.

first, one expects the appearance of a narrower J_1'' state Hanle effect, and then (at larger B) those of J_1' and J' states. For lower state J_1'' Hanle effect studies, $\hat{\mathbf{E}}_1 \perp \hat{\mathbf{E}}$ geometry is preferable, since in this case the signal amplitude (curves 4 and 6 in Fig. 4.15) is larger. In addition, the lower level and excited level signal (see curves 4 and 5) have different signs and are more easily separated. Note that, according to calculations, the minimum value of $_Q\mathcal{P}$, as determined by the superpositional curve 6, is smaller than the limiting value $_Q\mathcal{P} = -0.5$ for curves 4 and 5. The analysis presented demonstrates, that in the case of fluorescence population, Hanle signals permit us to investigate the magnetic properties and relaxation processes of level J_1''.

Similarly to the above case, signals of level crossing may also be obtained for (P, R)-transitions, as well as for excitation by circularly polarized light [30].

The example discussed considers the case of weak light excitation, where the first cycle $J'' \to J'$ (see Fig. 3.14) does not produce ground state optical polarization via depopulation of the J'' level. If this is not so, then the signal is described, accounting for depopulation effects, and naturally assumes a more complex shape [30].

4.4 Quantum beats

It follows, as we see, from (4.15), (4.17) and (4.35), that at stationary excitation the effect of a constant magnetic field is reduced to an irreversible destruction of the transversal components $_b\rho_Q^K$ and $_a\rho_q^\kappa$, where $Q, q \neq 0$, or to the destruction of coherence. In order to avoid this, i.e., in order to conserve (restore) coherence at an arbitrary magnitude of the magnetic field, it is necessary, arguing formally, to introduce *time dependence* into the right-hand side of (4.9) and (4.10). One usually organizes the time dependence of the pumping rate Γ_p either by means of a short *pulse*, when one may observe the so-called *quantum beats*, or in the form of a *harmonic* time function, when *beat resonance* emerges. We will now discuss both of these methods.

1 Excited state: weak excitation

At first let us revert to the case, already discussed in Section 4.1, where excitation takes place at time $t = 0$ and is affected by a weak light pulse of duration much below the characteristic relaxation time. Under conditions of independently proceeding precession (4.11) and decay (4.14) of the multipole moment $_b\rho_Q^K(t)$, affecting the signal (4.18), (4.19) and (4.28), the magnitude $\mathrm{Re}\,_b\rho_Q^K(t)$ is described by a very simple expression

$$\mathrm{Re}\,_b\rho_Q^K(t) = \mathrm{Mod}\left(_b\rho_Q^K(0)\right) e^{-\Gamma_K t} \cos(\psi_0 + Q\omega_{J'}t). \tag{4.40}$$

The basic result demonstrating (4.40) consists of the fact that after termination of excitation ($t > 0$) the transversal ($Q \neq 0$) components of the multipole moments (their real parts) *oscillate* with their own frequency $Q\omega_{J'}$, which coincides, according to (4.12), with a rate of phase change $\dot\psi = Q\omega_{J'}$, and with an amplitude fading exponentially at a rate Γ_K. It may be pointed out that (4.40) may be easily obtained from (4.9), assuming that $\Gamma_p = G\delta(t)$, G being the amplitude value of the exciting pulse, and that in the ground (initial) state only the population $_a\rho_0^0$ differs from zero.

Let us follow in time the intensity $I_{\varphi'}(t)$ of fluorescence linearly polarized at an angle φ', at linearly polarized excitation and with the geometry

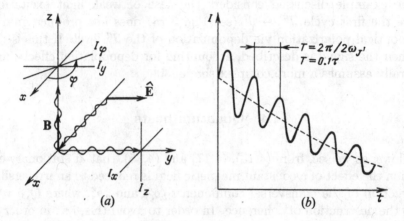

Fig. 4.16. (*a*) Geometry of the experiment. (*b*) Quantum beats in the kinetics of fluorescence.

shown in Fig. 4.16(*a*). Using (2.24), we obtain

$$I_{\varphi'}(t) = K_0 {}_b\rho_0^0(0)e^{-\Gamma_0 t} + K_2 \left[{}_b\rho_0^2(0)e^{-\Gamma_2 t} \right.$$
$$\left. - \sqrt{6}\text{Mod}\left({}_b\rho_2^2(0) \right) e^{-\Gamma_2 t} \cos(\psi_0 + 2\omega_{J'}t - 2\varphi') \right].$$

$$(4.41)$$

Assuming that $\Gamma_0 = \Gamma_2 = \Gamma$, we have

$$I_{\varphi'}(t) = \left[K_0 {}_b\rho_0^0(0) + K_2 {}_b\rho_0^2(0) \right] e^{-\Gamma t}$$
$$- \sqrt{6}K_2\text{Mod}\left({}_b\rho_2^2(0) \right) e^{-\Gamma t} \cos(\psi_0 + 2\omega_{J'}t - 2\varphi'),$$

$$(4.42)$$

where $K_0 = A/3$ and $K_2 = A(-1)^{\Delta'+1}C_{1-\Delta'1\Delta'}^{20}/\sqrt{6}$. The above shows that the oscillations of transversal alignment ('beats') *modulate* the exponential decay of the state at frequency $2\omega_{J'}$. The appearance of beats in the intensity of the radiation admits a very simple interpretation. Let us consider, for instance, a Q-transition. Then we have precession around the magnetic field of the angular momenta $\mathbf{J'}$, as well as that of the fading collinear dipoles $\mathbf{d}$. If we observe the evolution in time of the emission of the latter in a fixed direction, for instance along the field $\mathbf{B}$ direction, we will observe decay of the fluorescence, modulated by the double precession frequency $2\omega_{J'}$, or the *intensity beats*. The synchronism of the oscillations of different molecules is provided by the common excitation moment, $t = 0$. The concrete values of ${}_b\rho_0^0$, ${}_b\rho_0^2$ and ${}_b\rho_2^2$ at $t = 0$ may easily be found from (2.21) and (2.22), whilst the numerical value of the coefficient K_2 follows from Table C.1, Appendix C. It can easily be seen that the ratio between the *modulated* part and the *non-modulated* part of the intensity beats is simply equal to the degree of polarization $\mathcal{P}$ from Table 3.6. For instance, in the case of $Q \uparrow Q \downarrow$ transitions we have:

$$\frac{Mod}{Non\text{-}mod} = q\mathcal{P} = 0.5. \tag{4.43}$$

Examples, as described by Eqs. (4.41), (4.42) and (4.43) show what kind of information one may obtain directly by registering oscillations in the fluorescence decay. These are the *lifetime* $\tau = \Gamma^{-1}$ of the state, the factors affecting it, the *precession frequency* $\omega_{J'}$ and, consequently, the value of the Landé factor $g_{J'}$, as well as its sign (by the initial phase of oscillations), and, finally, the *degree of polarization* $\mathcal{P}$. A favorable condition for registration should be the validity of $T = 2\pi/(Q\omega_{J'}) \ll \tau, T$ being the period of oscillations. In Fig. 4.16(b) we assume that $T/\tau = 0.1$.

In spite of the apparent obviousness of the beat effect in optical radiation at pulsed excitation, it was only registered and studied comparatively recently. At the beginning of the 1960s Aleksandrov [3] and, independently, Dodd and coworkers [119] discovered beats in atomic emission. It may be pointed out that this, and the related phenomenon of beat resonance, was predicted by Podgoretskii [313], as well as by Dodd and Series [118]. The phenomenon was treated on the basis of well-known fundamental concepts on *coherent superposition* of states, and was named accordingly *quantum beats*. These ideas are amply expounded in reviews and monographs [4, 5, 6, 71, 96, 120, 146, 182, 188, 343, 348, 388].

Their essence is conveniently clarified if one continues the discussion begun in Fig. 4.5. Let the distance between the states $|M\rangle$ and $|M'\rangle$ be $\omega_{MM'}$, and let the optical transition proceed in such a way that at time $t = 0$ a coherent superposition of $|M\rangle$ and $|M'\rangle$ is excited from the same ground state $|\mu\rangle$. For the jth particle the superpositional state is

$$\begin{aligned}|\Psi_j(t_j)\rangle = & \; c_{jM}|M\rangle \exp\left[(i\omega_M - \Gamma_M/2)t_j\right] \\ & + c_{jM'}|M'\rangle \exp\left[(i\omega_{M'} - \Gamma_{M'}/2)t_j\right].\end{aligned} \tag{4.44}$$

Here c_{jM} and $c_{jM'}$ are the amplitudes depending on the conditions of excitation and Γ_M and $\Gamma_{M'}$ are the relaxation rates of corresponding magnetic sublevels. It can easily be seen that the probability density $\langle\Psi_j|\Psi_j\rangle$ of the superpositional state contains an *interference* ('crossing') term

$$c_{jM}c_{jM'}^* \exp\left[(i\omega_{MM'} - \Gamma_{MM'})t_j\right],$$

which shows the probability for oscillations to be in the states $|M\rangle$ and $|M'\rangle$, along with exponential decay at a rate $\Gamma_{MM'} = (\Gamma_M + \Gamma_{M'})/2$. They can be understood as beats between two fading oscillatory processes with different eigenfrequencies ω_M and $\omega_{M'}$. The phase difference of these oscillations $(\omega_M - \omega_{M'})t_j = \omega_{MM'}t_j$ changes over time following a linear law. In this connection one must emphasize in particular the decisive role of the phase of the wave function (4.44), which is frequently neglected.

The probability $W_j(t_j)$ of a radiative molecular transition from the $|\Psi_j\rangle$ state under the effect of a perturbation $\hat{V}$ into a certain state $|\mu'\rangle$ will also contain an interference term [4]:

$$\begin{aligned} W_j(t_j) &= \langle\mu'|\hat{V}|\Psi_j\rangle \\ &= [A + B\cos(\psi_j + \omega_{MM'}t_j)]\exp(-\Gamma_{MM'}t_j). \end{aligned} \quad (4.45)$$

It is this term that describes *magnetic* ('Zeeman') *quantum beats* in the form of intensity modulation with frequency $\omega_{MM'}$.

In order to describe the signal observed, it is necessary to average over the canonic ensemble of N identical particles j, or, in other words, to pass over to the density matrix $f_{MM'}$ [73]. The oscillatory term does not disappear after such a procedure, because at pulsed excitation all excitation moments are synchronized.

It is said that quantum beats are the result of *interference of non-degenerate sublevels* (not necessarily magnetic, in the general case), since for their observation the condition $\omega_{MM'} > \Gamma$ is necessary, which means the removal of degeneracy. Applying this terminology, *level crossing* ($\omega_{MM'} = 0$) is a special case of *degenerate state interference*. Coherence is automatically provided here, owing to the disappearance of the factor $\omega_{MM'}t_j$ in (4.45). The manifestation of such coherence leads to the specific angular distribution and polarization of fluorescence; see Section 4.2.

'Free' quantum beats, after excitation by a pulse of duration $\Delta t \ll \Gamma$, show the essence of the phenomenon clearly, since they appear in the fluorescence after termination of excitation and are determined only by the properties of the molecular ensemble itself. It is just this kind of beat which was observed in the form of the superposition of a harmonic component on the kinetics of fluorescence in the first experiments [3, 119] on mercury atoms $Hg(6^3P_1)$ and on cadmium atoms $Cd(5^3P_1)$ in the magnetic field. The 'magnetic' beats are connected with the interference of magnetic sublevels $m = \pm 1$ of the $^3P_1$ state at pulsed excitation by resonance lines of the same atoms. Subsequently, other kinds of coherent pulsed excitation were performed: in beam–foil collisions and in electron impact. New possibilities emerged with the appearance of pulsed tunable lasers. Such a technique was first demonstrated on ytterbium $Yb(6^3P_1)$ in [169]. 'Magnetic' beats in the system of Zeeman sublevels at laser excitation were observed with Ba and Ca atoms [338]. We limit ourselves to the above-cited works concerning atomic systems, referring to the above-mentioned reviews for more information. In order to follow the discussions it is essential to mention that there exists the possibility of observing beats on atoms after pulsed optical depopulation of the *ground state*. Thus, in [295] coherence of magnetic sublevels was created in $Na(3^2S_{1/2})$, whilst the beat effect was registered from absorption of the light beam, applying polarization spectroscopy (Section 3.5).

We will now consider diatomic molecules. The first experiment on observing 'free' magnetic quantum beats was performed on molecules in a magnetic field in 1974 by the Schawlow group [387] on iodine molecules $I_2(B^3\Pi_{0_u^+})$, and was used for the determination of the Landé factor. Considerably later the method of quantum beats was applied to the study of the magnetism of excited states of $Br_2(B^3\Pi_{0_u^+})$ [278]. An example of a highly effective combination of beat spectroscopy and selective excitation of rovibronic levels was achieved by Gouedard and Lehmann [174]. Scanning over the rovibronic levels they studied magnetism in the selenium molecule $^{80}Se_2(B0_u^+)$ in great detail. Quantum beats in fluorescence were obtained using LiH [84], using free OH radicals [266], and also using tellurium dimers $Te_2(A0_u^+)$ [233]. As an example we present data from a later work. Fig. 4.17 shows a beat signal in the fluorescence kinetics of $v' = 6, J' = 87$ level of $^{130}Te_2(A0_u^+)$, excited by the 514.5 nm line of an Ar^+-laser. The signal was obtained under conditions that were unfavorable for the observation of beats, where, unlike Fig. 4.16, the period of oscillations T was only about half as long as the lifetime, owing to limited magnetic field strength ($B \leq 1$ T). For this reason only some two oscillation periods took place, which, in addition to the high noise level of the low intensity signal and the comparatively small modulation amplitude (as follows from (4.43) and allowing for $_{P,R}\mathcal{P} = 1/7$), created difficulties in determining $\omega_{J'}$. However, even under such conditions the data processing method applied produced satisfactory results. The following procedure was used. At first the relaxation rate Γ was determined from curve 2 without a magnetic field. Then the values obtained in the presence of a field of 0.7 T, curve 3, were multiplied by $\exp(\Gamma t)$. The dependence obtained was approximated by a simple harmonic expression (curve 4):

$$\tilde{I} = C_1 \cos 2\omega_{J'}t + C_2 \sin 2\omega_{J'}t + C_3, \tag{4.46}$$

which permitted limitation to one non-linear parameter $\omega_{J'}$. The determination of the $\omega_{J'}$ dependence on B (see Fig. 4.17(b)) yields, from Eq. (4.2), the value $g_{J'} = (1.07 \pm 0.08) \cdot 10^{-4}$, which, even under such unfavourable circumstances, exceeds the accuracy of the result obtained by the Hanle method; see Fig. 4.6.

In many cases we obtain beats of various frequencies simultaneously which correspond to different values of hyperfine structure components, for instance with iodine I_2 [387] and bromine Br_2 [278]. In these cases we obtain a sum over the corresponding frequencies $\omega_{M_F M_{F'}}$ in (4.45). The problem of determining the set $\omega_{M_F M_{F'}}$ from the observed quantum beat signal is solved with the aid of Fourier transformation. It is more difficult to interpret the signals of magnetic quantum beats in the case of polyatomic molecules, such as in [389] with SO_2 and in [90] with $NO_2(\tilde{A}^2 B_2)$, where, as a general rule, one observes many perturbations between different states.

So far we have been discussing magnetic (Zeeman) quantum beats taking place at frequencies of coherent sublevel splitting in an external magnetic field.

There are no limitations, in principle, in applying the method of beats to molecular objects in the absence of a magnetic field; this is well known in the case of atomic systems, i.e. between different state *non-degenerate levels*; see previously cited reviews [5, 6, 71, 120, 146, 182, 188, 348, 388]. Indeed, instead of $|M\rangle$ and $|M'\rangle$ in (4.44) and (4.45) one may imagine any non-degenerate states which one must 'suddenly' prepare in a coherent way and provide for the possibility of registering oscillations on the time scale of the existence of coherence. Molecules, especially polyatomic molecules, ought to be favorable objects for the observation of beats between mixed electronic states, owing to manifold perturbations. Nevertheless, a demonstration of such a fundamental phenomenon encountered many difficulties, connected, in the first place, with simultaneous excitation of a large number of rotational states. The situation changed with the introduction of jet cooling and laser excitation and, beginning with the work of McDonald and his group [98], the beat method spread on quite a wide scale, and in its application to polyatomic molecules. We speak here of beats which are insensitive to the polarization of light. They may be classified in the following way: $a - (S_1-T_1)$-beats, as the result of excitation of superposition of closely situated singlet (S_1) and triplet (T_1) states, as performed, e.g., in [193] with biacetyl, and in [302] on pyrazine and pyramidine; $b - (S_1-S_0)$-beats between different singlet states, as, e.g., in [206], where beats were observed due to mixing levels $\tilde{C}(^1 B_2)$ with the ground state $X^1 A_1$ of an SO_2 molecule. The mixing mechanisms are essentially different. In the first case we have spin–orbital, whilst in second case we have vibronic mixing. The possibility of observing interference in large molecules [98, 193, 302], possessing a high density of states, is of considerable interest. On processing the signals one may determine the matrix elements of interaction.

Let us mention the 'rotational' beats which have recently been observed (with a *trans*-stilbene molecule in a gasodynamic jet) and interpreted by Felker and Zewail [54, 145]. The essence of this interesting phenomenon may be described in a simplified way as follows. Let a picosecond laser pulse be capable of coherently exciting from some rotational state J_0'' of a

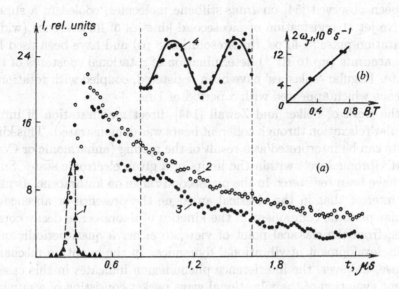

Fig. 4.17. (a) Signal of quantum beats of $^{130}Te_2(A0_u^+, v' = 6, J' = 87)$. (b) Field B dependence of beat frequency.

certain singlet ground state $|v_0 S_0\rangle$ transitions on to three rotational levels $J' = J_0'', J_0'' \pm 1$, which belong to some excited vibronic state $|v_1 S_1\rangle$. If we now observe the kinetics of radiation intensity $I_\parallel$ or $I_\perp$ at linearly polarized excitation (see Fig. 2.5) for a parallel transition, then the time dependence of fluorescence will be modulated by three beat components with differential frequencies $\Delta\omega_i/2\pi$, equalling $2J_0''B$ and $2(J_0''+1)B$, where B is the constant of rotation around the main axis. It is important, to the possibility of observing the effect, that the beats do not disappear at thermal averaging over a large number of rotational states J'' from which excitation takes place. This is due to the fact that the frequencies and amplitudes of rotational beats vary with changes in J'', K'', not at random, but following a certain rule, according to which $\Delta\omega_i/2\pi$ is a multiple of $2B$. Hence, after time $t = n/(2B)$, $n = 0, 1, 2, 3, \ldots$, phase synchronization will be observed, leading to a 'splash' of constructive interference. An analysis for the case of a classically rotating molecule leads to the conclusion that we have a case of direct manifestation of an evolution of *alignment of rotating dipole moments* (not of angular momenta); see Figs. 1.5 and 1.6. The beats are connected with the orientation of dipoles in excitation and in emission processes. They possess different signs for $I_\parallel$ and $I_\perp$, and disappear for $I_\parallel + 2I_\perp$ (2.26), as, for example, on observation in the direction of the 'magic angle' $\theta_0 = \arccos(1/\sqrt{3})$, which follows from the discussion on Eqs. (2.29) and (2.30). 'Purely rotational' beats have been observed [54] on *trans*-stilbene molecules, cooled in a super-sonic Ne jet at registration of picosecond kinetics of fluorescence (width of excitation pulse ≈ 40 ps, time resolution 8 ps) and have been used for highly accurate (up to 10^{-3}) determination of rotational constants of the S_1 state. Regular 'splashes' have been registered, coupled with rotational coherence which appeared with a period of 1 ns.

In the work of Felker and Zewail [144] direct manifestation of intra-molecular relaxation through coherent beats was demonstrated. This kind of beats can be interpreted as a result of the mixing (anharmonic or Cori-olis) of vibronic levels within the limits of a given electronic state. Such beats have been registered in the picosecond range on anthracene. It may be of interest that in the classical analogue the presence or absence of a regular periodic component in the kinetics of fluorescence décay corre-sponds, from the classical point of view, to either a quasi-periodic or a chaotic development of vibrational dynamics. In the quantum mechani-cal sense, however, the interference phenomenon indicates in this case a coherent evolution of the vibrational wave packet consisting of a number of states involved in the process.

The variety of manifestations in time of coherent development of molec-ular dynamics also includes such phenomena as mono- and bimolecular *chemical reactions*. Thus, Seideman *et al* [342] suggest the idea of 'gov-erning' the yield of a reaction by suddenly creating coherent superposition of two states of the transient complex and applying a second pulse with fixed delay for the dissociation of the complex. The appearance of co-herent beats in 'femtochemistry', in particular, at photodissociation, has been analyzed by Zewail (review [404]).

2   Depopulation quantum beats

Let us now revert to magnetic beats. We will consider the situation discussed in Section 4.3, when polarization and coherence of magnetic sublevels of the *ground state* was being created. Following the action of a short light pulse, the duration of which is much shorter then the characteristic relaxation time (δ-pulse), the angular momenta distribu-tion in the ground state will be in precession around the magnetic field **B** (z-axis) (see Fig. 4.18, bottom picture), whilst being 'filled' up to a spherical isotropic one at a rate γ, due to relaxation. A weak cw probe beam with the same polarization is applied in order to monitor the ground state angular momenta distribution. Let us assume that one may neglect precession of the angular momenta of the excited state in an external field and consider the excitation process by a probe beam as quasi-established on a γ^{-1} time scale, putting $\Gamma \gg \gamma$. Under these conditions the time de-pendence of the distribution $\rho_b(\theta, \varphi)$ for the excited state (see Fig. 4.18, top) will be governed, on the γ^{-1} scale, by a slower evolution of distribution $\rho_a(\theta, \varphi)$, owing to precession and relaxation. Thus, for instance, since the distribution $\rho_a(\theta, \varphi)$ in Fig. 4.18(b, d) is turned along $\mathbf{y} \parallel \hat{\mathbf{E}}$, i.e. in the most advantageous fashion for Q-absorption, the volumes of the upper figures $\rho_b(\theta, \varphi)$ in these positions will be larger than in neighboring positions (a, c, e). The amplitude of such 'pulsations' of the volumes will gradually decrease owing to isotropization of the ground state distribution. The pulsations of the $\rho_b(\theta, \varphi)$ volume must lead to pulsations in fluorescence during observation of the radiation from the excited state. The corresponding behavior of the time dependence of the fluorescence intensity is shown in Fig. 4.19. At first sight such *ground state quantum beats* are similar to quantum beats which take place, on a time scale Γ^{-1} in the excited state at weak pulsed excitation (see Fig. 4.16(b)). There is, however, a difference consisting of the fact that, since the lower state does not radiate, 'free' quantum beats cannot be observed in the fluorescence, and therefore a probe light beam is necessary.

For a description of the ground state magnetic quantum beats one might conveniently use the solution of Eq. (4.10) for multipole moments $_a\rho_q^\kappa$. Assuming that the excitation takes place by a δ-pulse at time $t = 0$, one may write its solution for $t > 0$ in the form:

$$_a\rho_q^\kappa(t) = \delta_{\kappa 0}\delta_{q 0}{}_a\rho_0^0(0) - \left[\delta_{\kappa 0}\delta_{q 0}{}_a\rho_0^0(0) - {}_a\rho_q^\kappa(0)\right]\exp\left[-(\gamma_\kappa - iq\omega_{J''})t\right].$$

(4.47)

The equation demonstrates the exponential behavior of $_a\rho_q^\kappa(t)$ in the transient process which differs from the one discussed in Section 3.6 (Eqs. (3.34) and (3.35)) by the presence of oscillations with frequency $q\omega_{J''}$ of the transversal components of the multipole moments $_a\rho_q^\kappa$, i.e., with frequency of splitting of the coherently depopulated magnetic sublevels μ, μ' ($q\omega_{J''} = (E_\mu - E_{\mu'})/\hbar$) in the ground state. We then substitute $_a\rho_q^\kappa$ from (4.47) into Eq. (4.15), which we will consider as a quasi-stationary solution of (4.9). The $_b\rho_Q^K(t)$ obtained can then be used in Eqs. (4.41) and (4.42) to calculate the intensity $I_{\varphi'}(t)$ of the radiation. A graphic illustration [23] of such signals in the geometry, as presented in Fig. 4.18(a), is shown in Fig. 4.19. It may be seen that in I_x and I_y *alignment beats* with frequency $2\omega_{J''}$ appear and dominate, whilst in I_z only *hexadecapole* beats with frequency $4\omega_{J''}$ appear (the latter are also present in I_y, but their relative contribution is negligible).

Fig. 4.20 shows an example of registered signals [20] for the ground state of $K_2(X^1\Sigma_g^+)$ under conditions following Table 3.7. The laser beam was modulated with depth 0.8–0.9 in the form of pulses of $\Delta t \cong 10^{-6}$ s duration. The kinetics of the transient process in the intensity of radiation (component I_y), excited by a weakened continuous wave 'background' after break-off of the pulse, is shown in Fig. 4.20(a). In Fig. 4.20(b), where

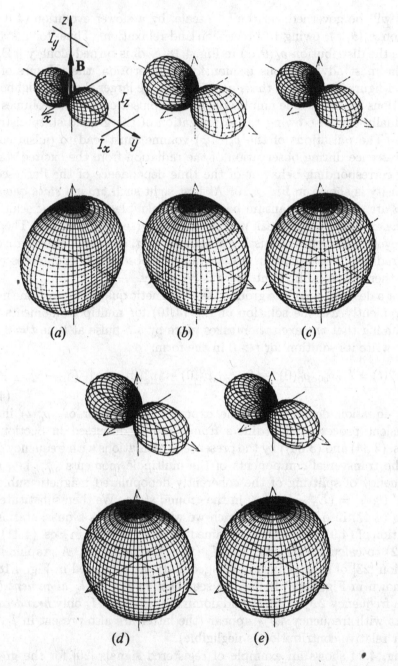

Fig. 4.18. Top diagrams – changes in time ('pulsations') of the angular momenta probability density $\rho_b(\theta, \varphi)$ of the upper level. Bottom diagrams – precession around **B**, at $\omega_{J''}/\gamma = 5$, of the angular momenta probability density $\rho_a(\theta, \varphi)$ of the lower level. Linearly polarized Q-excitation by strong δ-pulse: (a) $\gamma t = 0$; (b) $\gamma t = \pi/10$; (c) $\gamma t = \pi/5$; (d) $\gamma t = 3\pi/10$; (e) $\gamma t = 2\pi/5$.

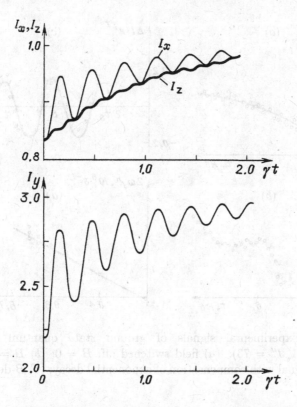

Fig. 4.19. Calculated signals of ground state quantum beats for $\omega_{J''}/\gamma = 10$. Geometry as in Fig. 4.18(a).

the magnetic field (0.816 T) is switched on, one can observe oscillations. The dependence in Fig. 4.20(c), obtained by subtraction of the above two signals and multiplication by the factor $\exp(\gamma t)$ compensating relaxation in (4.47), permits us to determine the modulation frequency $2\omega_{J''}$. Although here, as in Fig. 4.17, it is desirable to diminish the beat period in comparison with the mean relaxation time γ^{-1}, the B dependence of $\omega_{J''}$ in Fig. 4.20(d) makes it possible to find the Landé factor value presented in Table 4.2, Section 4.5. An increase in $\omega_{J''}/\gamma$ diminishes the error considerably, and such a method appears to be very fruitful for measuring the Landé factor in the electronic ground state. This is, in the first place, due to the simplicity of processing the signal, i.e. to the action of the strong light pulse being terminated. Thus, the approximation of the signal in Fig. 4.20(b) in the form

$$I(t) = I(t \to \infty) - [C_1 + C_2 \cos(2\omega_{J''}t + \psi_0)]\exp(-\gamma t) \qquad (4.48)$$

is perfectly satisfactory ($\gamma_\kappa = \gamma$ was supposed here).

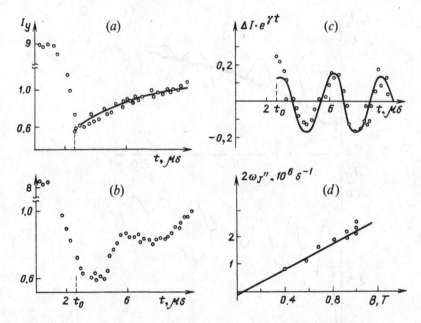

Fig. 4.20. Experimental signals of ground state quantum beats in $K_2(X^1\Sigma_g^+, v'' = 1, J'' = 73)$: (a) field switched off, $B = 0$; (b) $B = 0.816$ T; (c) differential signal with compensation of exponential decay; (d) B-dependence of $\omega_{J''}$.

3 Beats in laser-interrogated dichroism

Magnetic quantum beats in the transient process after pulsed depopulation of the ground state may be observed not only in fluorescence, but also in a more direct way, namely in *absorption*. In connection with what was discussed in Section 3.5, one must expect maximum sensitivity if the experiment is conducted according to the laser interrogated dichroism method; see Fig. 3.17. To this end it is convenient to direct the external magnetic field $\mathbf{B}$ along the z-axis as shown in Fig. 4.21 where the probe beam $\hat{\mathbf{E}}$-vector can be either in the xy plane ($\hat{\mathbf{E}}_{pr_1}$) or in the yz plane ($\hat{\mathbf{E}}_{pr_2}$).

Let us assume that the pumping beam is of the δ-pulse type, i.e. is switched on at time $t = 0$ for a short time as compared with the relaxation time, whilst the probe beam acts continuously. Substituting (3.31) into Eq. (3.32) in which only the first and the last terms are left and taking

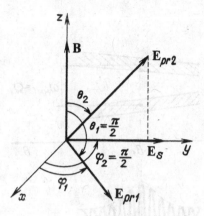

Fig. 4.21. Geometry of experiment on laser-interrogated dichroic spectroscopy.

(4.47) into account, we obtain, respectively, for Q- and (P, R)-absorption

$$\left. \begin{aligned}
_Q I_1 &\propto \frac{2\sqrt{2}}{\sqrt{3}} a\rho_2^2(0) \exp(-\gamma_2 t) l\varepsilon \sin(2\varphi_1 + 2\omega_{J''}t) + \frac{\varepsilon^2}{4}, \\
_Q I_2 &\propto -\left[\frac{\sqrt{2}}{\sqrt{3}} a\rho_2^2(0) \cos 2\omega_{J''}t + a\rho_0^2(0)\right] \\
&\quad \times \exp(-\gamma_2 t) l\varepsilon \sin 2\theta_2 + \frac{\varepsilon^2}{4}, \\
_{P,R} I_1 &\propto -\frac{\sqrt{2}}{\sqrt{3}} a\rho_2^2(0) \exp(-\gamma_2 t) l\varepsilon \sin(2\varphi_1 + 2\omega_{J''}t) + \frac{\varepsilon^2}{4}, \\
_{P,R} I_2 &\propto \left[\frac{1}{\sqrt{6}} a\rho_2^2(0) \cos 2\omega_{J''}t + \frac{1}{2} a\rho_0^2(0)\right] \\
&\quad \times \exp(-\gamma_2 t) l\varepsilon \sin 2\theta_2 + \frac{\varepsilon^2}{4},
\end{aligned} \right\} \quad (4.49)$$

where I_1 and I_2 correspond to $\hat{\mathbf{E}}_{pr_1}$ and $\hat{\mathbf{E}}_{pr_2}$ in Fig. 4.21. The mechanism of the appearance and the magnitude of the polarization moment $a\rho_q^\kappa$ during the action of the pumping beam is considered in detail in Section 3.5. The form of Eqs. (4.49) shows that, unlike quantum beats in the fluorescence from the ground state, laser interrogated dichroism in absorption reveals quantum beats only at one frequency $2\omega_{J''}$. This may simplify the treatment of the signals.

If the magnetic field is switched off, then, according to (4.49), one may observe relaxation of the anisotropy of the angular momenta, and the transient process will be determined by only one rate of relaxation of alignment γ_2, which is also an advantage against the situation that one faces in the case of registering the transient process in fluorescence (Section 3.6).

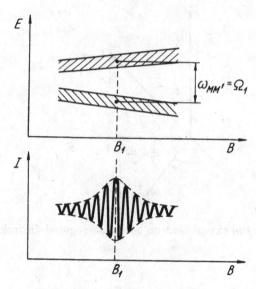

Fig. 4.22. Beat resonance.

4 Beat resonance

Among other methods of creating and observing optical manifestations of coherent superpositions of non-degenerate magnetic sublevels (an elegant analysis may be found in [4]), the most widely used method is that of *resonance of beats*. The effect appears when the frequency Ω_1 of *harmonic modulation* of the excitation rate

$$\Gamma_p = \Gamma_{p0}(1 - \varepsilon \cos \Omega_1 t) \qquad (4.50)$$

coincides with the splitting frequency of the coherent states, $\Omega_1 = Q\omega_{J'} = \omega_{MM'}$. Then the intensity of the spontaneous radiation will be modulated at frequency Ω_1 with maximum amplitude, whilst out of resonance modulation depth decreases (see Fig. 4.22). The first to register beat resonance in emission was Aleksandrov [2] in cadmium $Cd(5^3P_1)$ vapor, to be followed by Corney and Series on the same object [108]. The theory of the phenomenon was evolved in [133, 241]. It might be mentioned that the phenomenon of beat resonance actually had already taken place, essentially, in the experiment by Bell and Bloom [56] in the optical orientation of Rb in the ground state, when the signal was observed by the change in absorption of the orienting beam itself.

We will now try to present a comprehensible treatment of the phenomenon. To this end we turn again to Fig. 4.2. It is essential that the isotropy of angular momenta $\mathbf{J}'$ distribution in the xy plane in Fig. 4.2(e) actually is the result of the superposition of shapes similar to those in

Fig. 4.2(a), but in precession around the field $\mathbf{B}$ with a frequency $\omega_{J'}$, which considerably exceeds the relaxation rate, since $Q\omega_{J'}/\Gamma \gg 1$, but is in a different phase owing to excitation moments t distributed at random. Such distribution will be transformed into itself with a frequency $Q\omega_{J'}$ depending on the distribution symmetry as determined by Q. Let us now imagine that a regular excitation has been arranged, say one which is harmonically modulated. If the modulation frequency Ω_1 coincides with the frequency $Q\omega_{J'}$, we will observe the 'stroboscopic' effect of a quasi-static picture. This means partial return to the situation as shown in Fig. 4.2(a), or, in other words, *restitution of coherence*. One may also notice an analogy with resonance of oscillations in a system under the action of a periodic external force.

The above is confirmed by the solution of Eq. (4.9). Assuming that in the ground (initial) state there is only the population $_a\rho_0^0$, the solution for $_b\rho_Q^K$ at $t \gg \Gamma_K^{-1}$ (steady process) is obtained in the form of the expression presented in [133]:

$$_b\rho_Q^K(t) = \frac{(-1)^{\Delta+K}}{\sqrt{2K+1}} C_{1\Delta1-\Delta}^{K0} \Phi_Q^K(\hat{\mathbf{E}})\Gamma_{p0} \left\{ \frac{\Gamma_K - iQ\omega_{J'}}{\Gamma_K^2 + Q^2\omega_{J'}^2} \right.$$
$$\left. -\varepsilon \frac{[(\Gamma_K + iQ\omega_{J'})^2 + \Omega_1^2][(\Gamma_K + iQ\omega_{J'})\sin\Omega_1 t - \Omega_1\cos\Omega_1 t]}{[\Gamma_K^2 + (\Omega_1 - Q\omega_{J'})^2][\Gamma_K^2 + (\Omega_1 + Q\omega_{J'})^2]} \right\}.$$

$$(4.51)$$

The first term in curly brackets does not depend on t and represents the usual Hanle effect. The second term oscillates at modulation frequency Ω_1, with the modulation amplitude growing in resonance fashion if $Q\omega_{J'} = \Omega_1$ holds (see Fig. 4.22). For $\omega_{J'} \gg \Gamma_K$, when degeneracy is completely removed, we may write in the vicinity of resonance for $Q > 0$:

$$_b\rho_Q^K(t) = \frac{(-1)^{\Delta+K}}{\sqrt{2K+1}} C_{1\Delta1-\Delta}^{K0} \Gamma_{p0} \frac{\varepsilon}{2} \frac{\Omega_1 + Q\omega_{J'} + i\Gamma_K}{\Gamma_K^2 + (\Omega_1 - Q\omega_{J'})^2} \exp(i\Omega_1 t). \quad (4.52)$$

Resonance-shaped growth of modulation amplitude manifests itself in the intensity of fluorescence, as expressed through $_b\rho_Q^K(t)$, according to (2.24). It may be pointed out that, owing to the harmonic factor $\exp(i\Omega_1 t)$ the signal in (4.52) vanishes at registration averaged over the time span $\Delta t \gg \Omega_1^{-1}$. The position of resonance permits us to determine $Q\omega_{J'}$ (see Fig. 4.22), i.e. the Landé factor, whilst the shape (width) of the envelope yields the constant Γ_K.

The method of beat resonance has been used episodically to study the magnetism of *excited states of molecules*. One of the first papers of this kind appears to have been that by Lehmann *et al.* [89] on iodine $I_2(B^3\Pi_{0_u^+})$. Subsequently the same group [172, 173] used beat resonance for determining Landé factors for a number of rovibronic levels of

$^{80}\text{Se}_2(B1_u)$. It may be noted that the beat resonance method has certain advantages over the widely employed *magnetic resonance* method [4] since it can be worked at weak excitation without causing broadening and shift of the signal. Distortions of resonance signals appear due to the presence of an intense radiofrequency field in the magnetic resonance method.

We now pass on to discussing the possibility of extending this method to the *ground state* of molecules. Let the excitation be sufficiently effective with respect to the parameter $\chi = \Gamma_p/\gamma$ as to create noticeable optical polarization of the ground (initial) state. The essence of the phenomenon can be understood in a clear fashion if we continue the discussion on Figs. 4.9, 4.10 in Section 4.3. Let us consider the situation where the ground state Hanle effect has already been pronounced whilst the excited state effect has as yet not started; see cases (c) and (d). Let the modulation frequency of the exciting light Ω_1 coincide with the frequency $q\omega_{J''}$, $\Omega_1 = 2\omega_{J''}$ in the case of alignment. It is easy to understand that in this case the 'isotropic' distribution of angular momenta $\mathbf{J}_a$ in the xy plane in Figs. 4.9(c, d) and 4.10(c, d) (bottom) is now partially turned back to the 'anisotropic' situations in Figs. 4.9(a) and 4.10(a) due to the 'stroboscopic' synchronization of excitation and precession. Indeed, the distribution manages to turn into itself during the time period between the maximum excitations. This means that we again restore the ground state alignment and, consequently, the difference $I_\parallel - I_\perp$ and the degree of polarization $\mathcal{P}$ diminishes in the vicinity of $\Omega_1 = 2\omega_{J''}$. An important difference of non-linear ground state beat resonance from the linear signal of the excited state consists of the fact that the signal from the ground level does not vanish at stationary registration of the radiation averaging over the time interval $\Delta t \gg \Omega_1^{-1}$. If, for instance, one registers the degree of polarization of the fluorescence under the same conditions as in the lower state Hanle effect (see Fig. 4.11), but at modulated pumping (4.50), one ought to expect, in addition to a minimum at zero magnetic field (Hanle signal), the appearance of another minimum near such a value of the field B, where we have $\Omega_1 = 2\omega_{J''}$; see Fig. 4.23. But this is just the *beat resonance* of the *ground (initial) level*.

In order to perform quantitative analysis of such a non-linear effect of ground state beat resonance in terms of polarization moments, it is necessary to solve (4.9) and (4.10) with account being taken of (4.50). Here, direct numerical solution is possible, but its results are difficult to assess. The essence of the effect is easier to comprehend by means of expanding the polarization moments $_a\rho_q^\kappa$, $_b\rho_Q^K$ into series over powers of the pumping parameter Γ_p/γ. This method of obtaining approximations of different order is discussed in Section 5.5. Its application to the ground state resonance of beats is given in [153], where analytical expressions in the second order of expansion are obtained, leading to the terms in fluorescence intensity which contain the 'resonance-type' factor $[\gamma_\kappa^2 + (\Omega_1 - q\omega_{J''})^2]^{-1}$. Owing to the non-linear absorption dependent on Γ_p^2/γ^2, where Γ_p has the form (4.50) in the expression for fluorescence intensity, there remains the term proportional to $\varepsilon^2 \cos^2(\Omega_1 t)$. Therefore the resonance factor does not vanish in the time averaging process.

The experiment on ground state beat resonance was performed on tellurium $Te_2(X0_g^+)$ [153] and later on $K_2(X^1\Sigma_g^+)$ [14, 34].

Unfortunately, under realistic experimental conditions the second-order approximation is not sufficient for accurate description of the beat resonance signal, since it does not describe either non-linear shifts due to the light field, or shifts due to the effects of the magnetism of the excited state. Obtaining analytical expressions using further approximations is rather difficult, whilst for numerical calculations it was found to be more convenient to apply equations of the type (4.5) and (4.6) for the quasi-classical probability density $\rho_a(\theta, \varphi)$ and $\rho_b(\theta, \varphi)$ [21, 28, 34]. The numerical solution was found in [28, 34] using the finite difference method. The calculated steady ground state quantum beat signal in the degree of polarization for linearly polarized Q-excitation is presented in Fig. 4.23, where the modulation frequency Ω_1 was scanned at fixed $\omega_{J''}$ value. The problem of determining $\omega_{J''}$ at given field value **B**, yielding the Landé factor $g_{J''}$, must be solved not simply by finding the extremum position, but by approximating the whole dependence using $\omega_{J''}$ as one of the fitting parameters. In this process one must solve more complete equations than (4.5) and (4.6) (these are given in Chapter 5). In this condition, as in the case of the non-linear Hanle signal (Section 4.3), one must take into account the relation between the signs of $g_{J''}$ and $g_{J'}$ (curves 1 and 2 in Fig. 4.23). The $\omega_{J''}$ dependence on B (straight line 3) leads to $g_{J''} = (1.24 \pm 0.07) \cdot 10^{-5}$. We wish to point out an interesting peculiarity: in [14] a hexadecapole beat resonance signal of the lower state was registered in the form of a small peak in the region near $\Omega_1 = 4\omega_{J''}$.

Other possible types of resonance are discussed in [4, 100]. Non-linear parametric, phase and relaxation resonances of quantum beats and the possibilities of their observation in molecules are considered in [33, 37].

4.5 The molecular g-factor

As we have seen (Section 4.1), the behavior of the angular momentum **J** in a magnetic field is determined by the collinear *magnetic moment* μ_J coupled with it. It is therefore important to consider the relation between the magnitudes of the magnetic moment μ_J and the mechanical angular momentum **J**. This relation is, in its turn, determined by the *Landé factor* (the *g*-factor), g_J; see (4.1).

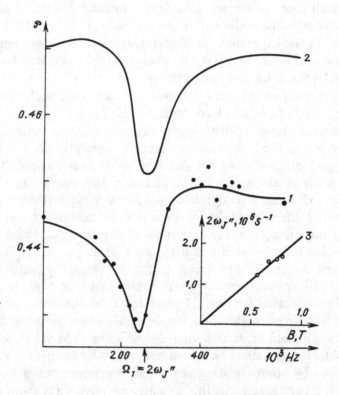

Fig. 4.23. Non-linear beat resonance signal of $K_2(X^1\Sigma_g^+, v'' = 1, J'' = 73)$ in the form of the dependence of the degree of polarization of radiation on modulation frequency Ω_1 in a constant magnetic field $B = 0.589$ T. The arrow indicates $\Omega_1 = 2\omega_{J''}$

The magnetic energy (4.7) may be expressed through the molecular g-factor g_J as

$$E_M^{mag} = -(\boldsymbol{\mu}_J \mathbf{B}) = -g_J \mu_B M_J B. \qquad (4.53)$$

This indicates the linear Zeeman effect with respect to the magnetic field, i.e. each energy level is split into $2J+1$ equidistant magnetic sublevels situated symmetrically with respect to the energy level for $B = 0$. There also exists a second-order Zeeman effect proportional to B^2. It is, however, small as a rule, and is thus neglected in (4.53), in which the g-factor functions as a dimensionless summary parameter determining the magnetic splitting of the level with a given value of the total angular momentum **J**. In this approximation the characteristic of molecular magnetism is determined solely by the molecular g-factor. The role of the quadratic Zeeman effect can be important in some cases and will be considered in the following chapter (see Section 5.4).

The problem of calculating g-factors for molecules is by no means trivial in the general case. Remaining within the scope of our discussion which is based on the maximum use of tangible vector schemes, we will first consider 'pure' Hund's coupling cases for diatomic molecules (see Section 1.2) when analytical expressions for g_J may be obtained. After that we will dwell briefly on the origins of the $^1\Sigma$ state magnetism in diatomic molecules including the usage of perturbation theory, mainly using an alkali dimer as an example. The basic theoretical concepts, as well as a more detailed discussion of the sources of molecular magnetism may be found, for instance, in monographs [155, 194, 267, 294, 319, 374] and papers [79, 87, 92, 290, 321, 378, 382, 395, 396, 399]. In our further explanations we will disregard nuclear spin effects.

1 'Pure' Hund's cases

Let the diatomic (or linear) molecule satisfy the conditions for the Hund's case (a) coupling scheme (see Section 1.2, Fig. 1.3(a)), where the electronic orbital and spin angular momenta are coupled with the internuclear axis. The magnetic moment μ_Ω, which is directed along the internuclear axis and corresponds to the projection Ω of the total angular momentum $\mathbf{J}$ upon the internuclear axis, has the value

$$\mu_\Omega = \mu_\Lambda + \mu_\Sigma = (g_l\Lambda + g_s\Sigma)\mu_B \cong -(\Lambda + 2.0023\Sigma)\mu_B. \qquad (4.54)$$

Account has been taken here of the fact that the Landé factor g_l which is connected with the electronic orbital momentum precisely equals unity $g_l = -1$, whilst the spin-connected one equals $g_s \cong -2.0023$. The recommended g_s and μ_B values can be found in Table 4.1. The discrepancy of the g_s value from two is due to quantum electrodynamical correction. It is important to mention that, in agreement with the definition of $\Omega = \Lambda + \Sigma$ as a positive value (see Section 1.2) and g_l, g_s as negative (see Section 4.1), we have from (4.54) that if $|\Sigma| < |\Lambda|$, but $|g_s\Sigma| > |g_l\Lambda|$, μ_Ω can take a positive value because μ_Ω and Ω possess the same direction (positive Landé factor g_Ω). In other cases μ_Ω is negative, μ_Ω being directed opposite to Ω as shown in Fig. 4.24 (negative g_Ω).

Let us consider some simple examples. For a singlet state $^1\Pi_{\Omega=1}$ with $\Lambda = 1$, $S = \Sigma = 0$ we have a negative moment value $\mu_\Omega = g_l\mu_B$ determined only by the orbital motion of electrons. It thus precisely equals the Bohr magneton. For the doublet state $^2\Pi_\Omega$, where Ω can be $3/2$ or $1/2$, μ_Ω is determined both by the orbital and by the spin contribution. Thus, for $\Omega = 3/2$, $\Lambda = 1$, $\Sigma = 1/2$ according to (4.54) we obtain $\mu_\Omega = (g_l + g_s/2)\mu_B \cong= -2.00116\mu_B$. Of particular interest is the state $^2\Pi_{1/2}$ with $\Omega = 1/2$, $\Lambda = 1$ and $\Sigma = -1/2$ where $\mu_\Omega = (g_l - g_s/2)\mu_B$, and μ_Ω is determined only by the 'anomalous' quantum electrodynamical correction $\mu_\Omega \cong 0.00116\mu_B$. Thus, as pointed out in [113], the $^2\Pi_{1/2}$ state of non-rotating ($R = 0$, $J = \Omega$) light molecules with one unpaired electron and without nuclear spin, such as $^{12}C^2D$, $^{14}N^2D$, $^{16}O^2D$, provides the possibility of measuring in a 'pure' way the quantum electrodynamical correction only, and the degree of agreement with the theoretical value will characterize the validity of the Hund's (a) case coupling of angular momenta. Other convenient cases are the $^3\Delta_1$ and $^4\Phi_{3/2}$ states [113, 374]. For the triplet state $^3\Pi_\Omega$ we have after substituting $\Lambda = 1$ in (4.54) $\mu_\Omega(g_l - g_s) \cong 1.0023\mu_B$ at $\Omega = 0$, $\Sigma = -1$, $\mu_\Omega = -\mu_B$ at $\Omega = 1$, $\Sigma = 0$, and $\mu_\Omega = (g_l + g_s)\mu_B \cong -3.0023\mu_B$ for $\Omega = 2$, $\Sigma = 1$.

In a non-rotating molecule ($R = 0$) the angular momentum $\mathbf{J}$ and the corresponding magnetic moment $\boldsymbol{\mu}_J = \boldsymbol{\mu}_\Omega$ are directed along the molecular axis. If, however, $R \neq 0$ (see Fig. 1.3(a)), and if we have a weak magnetic field, i.e. we have faster nutation of $\boldsymbol{\Omega}$ (and of magnetic moment $\boldsymbol{\mu}_\Omega$, coupled with it) around $\mathbf{J}$ than that of the precession of $\boldsymbol{\mu}_J$ around $\mathbf{B}$ (see Fig. 4.24), then the $\boldsymbol{\mu}_\Omega$ component, which is orthogonal with respect to $\mathbf{J}$, is averaged, and only the projection of $\boldsymbol{\mu}_\Omega$ upon $\mathbf{J}$ is left, equalling

$$\langle \mu_\Omega \rangle = \mu_J = \mu_\Omega \cos(\boldsymbol{\Omega}, \mathbf{J}) = \frac{\mu_\Omega \Omega}{\sqrt{J(J+1)}}. \tag{4.55}$$

Hence, following (4.1) and (4.54), the desired g-factor is

$$g_J = \frac{\mu_J}{\mu_B \sqrt{J(J+1)}} \approx \frac{(\Lambda + \Sigma)(g_l\Lambda + g_s\Sigma)}{J(J+1)}. \tag{4.56}$$

For instance, for the $^1\Pi$ state we have $g_J = -1/[J(J+1)]$, as was immediately confirmed for the case of alkali dimers and applied, in particular, to $Na_2(B^1\Pi_u)$ [290] and to $K_2(B^1\Pi_u)$ [312] in order to determine the lifetime from the Hanle effect.

Let us now consider *Hund's case* (b) coupling scheme. Our argument will be similar to the preceding one, based on the vector model for the case shown in Fig. 1.3(b). Here the magnetic moment along the internuclear axis equals $\mu_\Lambda = g_l|\Lambda|\mu_B$. Next in the hierarchy of momenta interaction will be the nutation of $\boldsymbol{\Lambda}$ around $\mathbf{N}$, then the precession of $\mathbf{N}$ and $\mathbf{S}$ around $\mathbf{J}$, determining the magnetic moment $\boldsymbol{\mu}_J$ and the molecular g-factor. Applying, within the framework of the vector scheme, interaction of the vectors in the above sequence, we obtain, for instance for a linear molecule in the $^{2S+1}\Sigma$ state with $\Lambda = 0$, $S \neq 0$, $R = N$ [374]:

$$g_J \approx g_s[J(J+1) + S(S+1) - N(N+1)]/[2J(J+1)]. \tag{4.57}$$

In particular, for the $^2\Sigma$ state with $S = 1/2$ we have

$$g_{J=N+1/2} = g_s/(2J) \cong -2.0023/(2J), \tag{4.58}$$

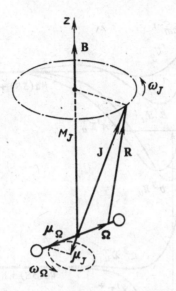

Fig. 4.24. Vector scheme for the interaction of momenta of a diatomic molecule in a magnetic field in Hund's case (a) coupling scheme. The frequency ω_Ω of nutation of Ω and μ_Ω around J considerably exceeds the frequency ω_J of the precession of J around B.

$$g_{J=N-1/2} = -g_s/(2J+2) \cong 2.0023/(2J+2). \qquad (4.59)$$

The case of intermediate coupling of momenta (between Hund's cases (a) and (b)), as well as that of breaking weak field approximation are discussed in [294]. The molecular g-factors for *Hund's case (c)* coupling are discussed in [92, 364].

2 Landé factors for the $^1\Sigma$ states

The above approach is, however, not applicable to the most frequent case, namely to that of non-excited molecules in the $^1\Sigma$ state, in which both the electronic and spin magnetic moments are absent. The reasons for the appearance of a non-zero magnetic moment of the rotating $^1\Sigma$ molecule, disregarded in (4.55)–(4.59) are as follows. Firstly, it is the *contribution of nuclear g_J^n*, connected with the rotation of the nuclear core of the linear molecule [80, 87, 294, 319, 321, 322, 395, 396]. For a diatomic molecule the contribution of nuclear rotation can be estimated from the simple expression:

$$g_J^n = \frac{m_e(Z_1^* M_1^2 + Z_2^* M_2^2)}{M_1 M_2(M_1 + M_2)}, \qquad (4.60)$$

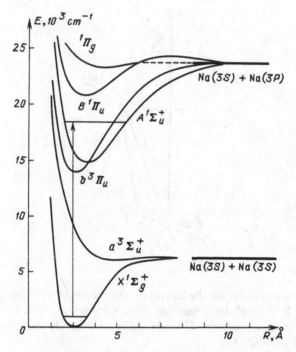

Fig. 4.25. Potential curves of some electronic states of Na_2.

where M_1, M_2 are the mass of the nuclei of atoms forming the molecule, m_e is the electron mass and Z_1^* and Z_2^* are the effective charge numbers of the atomic core, screened off by the internal ('non-valent') atomic shell. The value of Z^* is not known *a priori*, and its calculation requires the use of the electronic wave functions of the molecule. Secondly, it is the *electronic state perturbation* contribution g_J^e, created by the electronic shell of the rotating molecule due to the interaction of the given $^1\Sigma$ state with other electronic states possessing non-zero magnetic moment. This problem is dealt with in parts of the monographs [267, 294] and in sources quoted therein, as well as in, for instance, such papers as [80, 174, 175, 234, 364, 365, 378].

In order to give an idea of these effects, let us consider in greater detail a concrete case of a hydrogen-like molecule, in particular the 'test' molecule Na_2 (see Fig. 4.25). In order to evaluate g_J^n from (4.60) we will consider that the structure of an alkali dimer is completely hydrogenlike, in other words, that only the valence electron participates in the formation of the molecule, and one may assume, following [87], that Z^* is close to unity. Thus, g_J^n is always positive and does not, in such approximations, depend on v, J and on the electronic term. For alkali dimers the values of g_J^n, as calculated from (4.60) and accounting for $M_1 = M_2 = M$, are presented in Table 4.2. A comparison with experimental results shows that they produce the main contribution to the g-factor of the electronic ground state $X^1\Sigma_g^+$ of the alkali dimer.

Let us now consider the second mechanism, namely, the appearance of the electronic contribution g_J^e due to the interaction with the paramagnetic electronic states. In particular, the singlet terms $^1\Pi$ and $^1\Sigma$ of one parity (either $u \leftrightarrow u$ or $g \leftrightarrow g$) interact because of the non-zero matrix elements of the *electron–rotation* operator $[-1/(2\mu r_0^2)](J^+L^- + J^-L^+)$, where μ is the reduced mass, r_0 is the internuclear distance (in atomic units) and the cyclic components of the vectors are defined in the same way as in [267]: $L^\pm = L_x \pm iL_y$, $J^\pm = J_x \pm iJ_y$ connecting the x and y components in the molecular coordinate system. Such electron–rotation (or gyroscopic) interaction tends to sever the bond between $\mathbf{L}$ and the internuclear axis responsible for Λ-doubling of the $^1\Pi$ state (the q-factor) and is thus responsible for the appearance of some magnetism in a diamagnetic $A^1\Sigma_u^+$ state. For sufficiently far states, such as $A^1\Sigma_u^+$ and $B^1\Pi_u$ of the Na$_2$ molecule (see Fig. 4.25), the potential curve shapes of which are similar and are not shifted too far from each other, and the values of the vibrational and rotational constants of which are close to each other, the order of magnitude of the $B^1\Pi_u$ state contribution $g_J^{^1\Pi_u}(^1\Sigma_u^+)$ to the g-factor of the $A^1\Sigma_u^+$ state may be estimated very simply [365]:

$$g_J^{^1\Pi_u}(^1\Sigma_u^+) \cong 2|L^+|^2 B_e/\Delta T_e \cong g_l q(^1\Pi)/B_e, \qquad (4.61)$$

where B_e is a rotation constant and $q(^1\Pi)$ is the constant of Λ-doubling of the $B^1\Pi_u$ state. It is assumed here that the matrix element $|L^+| = \langle ^1\Pi|L^+|^1\Sigma\rangle$ and the energy difference between the interacting levels ΔT_e are independent of internuclear distance and that we have $\Delta T_e \gg B_e$. It is essential to stress that the g-factor, following (4.61), does not depend on J. This is due to the fact that the interaction matrix element includes $\sqrt{J(J+1)}$, which compensates for the diminution of g_J, after (4.56). If the approximation of 'pure precession' after van Vleck [378] holds, when $\langle ^1\Pi|L^+|^1\Sigma\rangle = \sqrt{l(l+1)} = \sqrt{2}$, we may obtain the estimate $g_J^{^1\Pi_u}(^1\Sigma_u^+) \cong -6 \cdot 10^{-5}$ for (4.61). Although the estimate is rough, the sum of this value with the nuclear contribution g_J^n nevertheless permits us to interpret the order of magnitude and the sign of the g-factor of the level Na$_2(A^1\Sigma_u^+, v = 16, J = 17)$ (see Table 4.2), the measured values of which can be obtained from the results presented in Fig. 4.7. More accurate calculations are given in [365, 366]. It is also possible that it is just the rotational interaction of the $X^1\Sigma_g^+$ state with the $^1\Pi_g$ term of Na$_2$ (see Fig. 4.25) which produces a negative electronic contribution to the g-factor of the ground state $X^1\Sigma_g^+$, which is responsible for the discrep-

Table 4.2. Measured Landé factor g_J values for diamagnetic electronic states of alkali dimers and for the tellurium dimer

Dimer	State	Measured g_J, 10^{-5}		g_J^n, 10^{-5}
$^7\text{Li}_2$	$X^1\Sigma_g^+$	5.9170	[87]	7.80
$^{23}\text{Na}_2$	$X^1\Sigma_g^+$	2.1329	[87]	2.37
	$A^1\Sigma_u^+$			
	$v' = 16, J' = 17$	-5.3	[365]	1.40
$^{39}\text{K}_2$	$X^1\Sigma_g^+$	1.1854	[87]	
	$v'' = 1, J'' = 73$	1.24 ± 0.07	[34]	
		1.30 ± 0.27	[23]	
$^{85}\text{Rb}_2$	$X^1\Sigma_g^+$	0.522	[87]	0.642
$^{133}\text{Cs}_2$	$X^1\Sigma_g^+$	0.298	[87]	0.410
$^{130}\text{Te}_2$	$X0_g^+$			
	$v'' = 6, J'' = 52$	-19.6	[34]	

ancy between the g_J^n value, as calculated from (4.60), and the measured g-factor (see Table 4.2). However, in this case the approximation of 'pure precession' clearly does not hold.

Using the same Na$_2$ molecule as an example, we will now discuss the effect of *spin–orbit* interaction. A glance at Fig. 4.25 might, naturally, suggest that a large part of the appearance of magnetism in the $A^1\Sigma_u^+$ state is due to interaction with the intercrossing and strongly displaced term $b^3\Pi_u$. One must consider, however, that the strongest $^3\Sigma_u \sim {}^3\Pi_u$ interaction in first-order perturbation theory takes place owing to the spin–orbit operator $\sum_k a_k \mathbf{l_k s_k}$, where summation is performed over the orbital $\mathbf{l_k}$ and spin $\mathbf{s_k}$ momenta of all valence electrons. Owing to the selection rule $\Delta\Omega = 0$ this interaction connects $^1\Sigma(\Omega = 0)$ only with the component $^3\Pi(\Omega = 0) \equiv {}^3\Pi_0$, which is itself diamagnetic. Interaction does not explicitly depend on J, and the correspondent matrix element $\mathcal{H}_{\Sigma\Pi_0}^{SO} \cong \xi\langle v_\Sigma | v_{\Pi_0}\rangle$ contains the overlap integral between the vibrational wave functions $|v_\Sigma\rangle$, $|v_{\Pi_0}\rangle$. It may seem, at first sight, that this does not lead to any magnetism, since according to (4.56), $g_J^e = 0$ for the $^3\Pi_0$ component of the non-perturbed $b^3\Pi_u$ state on the assumption of Hund's case (a) coupling. However, the rotational perturbation discussed above alters the g-factors of the $^3\Pi_\Omega$ components. The dependence of $g_J^e(b^3\Pi_u)$ on J for Na$_2$, as calculated in [365] using the basis function set from [244], is shown in Fig. 4.26. The values of g_J are practically independent of the vibrational quantum number. If we compare g_J in Fig. 4.26 with the result obtained from (4.56), one may see that the component $^3\Pi_0$ obtains a certain magnetism, 'borrowing' it from other components. Hence, the

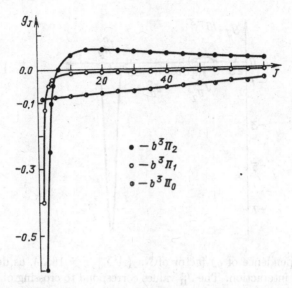

Fig. 4.26. J-dependence of the g-factors of the $^3\Pi_\Omega$ components of the non-perturbed state $Na_2(b^3\Pi_u)$ with $v = 0$ [365].

contribution of the $b^3\Pi_u$ state to the $A^1\Sigma_u^+$ magnetism is due to indirect interaction with the paramagnetic components $^3\Pi_1$ and $^3\Pi_2$ through the component $^3\Pi_0$. This effect is similar to the well-known phenomenon of 'accidental' predissociation [267]. The aforesaid is confirmed by the example of calculation from [365] and shown in Fig. 4.27. As can be seen, the largest contribution to the Landé factor g_J for Na_2 ($A^1\Sigma_u^+, v = 16, J$) from the $b^3\Pi_u$ state emerges at the point of intersection ($J = J_\Pi^c = 37$) with the component $^3\Pi_0$ where $v_{\Pi_0} = 20$. It is of very considerable value, exceeding by two orders of magnitude the contribution from other sources of magnetism (Table 4.2). Measurements of such 'resonance' values of g_J-factors present a singular possibility for determining the matrix element of electronic interaction and vibrational and rotational constants of the interacting electronic states. Far from resonance the contribution of $^3\Pi_u \sim {}^1\Sigma_u$ interaction may be negligibly small. Thus, for the $A^1\Sigma_u^+$ state level with $v_\Sigma = 16, J = 17$ (Table 4.2) it may be as low as $\approx 10^{-7}$.

Let us imagine an example of a heavy molecule, say, Te_2, where the electronic term notation is given according to Hund's case (c) coupling scheme. The g-factor for the electronic ground state $Te_2(X0_g^+)$ is determined by electron–rotation interaction with the $X1_g^+$ state which has a shape similar to that of a potential curve, being positioned higher at $\Delta T_e = 1987.5$ cm^{-1}. Accounting for $B_e = 0.03967$ cm^{-1} and for the fact that $X0_g^+$ is a $\Omega = 0$ component of the $^3\Sigma_g^-$ state, we obtain the value

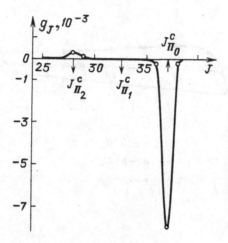

Fig. 4.27. J-dependence of g_J-factor of $Na_2(A^1\Sigma_u^+, v = 16, J)$, as determined by $b^3\Pi_u \sim A^1\Sigma_u^+$ interaction. The J_Π^c values correspond to crossing of the terms with $v_\Sigma = 16$ and $v_{\Pi_0} = 20$. The overlap integral $\langle v_\Sigma | v_{\Pi_0} \rangle = 0.0364$ [365].

$g_J^{1_g^+}(0_g^+) \cong -2q/B_e = -8B_e/\Delta T_e = -16.0 \cdot 10^{-5}$, which is in fair agreement with experiment [34] performed by the beat resonance method, see Section 4.4 (the contribution of g_J^n due to nuclear rotation is small for Te_2); q here denotes the Ω-doubling factor of the $X1_g$ state.

In the above situation one may, to a certain extent, speak of a characteristic value (and sign) of the g-factor for the electronic state as a whole. This is not quite so if the terms cross and interaction is of an irregular nature. Results of such irregular electron–rotation interaction demonstrating the drastic dependence of the value and sign of the magnitude on v_0, J are, for instance, the g-factors of the diamagnetic excited state $B0_u^+$ of $^{130}Te_2$ as mentioned in Section 4.2. This is so because, for the concrete level $B0_u^+, v_0, J$, one may separate, as a rule, one most closely (in energy) positioned level of the paramagnetic state 1_u with a certain vibrational quantum number v_1 and a coinciding rotational number J. In fact, the sign of the energy difference $\Delta T_{v_0 v_1 J}$ mainly determines the sign of the g-factor. In this case one cannot really speak of any characteristic value or of a simple rule of the $B0_u^+$ state Landé factor behavior. A similar situation takes place for Landé factors of the states $B0_u^+$ and 1_u^+ in $^{80}Se_2$, which have been studied in detail in [174] by measuring g_J values using the magnetic quantum beat method.

Thus, in spite of the large variety of situations, we hope that the above brief information on molecular magnetic moments, as well as the concrete examples discussed, might be of use in estimating the nature and value of the changes in optical polarization of angular momenta **J** connected with precession of the latter in an external magnetic field. Some more information about the magnetic field effect on molecular angular momenta distribution, in particular that connected with the quadratic Zeeman effect, will be given in Chapter 5.

5

General equations of motion
for arbitrary J values

In the three preceding chapters we discussed various manifestations of light-created alignment and orientation of both the excited and the ground state of molecules. In those chapters we took into consideration only the basic processes determining the phenomena, neglecting processes of secondary importance, such as stimulated and reverse spontaneous transitions. For detailed understanding of the phenomenon and for quantitative treatment of the observed signals it is necessary to account for all the processes taking place in the interaction between sufficiently intensive radiation and the ensemble of the molecules in the gas phase. We will therefore attempt to present a comprehensive system of equations describing the optical polarization of molecular angular momenta in an external field.

5.1 Equation of motion for the density matrix

We will begin our discussion by considering the most general case of optical polarization of angular momenta in the lower and upper states as a result of $J'' \rightarrow J'$ transitions with arbitrary quantum numbers J'', J'. As a basis for further analysis we take a system of equations for the *density operator* [104] in matrix representation. Assuming broad line approximation, and keeping, for matrix elements of the excited state density matrix, the notation $f_{MM'}$, and for those of the ground state $\varphi_{\mu\mu'}$, we obtain the following system [104, 242]:

$$
\begin{aligned}
\dot{f}_{MM'} = \ & \tilde{\Gamma}_p \sum_{\mu\mu'} \langle M|\hat{\mathbf{E}}^*\mathbf{d}|\mu\rangle \langle M'|\hat{\mathbf{E}}^*\mathbf{d}|\mu'\rangle^* \varphi_{\mu\mu'} \\
& - \left(\frac{\hat{\Gamma}_p}{2} + i\hat{\omega}_S\right) \sum_{M''\mu} \langle M|\hat{\mathbf{E}}^*\mathbf{d}|\mu\rangle \langle M''|\hat{\mathbf{E}}^*\mathbf{d}|\mu\rangle^* f_{M''M'} \\
& - \left(\frac{\hat{\Gamma}_p}{2} - i\hat{\omega}_S\right) \sum_{M''\mu} \langle M''|\hat{\mathbf{E}}^*\mathbf{d}|\mu\rangle \langle M'|\hat{\mathbf{E}}^*\mathbf{d}|\mu\rangle^* f_{MM''} \\
& - \sum_{M_1 M_1'} \Gamma_{MM'}^{M_1 M_1'} f_{M_1 M_1'} - i_{J'}\omega_{MM'} f_{MM'}, \qquad (5.1)
\end{aligned}
$$

160

$$\dot{\varphi}_{\mu\mu'} = -\left(\frac{\tilde{\Gamma}_p}{2} + i\tilde{\omega}_S\right) \sum_{\mu''M} \langle\mu|\hat{\mathbf{E}}^*\mathbf{d}|M\rangle\langle\mu''|\hat{\mathbf{E}}^*\mathbf{d}|M\rangle^* \varphi_{\mu''\mu'}$$

$$-\left(\frac{\tilde{\Gamma}_p}{2} - i\tilde{\omega}_S\right) \sum_{\mu''M} \langle\mu''|\hat{\mathbf{E}}^*\mathbf{d}|M\rangle\langle\mu'|\hat{\mathbf{E}}^*\mathbf{d}|M\rangle^* \varphi_{\mu\mu''}$$

$$+\tilde{\Gamma}_p \sum_{MM'} \langle\mu|\hat{\mathbf{E}}^*\mathbf{d}|M\rangle\langle\mu'|\hat{\mathbf{E}}^*\mathbf{d}|M'\rangle^* f_{MM'} - \sum_{\mu_1\mu_1'} \gamma_{\mu\mu'}^{\mu_1\mu_1'} \varphi_{\mu_1\mu_1'}$$

$$-i_{J''}\omega_{\mu\mu'}\varphi_{\mu\mu'} + \sum_{MM'} \Gamma_{\mu\mu'}^{MM'} f_{MM'} + \lambda\delta_{\mu\mu'}. \tag{5.2}$$

The first term on the righthand side of Eq. (5.1) describes the effect of light absorption on $f_{MM'}$ at reduced absorption rate

$$\tilde{\Gamma}_p = \frac{8\pi^3}{\hbar^2}i(\omega_0), \tag{5.3}$$

where $i(\omega_0)$ is the spectral density of the exciting radiation at molecular transition frequency ω_0. The matrix elements of the form $\langle\mu|\hat{\mathbf{E}}^*\mathbf{d}|M\rangle$ account for the conservation of the angular momentum at photon *absorption* from the light wave with unit polarization vector $\hat{\mathbf{E}}$. The second and third terms together describe the *stimulated emission* of light and the shift in the reduced resonance transition frequency by a value of $\tilde{\omega}_S$ owing to the *dynamic Stark effect*. In broad excitation line approximation this value is determined by mutual spectral positioning of the broad contour of the exciting radiation $i(\omega_l)$ and of the resonance frequency of the absorbing transition ω_0, i.e.

$$\tilde{\omega}_S = \frac{1}{\hbar^2}\text{v.p.} \int \frac{i(\omega_l)}{\omega_l - \omega_0} d\omega_l. \tag{5.4}$$

The fourth term characterizes the relaxation of the density matrix $f_{MM'}$, both in 'purely depolarizing' *elastic* collisions with rate constants $\Gamma_{MM'}^{M_1M_1'}$ and differing indices M, M', and M_1, M_1' and in *inelastic* collisions with coinciding indices $M = M_1$ and $M' = M_1'$. In the latter case the $\Gamma_{MM'}^{MM'}$ values also include radiational decay of the matrix element $f_{MM'}$. Finally, the fifth term on the righthand side of Eq. (5.1) describes the effect of the *external field* producing splitting of the magnetic sublevels M and M' by a value of $_{J'}\omega_{MM'} = (E_M - E_{M'})/\hbar$.

In the righthand part of Eq. (5.2) the first and second terms together describe light absorption and the effect of the dynamic Stark effect, the third term describes stimulated light emission, the fourth term describes relaxation processes in the ground state, the fifth term describes the external field effect causing splitting of neighboring magnetic sublevels of the ground state J'' by a value $_{J''}\omega_{\mu\mu'}$, the sixth term describes the reverse spontaneous transitions at a rate $\Gamma_{\mu\mu'}^{MM'}$, and, finally, the seventh term describes the relaxation of the density matrix $\varphi_{\mu\mu'}$ at the interaction between the molecule with the isotropic thermostat, the role of which has been discussed in Section 3.1.

The diagonal elements of density matrices f_{MM} and $\varphi_{\mu\mu}$, as already mentioned in Section 3.1, characterize the population of the magnetic sublevels M and μ for the states J' and J'', whilst the non-diagonal elements $f_{MM'}$ and $\varphi_{\mu\mu'}$ describe the phase coherence of the corresponding wave functions $|J'M\rangle$ and $|J'M'\rangle$, or of $|J''\mu\rangle$ and $|J''\mu'\rangle$. Sometimes even special normalized magnitudes

$$0 \leq h'_{MM'} = \sqrt{\frac{f_{MM'}f_{M'M}}{f_{MM}f_{M'M'}}} \leq 1, \quad 0 \leq h''_{\mu\mu'} = \sqrt{\frac{\varphi_{\mu\mu'}\varphi_{\mu'\mu}}{\varphi_{\mu\mu}\varphi_{\mu'\mu'}}} \leq 1 \quad (5.5)$$

are introduced [240]. One may consider them as a measure of the *degree of coherence*. Eq. (5.5) is close to the definition of the *correlation coefficient* through elements of the covariation matrix, as accepted in mathematical statistics, see, for instance, [77].

The intensity of the fluorescence appearing in the radiational transition of the molecule from level J' upon level J_1'' (see Fig. 3.14) may be calculated by means of the formula [133]

$$I(\hat{\mathbf{E}}') = \tilde{I}_0 \sum_{MM'\mu} \langle M|(\hat{\mathbf{E}}')^*\mathbf{d}|\mu\rangle\langle M'|(\hat{\mathbf{E}}')^*\mathbf{d}|\mu\rangle^* f_{MM'}, \quad (5.6)$$

where $\hat{\mathbf{E}}'$ is the unit polarization vector of fluorescence, $\tilde{I}_0$ is a proportionality factor, and μ denotes the magnetic sublevels of the final state J_1'' of the molecular transition.

In the analysis of optical pumping of atoms, when the values of the quantum numbers J'', J' and J_1'' are, as a rule, of the order of unity, and hence the number of magnetic sublevels is small, the system of equations (5.1) and (5.2) may be solved directly, as, for instance, in [105].

If the quantum numbers J'', J' and J_1'' considerably exceed unity, as frequently occurs in the case of molecules, then such a system can easily be solved only in situations where excitation takes place in weak light. At such excitation light does not affect the ground state density matrix, and the equations of motion of the excited state density matrix elements (5.1) assume a comparatively simple form:

$$\dot{f}_{MM'} = \tilde{\Gamma}_p \sum_\mu \langle M|\hat{\mathbf{E}}^*\mathbf{d}|\mu\rangle\langle M'|\hat{\mathbf{E}}^*\mathbf{d}|\mu\rangle^* \varphi_{\mu\mu}$$

$$-\Gamma_{MM'}f_{MM'} - i_{J'}\omega_{MM'}f_{MM'}, \quad (5.7)$$

where all $\varphi_{\mu\mu}$ with different μ coincide and, for the sake of simplicity, will be further assumed to be equal to 1. This equation is a generalization of

Eq. (3.22) previously given in Section 3.2. We will now analyze in greater detail the methods of solving this equation. As is known, the product $\hat{\mathbf{E}}^*\mathbf{d}$ has to be considered as a *hermitian product* of $\hat{\mathbf{E}}$ and $\mathbf{d}$ [140]. It is just such a product which represents the amplitude with which the vector $\hat{\mathbf{E}}$ is 'contained' in the vector $\mathbf{d}$. As a result we have for the matrix elements of the electric dipole transition

$$\langle M|\hat{\mathbf{E}}^*\mathbf{d}|\mu\rangle = \sum_q (E^q)^*\langle M|d^q|\mu\rangle. \tag{5.8}$$

For the matrix element $\langle M|d^q|\mu\rangle$ entering into the righthand part of Eq. (5.8) we obtain, applying the Wigner–Eckart theorem (D.7),

$$\langle M|d^q|\mu\rangle = \frac{1}{\sqrt{2J'+1}}C^{J'M}_{J''\mu 1q}(J'\|d\|J''), \tag{5.9}$$

where $(J'\|d\|J'')$ is the reduced matrix element; see Appendix D. With the aid of (5.8) and (5.9) a stationary solution for (5.7) may be obtained in the form

$$f_{MM'} = \frac{\tilde{\Gamma}_p|(J'\|d\|J'')|^2}{2J'+1}\frac{1}{\Gamma_{MM'}+i_{J'}\omega_{MM'}}$$
$$\times \sum_{\mu q_1 q_2}(E^{q_1})^*(E^{q_2})C^{J'M}_{J''\mu 1q_1}C^{J'M'}_{J''\mu 1q_2}. \tag{5.10}$$

Here the proportionality factor $\tilde{\Gamma}_p|(J'\|d\|J'')|^2/(2J'+1)$ coincides with the *dynamic part* Γ_p of the absorption probability introduced in Chapter 2. The sum in (5.10) presents the *angular part* of the absorption probability, whilst the factor $(\Gamma_{MM'}+i_{J'}\omega_{MM'})^{-1}$ describes the effect upon $f_{MM'}$ both on the part of perturbation by the *external field*, causing splitting of the magnetic sublevels $_{J'}\omega_{MM'}$, and by *spontaneous decay* and *collisions* (anisotropic in the general case), together described by a set of relaxation rates $\Gamma_{MM'}$. Applying similar manipulations as for (5.7) to Eq. (5.6), we obtain an equation for fluorescence *intensity*:

$$I(\hat{\mathbf{E}}') = I_0 \sum_{\mu MM' q_1 q_2} f_{MM'}(-1)^{q_1+q_2}(E'^{-q_1})^*(E'^{-q_2})C^{J'M}_{J_1''\mu 1q_1}C^{J'M'}_{J_1''\mu 1q_2}. \tag{5.11}$$

In the calculation of circularly or elliptically polarized fluorescence it is necessary to keep in mind the fact that Eq. (5.11) yields the intensity of light with polarization $(\hat{\mathbf{E}}')^*$ [6, 192].

5.2 Electric field effect on fluorescence of molecules

One of the most interesting effects that can be analyzed by applying the formulae obtained in the preceding section is the influence of an external *electric field* on the fluorescence of molecules. Such an influence cannot be described by means of classical probability density $\rho(\theta, \varphi)$, unlike the case of an external magnetic field, which was considered in Chapter 4.

Let us assume the geometry of excitation and observation (see Fig. 5.1) as being similar to that used in the registration of traditional Hanle effect signals (Chapter 4). Substituting into Eq. (5.10) the cyclic components E^q for the $\hat{\mathbf{E}}$-vector of the exciting light (see (A.4)) we obtain expressions for the non-zero elements $f_{MM'}$ of the density matrix of the excited state J':

$$f_{MM} = \frac{\Gamma_p}{\Gamma}\frac{1}{2}\left[\left(C^{J'M}_{J''M-111}\right)^2 + \left(C^{J'M}_{J''M+11-1}\right)^2\right], \qquad (5.12)$$

$$f_{M-1M+1} = f^*_{M+1M-1} = \frac{\Gamma_p}{\Gamma + i_{J'}\omega_{M-1M+1}}\frac{1}{2}C^{J'M-1}_{J''M1-1}C^{J'M+1}_{J''M11}, \qquad (5.13)$$

where, as a further simplification, it is assumed that the relaxation of state J' is described with the aid of one constant Γ. It means that only radiational decay and isotropic inelastic collisional relaxation of state J' is considered.

For two components of the fluorescence intensity I_y and I_x we may write, according to the expressions obtained, Eqs. (5.12), (5.13) and (5.11):

$$I_{x,y} = I_0 \sum_M \left\{ \frac{\Gamma_p}{\Gamma}\frac{1}{4}\left[\left(C^{J'M}_{J''M-111}\right)^2 + \left(C^{J'M}_{J''M+11-1}\right)^2\right] \right.$$
$$\times \left[\left(C^{J'M}_{J_1''M-111}\right)^2 + \left(C^{J'M}_{J_1''M+11-1}\right)^2\right]$$
$$\left. \mp\frac{1}{2}\frac{\Gamma_p\Gamma}{\Gamma^2 + {}_{J'}\omega^2_{M-1M+1}}C^{J'M+1}_{J''M11}C^{J'M-1}_{J''M1-1}C^{J'M+1}_{J_1''M11}C^{J'M-1}_{J_1''M1-1}\right\}, \qquad (5.14)$$

where the upper sign (minus) refers to I_x, but the lower (plus) refers to I_y. Eqs. (5.10) and (5.11), as well as their special cases (5.12), (5.13) and (5.14) make it possible, in particular, to describe the Hanle effect at weak excitation; see Section 4.2. In reality, however, these formulae are of a much more general nature, and permit much wider application, as compared to the description of molecular state in terms of polarization moments, as used in Section 4.2. Indeed, the formulae obtained in the present section make it possible to calculate the fluorescence intensity for arbitrary splitting ${}_{J'}\omega_{MM'}$ of magnetic sublevels of the excited state. It makes no difference what the nature of this splitting is. At the same time, the method of polarization moments assumes that at a given field value the separation between any adjacent magnetic sublevels ${}_{J'}\omega_{MM\pm1} = \omega_{J'}$

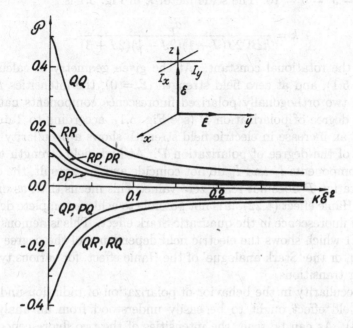

Fig. 5.1. Hanle effect analogue in the case of the quadratic Stark effect.

is constant and does not depend on M as is the case in the linear Zeeman effect. Such an assumption corresponds, in the classical model, to the picture when all the angular momenta of the molecular ensemble are in precession around the direction of the external field with equal angular velocity $\omega_{J'}$, independently of their orientation with respect to this field, as discussed in detail in Chapter 4.

As example of the employment of the density matrix method for calculation of observable signals, we propose to consider the *quadratic Stark effect* in the simplest case of the $^1\Sigma$ states of molecules possessing a constant electric dipole moment $\mathbf{d}_p$. In an electric field $\mathcal{E}$ the magnetic sublevel M of a rotational state J of such a molecule acquires additional energy [374]:

$$\Delta E_M^{el} = \frac{d_p^2 \mathcal{E}^2}{hB} \left[\frac{J(J+1) - 3M^2}{2J(J+1)(2J-1)(2J+3)} \right]. \qquad (5.15)$$

Using this expression for calculating the value of $_{J'}\omega_{M-1M+1}$, then summing up Eqs. (5.14), we obtain the dependence of fluorescence intensity and degree of polarization $\mathcal{P}$ on the strength $\mathcal{E}$ of the external electric field $\mathcal{E}$. Fig. 5.1 shows such a dependence, as calculated for the rotational state $J'' = J' = J = 10$. The scale factor k in Fig. 5.1 is

$$k = \frac{d_p^2}{hB\Gamma 2J(J+1)(2J-1)(2J+3)}, \tag{5.16}$$

B being the rotational constant. At the given geometry of calculation (see Fig. 5.1), and at zero field strength ($\mathcal{E} = 0$), the intensities I_x and I_y of the two orthogonally polarized fluorescence components naturally yield the degree of polarization $\mathcal{P}$ (see Fig. 5.1), according to Table 3.6. However, an increase in electric field strength shows a peculiarity in the behavior of the degree of polarization $\mathcal{P}$. At high field strength the intensity components I_x and I_y do not coincide and, as a result, the degree of polarization $\mathcal{P}$ does not reach zero value. This means that, as distinct from the Hanle effect (4.22), it is not possible to achieve complete depolarization of fluorescence in the quadratic Stark effect. This is demonstrated in Fig. 5.1 which shows the electric field dependence of the degree of polarization, or the 'Stark analogue' of the Hanle effect, for various types of molecular transitions.

This peculiarity in the behavior of polarization of radiation under the electric field effect ought to be easily understood from an analysis of Eq. (5.14). As can be seen, the intensities of the two fluorescence components differ in the sign of the second term, which is proportional to $(\Gamma^2 + {}_{J'}\omega_{M-1M+1}^2)^{-1}$. In the case of the Hanle effect at increase in magnetic field strength all ${}_{J'}\omega_{M-1M+1}^2$ increase, and the second term becomes zero at very strong magnetic field. It is just this circumstance which implies complete depolarization of fluorescence in the Hanle effect. The situation changes in the case of the quadratic Stark effect. Here the external electric field does not remove the degeneracy of all magnetic sublevels, since, according to Eq. (5.15), the sublevels $+M$ and $-M$ remain degenerate. As a result, all ${}_{J'}\omega_{M-1M+1}$ increase in an electric field, except for one, namely ${}_{J'}\omega_{-11}$, which equals zero at all electric field values. The presence of this term in (5.14) leads to different values of the two intensity components I_x and I_y, and this difference remains constant up to electric field strength as large as desired. In addition, this difference is larger the smaller the value of the angular momentum J'. Thus, in the geometry shown in Fig. 5.1 and at $J' = 1$, it is obvious that the term with ${}_{J'}\omega_{-11}^2$ is the only one in the last term of Eq. (5.14). But this means that the quadratic Stark effect will not affect the radiation from the $J' = 1$ state at all, at the given geometry. With increase in J' the number of terms in (5.14) increases, and the relative contribution of ${}_{J'}\omega_{-11}$ diminishes. At the limit $J' \to \infty$ the electric field is, obviously, capable of depolarizing fluorescence completely. The zero field level crossing in an external static electric field, was measured by Hese and coworkers [121] in the excited $A^1\Sigma^+$ state of BaO with $v' = 1$ and 3, $J' = 1$.

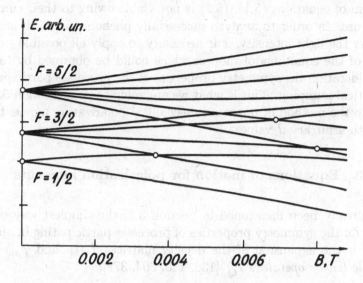

Fig. 5.2. Zeeman energy level pattern for the OD $A^2\Sigma^+, v' = 0, N = 1, J' = 3/2$ level as a function of magnetic field.

Eq. (5.14) can also be used for the quantitative description of signals of *magnetic sublevel crossing* in non-zero external fields. This phenomenon was mentioned in Chapter 4, if only qualitatively; see Fig. 4.5. Clearly, if two magnetic sublevels $M - 1$ and $M + 1$ intercross as the result of some effect, i.e. produce a zero value of $_{J'}\omega_{M-1M+1}$, then resonance increase or decrease of the signal in fluorescence must appear. In addition, these resonances, connected with 'nullification' of $_{J'}\omega_{M-1M+1}$, must be the more pronounced the fewer terms there will be in the sum (5.14) over M from $-J$ to J. This means that such resonance signals may be better pronounced in the case of states with small angular momenta.

Fig. 5.2 shows the external magnetic field dependence of the energies of hyperfine sublevels of the OD molecule in the $A^2\Sigma^+, v' = 0, N = 1, J' = 3/2$ state [393]. Three intercrossing points with $\Delta M = 2$ are shown. The authors of the paper [393] observed resonance changes in fluorescence intensity at magnetic field values corresponding to these intercrossings. From the results of these measurements and from measurements of additional shifts of magnetic sublevels in cases where, simultaneously with the magnetic field, an electric field was superimposed upon an OD molecule in the previously mentioned state, the hyperfine splitting constants and the dipole moment of the molecule were determined.

We thus see that Eqs. (5.10) and (5.11) make it possible to analyze many sufficiently interesting problems, as we have been trying to show. However, in order to approach the case of excitation by intense radiation the system of equations (5.1), (5.2) is not viable owing to their cumbersome nature. In order to analyze successfully phenomena which are not linear over the light intensity, it is necessary to apply all possible simplifications of the equations of motion which could be obtained by taking into consideration the symmetry properties of the processes participating in the optical pumping. This is what we are going to do in Section 5.3 by turning to the method of quantum mechanical polarization moments for states with arbitrary J values.

5.3   Equations of motion for polarization moments

As has already been mentioned in Section 3.1, the simplest way of accounting for the symmetry properties of processes participating in optical pumping is the expansion of the density matrices $f_{MM'}$ and $\varphi_{\mu\mu'}$ over *irreducible tensor operators* T_Q^K [136, 140, 304, 379]:

$$f_{MM'} = \sum_{K=0}^{2J'} \sum_{Q=-K}^{K} (-1)^Q f_Q^K \left(T_{-Q}^K\right)_{MM'},$$

$$\varphi_{\mu\mu'} = \sum_{\kappa=0}^{2J''} \sum_{q=-\kappa}^{\kappa} (-1)^q \varphi_q^\kappa \left(T_{-q}^\kappa\right)_{\mu\mu'}, \tag{5.17}$$

where the expansion coefficients f_Q^K and φ_q^κ are *quantum mechanical polarization moments*. The expansion (5.17) is, in essence, a full quantum mechanical analogue of the expansion (2.14) of the probability density $\rho(\theta, \varphi)$ over the spherical functions $Y_{KQ}(\theta, \varphi)$, yielding classical polarization moments ρ_Q^K and ρ_q^κ, as used by us previously in Chapter 2. This statement becomes obvious if one keeps in mind [379] the fact that from the mathematical point of view the spherical functions Y_{KQ} are irreducible tensors of rank K. The chief distinction between classical and quantum mechanical expansion consists of the fact that summation in (5.17) over the rank K or κ is performed up to the values $2J'$ and $2J''$, whilst in the classical expansion the upper summation limit is infinity. This means that for a full description of the polarization of a molecular state, in the general case of the quantum mechanical approach, a finite number of polarization moments is required, whilst in the classical case their number is infinite. This is obvious: in the quantum mechanical description of the angular momentum only a finite number of discrete orientations is possible with respect to the quantization axis z, which are determined by the magnetic quantum number M. In the classical case, on the other hand, the orientation of $\mathbf{J}(\theta, \varphi)$ may change continuously, and there exists an infinite number of possible directions of the angular momentum.

The choice of the phase and the normalization of the irreducible tensor operators is somewhat arbitrary [379]. Following [133], we will employ the following definition of matrix elements of tensor operators (other existing methods are discussed in Appendix D):

$$\left(T_Q^K\right)_{MM'} = \frac{2K+1}{\sqrt{2J'+1}}(-1)^{J'-M'}\begin{pmatrix} J' & K & J' \\ -M & Q & M' \end{pmatrix} \tag{5.18}$$

and a similar expression for $\left(T_q^\kappa\right)_{\mu\mu'}$. In the righthand part of (5.18) the magnitude in round brackets is a $3j$-symbol connected with the Clebsch–Gordan coefficient through Eq. (C.6) of Appendix C. In certain cases it is more convenient to use $3j$-symbols instead of Clebsch–Gordan coefficients, as we are sometimes going to do in the present chapter.

The determination of $\left(T_Q^K\right)_{MM'}$ by means of Eq. (5.18) is, to a certain extent, untraditional: the tensor operators are not normalized to unity. The orthogonality relations, also reflecting normalization, are of the following form:

$$\sum_{MM'} \left(T_Q^K\right)_{MM'}\left(T_{Q_1}^{K_1}\right)_{MM'} = \frac{2K+1}{2J'+1}\delta_{KK_1}\delta_{QQ_1}. \tag{5.19}$$

Other forms of normalization, as well as forms denoting irreducible tensor operators may be found in [304] and in Appendix D. With the aid of the orthogonality relation one may easily express the quantum mechanical polarization moments f_Q^K and φ_q^κ through the elements of the density matrix $f_{MM'}$ and $\varphi_{\mu\mu'}$

$$f_Q^K = (-1)^Q\frac{2J'+1}{2K+1}\sum_{MM'} f_{MM'}\left(T_{-Q}^K\right)_{MM'},$$

$$\varphi_q^\kappa = (-1)^q\frac{2J''+1}{2\kappa+1}\sum_{\mu\mu'}\varphi_{\mu\mu'}\left(T_{-q}^\kappa\right)_{\mu\mu'}. \tag{5.20}$$

Since the density matrix is Hermitian, we obtain the property of polarization moments which is analogous to the classical relation (2.15): $f_Q^K = (-1)^Q(f_{-Q}^K)^*$ and $\varphi_q^\kappa = (-1)^q(\varphi_{-q}^\kappa)^*$. The adopted normalization of the tensor operators (5.19) yields the most lucid physical meaning of quantum mechanical polarization moments f_Q^K and φ_q^κ which coincides, with accuracy up to a normalizing coefficient that is equal for polarization moments of all ranks, with the physical meaning of classical polarization moments ρ_Q^K, as discussed in Chapter 2. For a comparison between *classical* and *quantum mechanical* polarization moments of the lower ranks see Table 5.1.

Applying the Wigner–Eckart theorem [136, 140, 304, 379] to the matrix elements of form $\langle M|\hat{\mathbf{E}}^*\mathbf{d}|\mu\rangle$ (see (5.9)), the latter may be written through

Table 5.1.   Comparison between classical ρ_Q^K and quantum mechanical f_Q^K polarization moments

f_Q^K	ρ_Q^K
$f_0^0 = \sum_M f_{MM}$ – population of rotational level J'	ρ_0^0 – probability of detecting the molecule at level J'
$f_Q^1 = \dfrac{\langle J_Q^\Sigma \rangle (-1)^Q}{\sqrt{J'(J'+1)}}$, where $\langle J_Q^\Sigma \rangle$ is the Q-component of the total angular momentum of the ensemble of particles	$\rho_Q^1 = \dfrac{(-1)^Q \langle J_Q \rangle_J \rho_0^0}{|J'|}$, where $\langle J_Q \rangle$ is the Q-component of the angular momentum of the molecule, averaged over the ensemble

the 3j-symbols

$$\langle M|\hat{\mathbf{E}}^* \mathbf{d}|\mu \rangle = \sum_q (E_q)^* (-1)^{J'-M} (J'\|d\|J'') \begin{pmatrix} J' & 1 & J'' \\ -M & q & \mu \end{pmatrix}. \quad (5.21)$$

Substituting (5.17) and (5.21) into the system of equations (5.1) and (5.2), using the properties of tensor operators (5.19), and some properties of the sums of 3j-symbols (C.7), (C.8), (C.9) and (C.10), we obtain a system of equations of motion for quantum mechanical polarization moments f_Q^K and φ_q^κ [36, 39]:

$$\dot{f}_Q^K = \Gamma_p \sum_{X\kappa} {}^K F^{X\kappa} \left\{ \Phi^{(X)} \otimes \varphi^{(\kappa)} \right\}_Q^K$$

$$+ 2i\omega_S \sum_{XK'} {}^K A_{1-}^{XK'} \left\{ \Phi^{(X)} \otimes f^{(K')} \right\}_Q^K$$

$$- \Gamma_p \sum_{XK'} {}^K A_{1+}^{XK'} \left\{ \Phi^{(X)} \otimes f^{(K')} \right\}_Q^K - (\Gamma_K - iQ\omega_{J'}) f_Q^K, \tag{5.22}$$

$$\dot{\varphi}_q^\kappa = -\Gamma_p \sum_{X\kappa'} {}^\kappa A_+^{X\kappa'} \left\{ \Phi^{(X)} \otimes \varphi^{(\kappa')} \right\}_q^\kappa$$

$$+ 2i\omega_S \sum_{X\kappa'} {}^\kappa A_-^{X\kappa'} \left\{ \Phi^{(X)} \otimes \varphi^{(\kappa')} \right\}_q^\kappa$$

$$+ \Gamma_p \sum_{XK} {}^\kappa F_1^{XK} \left\{ \Phi^{(X)} \otimes f^{(K)} \right\}_q^\kappa - (\gamma_\kappa - iq\omega_{J''}) \varphi_q^\kappa$$

$$+ \Gamma_{J'J''} C_\kappa \delta_{K\kappa} \delta_{Qq} f_Q^K + \lambda_q^\kappa \delta_{\kappa 0} \delta_{q0}. \tag{5.23}$$

On the righthand side of Eqs. (5.22) and (5.23) the first term describes the *absorption* of light at a rate

$$\Gamma_p = \tilde{\Gamma}_p \frac{|(J'\|d\|J'')|^2}{2J'+1} = \frac{2\pi|(J'\|d\|J'')|^2}{\hbar^2(2J'+1)} i(\omega_0). \qquad (5.24)$$

The second term describes the influence of the *dynamic Stark effect* leading to a shift in the resonance frequency of the molecular transition by the value

$$\omega_S = \tilde{\omega}_S \frac{|(J'\|d\|J'')|^2}{2J'+1} = \frac{|(J'\|d\|J'')|^2}{\hbar^2(2J'+1)} \text{v.p.} \int \frac{i(\omega_l)}{\omega_l - \omega_0} d\omega_l. \qquad (5.25)$$

The third term describes the *stimulated transitions*, the fourth term describes the *relaxation* of the polarization moments of corresponding rank, together with the effect of the *external field*.

In the general case, the terms responsible for relaxation ought to be written as sums of the form $\sum_{K_1 Q_1} \Gamma^{QQ_1}_{KK_1} f^{K_1}_{Q_1}$. Then we have [133]

$$\Gamma^{QQ_1}_{KK_1} = (2K+1) \sum_{MM'M_1M'_1} (-1)^{M-M_1} \begin{pmatrix} J' & J' & K \\ M & -M' & Q \end{pmatrix} \Gamma^{M_1M'_1}_{MM'}$$

$$\times \begin{pmatrix} J' & J' & K_1 \\ M_1 & -M'_1 & Q_1 \end{pmatrix}. \qquad (5.26)$$

A similar formula describes relaxation in the ground state. However, in obtaining the Eqs. (5.22) and (5.23) we assumed that relaxation takes place in isotropic processes with respect to directions in space. In this case the relaxation matrix is diagonal and does not depend on Q, Q_1, i.e. $\Gamma^{QQ_1}_{KK_1} = \Gamma_K \delta_{KK_1} \delta_{QQ_1}$ [132, 134]. This is the case, e.g., of radiational decay, of relaxation as the result of isotropic pair collisions [133], of fly-through relaxation (which may be described by means of relaxation constants Γ_K, γ_κ, as long as we consider it as proceeding according to the exponential law; see Section 3.6). This means that the relaxation of different rank K, κ polarization moments takes place independently. Some more subtle aspects of this question are discussed in Section 5.8. In the particular case of 'pure' radiational or fly-through relaxation, the dependence of the relaxation rates on the rank of the polarization moment disappears: $\Gamma_K = \Gamma$, $\gamma_\kappa = \gamma$.

Further, the fifth term on the righthand side of Eq. (5.23) describes *reverse spontaneous transitions* at the rate $\Gamma_{J'J''}$. In the analysis of reverse spontaneous transitions we make use of the fact that we have, as shown in [104],

$$\Gamma^{MM'}_{\mu\mu'} = \Gamma_{J'J''}(-1)^{-M-\mu} \sum_q \begin{pmatrix} J' & 1 & J'' \\ -M & q & \mu \end{pmatrix} \begin{pmatrix} J'' & 1 & J' \\ -\mu' & q & M' \end{pmatrix}, \qquad (5.27)$$

as well as the properties of the sums of $3j$-symbols (C.9). Finally, the sixth term in (5.23) describes relaxation in *isotropic processes* increasing the population of the lower level J''. In the case of restitution of the ground state population in interaction with an isotropic thermostat (Section 3.1), the constant $\lambda_0^0 = \gamma_0 \overset{(0)}{\varphi_0^0}$, where $\overset{(0)}{\varphi_0^0}$ is the population of level J'' at switch off of excitation, i.e. the thermally equilibrated population.

The coefficients accounting for the conservation of the angular momentum in optical transitions are of the form [36, 39, 153]

$$
{}^K F^{X\kappa} = \frac{(2J'+1)^{3/2}(2X+1)(2\kappa+1)}{(2J''+1)^{1/2}(2K+1)^{1/2}}(-1)^{X+1}
$$
$$
\times \left\{ \begin{array}{ccc} K & J' & J' \\ X & 1 & 1 \\ \kappa & J'' & J'' \end{array} \right\}, \tag{5.28}
$$

$$
{}^\kappa F_1^{XK} = \frac{(2J'+1)^{1/2}(2J''+1)^{1/2}(2X+1)(2\kappa'+1)}{(2\kappa+1)^{1/2}}(-1)^{X+1}
$$
$$
\times \left\{ \begin{array}{ccc} \kappa & J'' & J'' \\ X & 1 & 1 \\ K & J' & J' \end{array} \right\}, \tag{5.29}
$$

$$
{}^\kappa A_\pm^{X\kappa'} = \frac{1 \pm (-1)^{\kappa+X+\kappa'}}{2} \frac{(2J'+1)(2X+1)(2\kappa'+1)}{(2\kappa+1)^{1/2}}(-1)^{J'-J''+\kappa'}
$$
$$
\times \left\{ \begin{array}{ccc} \kappa & X & \kappa' \\ J'' & J'' & J'' \end{array} \right\} \left\{ \begin{array}{ccc} 1 & 1 & X \\ J'' & J'' & J' \end{array} \right\}, \tag{5.30}
$$

$$
{}^K A_{1\pm}^{XK'} = \frac{1 \pm (-1)^{K+X+K'}}{2} \frac{(2J'+1)(2X+1)(2K'+1)}{(2K+1)^{1/2}}(-1)^{J'-J''+K'}
$$
$$
\times \left\{ \begin{array}{ccc} K & X & K' \\ J' & J' & J' \end{array} \right\} \left\{ \begin{array}{ccc} 1 & 1 & X \\ J' & J' & J'' \end{array} \right\}, \tag{5.31}
$$

$$
C_\kappa = (-1)^{J'-J''+\kappa+1}(2J'+1)^{1/2}(2J''+1)^{1/2} \left\{ \begin{array}{ccc} J'' & J'' & \kappa \\ J' & J' & 1 \end{array} \right\}. \tag{5.32}
$$

where the quantities in curly brackets represent $9j$- and $6j$-symbols; see Appendix C and [379].

According to the definition in [379] the irreducible tensor product is

$$
\left\{ \Phi^{(X)} \otimes \varphi^{(\kappa)} \right\}_Q^K = \sum_\xi C_{X\xi\kappa q}^{KQ} \Phi_\xi^X \varphi_q^\kappa. \tag{5.33}
$$

For analysis of the optical pumping processes it is important to keep in mind the fact that the coefficients ${}^K F^{X\kappa}$, ${}^\kappa F_1^{XK}$, ${}^\kappa A_+^{X\kappa'}$ and ${}^K A_{1+}^{XK'}$ differ from zero only in those cases where the sum of the top indices is an even number. For the first two quantities this follows from the properties of the 9j-symbols [379], and for the second quantities this is indicated by the first factor in Eqs. (5.30) and (5.31). The coefficients $^{\kappa}A_{-}^{X\kappa'}$ and $^{K}A_{1-}^{XK'}$, on the other hand, differ from zero only in the case of an odd sum of the top indices. All the above-mentioned coefficients differ from zero only if the triangle rule holds for the top indices. For instance, the relation $|K - \kappa| \leq X \leq K + \kappa$ has to be valid for $^{K}F^{K\kappa}$. The Dyakonov tensor Φ_{ξ}^{X} is defined in the same way as in the classical description of the optical pumping process in Chapter 2; see Eq. (2.22).

We will now discuss polarization moments in the expression for fluorescence intensity in the $J' \to J_1''$ transition with polarization $\hat{\mathbf{E}}'$. To this end we will use the expansions (5.17) and (5.21), as well as the relation (C.8) from Appendix C. After some simple manipulations we obtain [96, 133]

$$I(\hat{\mathbf{E}}') = I_0 |(J'\|d\|J_1'')|^2 (-1)^{J'+J_1''} \frac{1}{\sqrt{2J'+1}}$$

$$\times \sum_{K=0}^{2} (2K+1) \left\{ \begin{matrix} 1 & 1 & K \\ J' & J' & J_1'' \end{matrix} \right\} \sum_{Q=-K}^{K} (-1)^{Q} f_Q^K \Phi_{-Q}^K(\hat{\mathbf{E}}').$$

$$(5.34)$$

Let us consider the *hyperfine depolarization* effect caused by the presence of *nuclear spin* that is not oriented or aligned by light but acts as a randomly oriented 'flywheel' to reduce the average polarization of fluorescence. The emission process can be effected by the presence of nuclear spin I, even if the hyperfine structure is unresolved by the detection apparatus. For this purpose one has to multiply each polarization moment f_Q^K in Eq. (5.34) by a coefficient $\overline{g}^{(K)}$ which is equal to [177, 402]

$$\overline{g}^{(K)} = \sum_{FF'} \frac{(2F+1)(2F'+1)}{2I+1} \left\{ \begin{matrix} F & F' & K \\ J' & J' & I \end{matrix} \right\}^2 \frac{1}{1 + \omega_{FF'}^2 \tau^2}, \quad (5.35)$$

where τ is the radiative lifetime of excited state and $\omega_{FF'}$ is the frequency splitting of excited state hyperfine levels. Note that $\overline{g}^{(0)} = 1$ and hence the population of the level is not effected by hyperfine depolarization. In the limit $|\omega_{FF'}| \ll \tau^{-1}$ one arrives at $\overline{g}^{(K)} \cong 1$ which means that no hyperfine precession occurs before the light is emitted. In the opposite extreme $|\omega_{FF'}| \gg \tau^{-1}$ many periods of precession occur before light is emitted and Eq. (5.35) is reduced to [177, 249, 402]

$$\overline{g}^{(K)} \approx \sum_{F} \frac{(2F+1)^2}{2I+1} \left\{ \begin{matrix} F & F & K \\ J' & J' & I \end{matrix} \right\}^2. \quad (5.36)$$

Since $\overline{g}^{(K)} \cong 1$ for $J' \gg I$ the hyperfine depolarization must be considered only if J' is comparable to I.

Precise analytical solutions of the equation system (5.22) and (5.23) may be found comparatively simply for weak excitation when the approximation $\Gamma_p/\gamma_\kappa, \Gamma_p/\Gamma_K, \omega_S/\gamma_\kappa, \omega_S/\Gamma_K \to 0$ is valid. In this case the system of equations (5.22), (5.23) is of much simpler form [303]. Examples of such solutions may be found in [96, 133, 303]. For strong excitation, when the interaction between the molecular ensemble and light becomes non-linear, whilst the above parameters still remain smaller than unity, the solution for polarization moments may be obtained in the form of an expansion over the powers of these parameters. Finally, at excitation by very strong irradiation, when non-linearity is considerable, the determination of polarization moments f_Q^K and φ_q^κ requires the application of numerical methods for solving Eqs. (5.22) and (5.23).

In all cases of non-linear light absorption the amount of calculations can be reduced if one applies the symmetry properties of the coefficients $^\kappa A_\pm^{X\kappa'}$, $^K A_{1\pm}^{XK'}$, $^K F^{X\kappa}$, $^\kappa F_1^{XK}$. In the most simple way these properties can be written for P, Q and R transitions in the case of the first two coefficients which do not depend on the value of the lower index. Thus omitting this index, from the properties of the $6j$-symbols [379] we obtain [27]

$$^{L_1}A^{L_2 L_3} = {}^{L_3}A^{L_2 L_1}(-1)^{L_3+L_1}\left(\frac{2L_3+1}{2L_1+1}\right)^{3/2}. \qquad (5.37)$$

For the coefficients $^K F^{X\kappa}$, $^\kappa F_1^{XK}$ the corresponding symmetry properties are of a more complex form [27]:

$$^{L_1}F^{L_2 L_3} = {}^{L_3}F^{L_2 L_1}(-1)^{L_3+L_1}\left(\frac{2L_3+1}{2L_1+1}\right)^{3/2}C, \qquad (5.38)$$

where, for Q-type transitions $C = 1$, for $^K F^{X\kappa}$ as well as for $^\kappa F_1^{XK}$. For P-type transitions, in the case of $^K F^{X\kappa}$ we have

$$C = (-1)^X \sqrt{\frac{(2J'+3+L_3)!(2J'+2-L_3)!(2J'+1+L_3)!(2J'-L_3)!}{(2J'+3+L_1)!(2J'+2-L_1)!(2J'+1+L_1)!(2J'-L_1)!}}. \qquad (5.39)$$

In the case of R-transitions it is necessary for $^K F^{X\kappa}$ in (5.39) to interchange L_1 and L_3, and substitute J' for J''. For $^\kappa F_1^{XK}$ in the case of P-transitions it is necessary to interchange L_1 and L_3 in (5.39), but for R-transitions it is necessary to substitute J' for J'' in (5.39).

As a model case for discussing optical alignment and orientation of molecules possessing low values of the angular momentum quantum number one may consider the transition $(J'' = 1) - (J' = 1)$. With such a choice of transition the values of the quantum numbers are the smallest ones at which both orientation $(K, \kappa = 1)$ and alignment $(K, \kappa = 2)$ can exist on both levels, since $K \leq 2J' = 2$, $\kappa \leq 2J'' = 2$; see (5.17). A

Table 5.2. Coefficients $^\kappa A_\pm^{X\kappa'}$, $^K A_1^{XK'}$, $^K F_1^{X\kappa}$, $^\kappa F_1^{XK}$ and C_κ for the transition $(J'' = 1) - (J' = 1)$. The upper indices are numbered from left to right

Indices			$^\kappa A_+^{X\kappa'}$	$^\kappa A_-^{X\kappa'}$	$^K A_{1+}^{XK'}$	$^K A_1^{XK'}$	$^K F^{X\kappa}$	$^\kappa F_1^{XK}$
1st	2nd	3rd						
0	0	0	$-\frac{1}{\sqrt3}$	0	$-\frac{1}{\sqrt3}$	0	$-\frac{1}{\sqrt3}$	$-\frac{1}{\sqrt3}$
0	2	2	$\frac{5\sqrt5}{2\sqrt3}$	0	$\frac{5\sqrt5}{2\sqrt3}$	0	$\frac{5\sqrt5}{2\sqrt3}$	$\frac{5\sqrt5}{2\sqrt3}$
1	0	1	$-\frac{1}{\sqrt3}$	0	$-\frac{1}{\sqrt3}$	0	$-\frac{1}{2\sqrt3}$	$-\frac{1}{2\sqrt3}$
1	2	1	$-\frac{5}{4\sqrt3}$	0	$-\frac{5}{4\sqrt3}$	0	$-\frac{5}{2\sqrt3}$	$-\frac{5}{2\sqrt3}$
1	2	2	0	$-\frac{5\sqrt5}{2\sqrt3}$	0	$-\frac{5\sqrt5}{2\sqrt3}$	0	0
2	0	2	$-\frac{1}{\sqrt3}$	0	$-\frac{1}{\sqrt3}$	0	$\frac{1}{2\sqrt3}$	$\frac{1}{2\sqrt3}$
2	2	0	$\frac{1}{2\sqrt3}$	0	$\frac{1}{2\sqrt3}$	0	$\frac{1}{2\sqrt3}$	$\frac{1}{2\sqrt3}$
2	2	1	0	$\frac{3}{2}$	0	$\frac{3}{2}$	0	0
2	2	2	$\frac{\sqrt{35}}{4\sqrt3}$	0	$\frac{\sqrt{35}}{4\sqrt3}$	0	$-\frac{\sqrt{35}}{2\sqrt3}$	$-\frac{\sqrt{35}}{2\sqrt3}$
	$C_0 = 1$				$C_1 = \frac{1}{2}$		$C_2 = -\frac{1}{2}$	

full set of numerical values for all coefficients entering into the system of Eqs. (5.22) and (5.23) in transitions $(J'' = 1) - (J' = 1)$ is presented in Table 5.2.

The system of equations obtained, (5.22) and (5.23), in broad line approximation in many cases allows us to carry out the analysis of non-linear optical pumping of both atoms and molecules in an external magnetic field. Some examples will be considered in Section 5.5, among them the comparatively unexplored problem of transition from alignment to orientation under the influence of the dynamic Stark effect. But before that we will return to the weak excitation and present, as examples, some cases of the simultaneous application of density matrix equations (5.7) and expansion over state multipoles (5.20).

5.4 Transition from alignment to orientation. Weak excitation

We will consider the kind of external action that can cause the transition from alignment to orientation in the excited state of molecules at weak excitation. It follows from all our discussions in the preceding chapters that excitation by linearly polarized light is capable of creating alignment of the angular momenta of the molecular ensemble only, and is not capable of producing orientation of the angular momenta. This follows clearly from symmetry considerations, since the light vector $\hat{\mathbf{E}}$ of linearly polarized light determines only the axis of cylindrical symmetry ($\leftrightarrow$), without singling out a definite direction ($\Leftarrow$). In other words, it does not create any orientation. For a considerable time there has been major interest in the question of the conditions under which these strict symmetry rules may be broken. In particular, when, for instance, does plane polarized light-created alignment become transformed into orientation under the effect of some external perturbation?

1   General consideration

Let the alignment emerge in weak optical excitation. Two basic directions exist for approaching this problem. One consists of studying the perturbations in the form of *anisotropic collisions* the other; considers the effects of an *external field*. This becomes clear from Eq. (5.10) for the excited state matrix element $f_{MM'}$, in which the righthand side contains a factor into which the relaxation term $\Gamma_{MM'}$ and the term $i_{J'}\omega_{MM'}$ which describes the effect of an external field enter symmetrically. In conditions of broad line excitation only these two quantities can reflect collisional and external field influence on the molecules. Neither collisions, nor external field effects can change either the Clebsch–Gordan coefficients, or the unit vectors E^q describing the exciting light. It ought to be borne in mind that since in this book we use the approximation of broad line excitation throughout we are not going to consider the most trivial case here, when, at excitation of molecules by monochromatic light, some of the magnetic sublevels are brought out of resonance by the magnetic field. By this we exclude the so-called 'magnetic scanning' which leads to population through absorption of mainly positive or negative magnetic sublevels and would lead to the creation of longitudinal orientation of the state.

In order to determine whether orientation has taken place or not, it is convenient to expand the matrix elements $f_{MM'}$ over irreducible tensor operators according to (5.17). As follows from Appendix D, reverse expansion (5.20) may be written conveniently in the following way:

$$f_Q^K = (-1)^Q \sum_{MM'} C_{J'MKQ}^{J'M'} f_{MM'}. \tag{5.40}$$

Since longitudinal orientation f_0^1 can arise only from diagonal matrix elements f_{MM}, which are not affected by external perturbation in the form of anisotropic collisions or in the form of an external field, at linearly polarized excitation we have $f_0^1 = 0$ irrespective of the type of perturbation. This means that the orientation which may emerge must be transversal, i.e. the corresponding components $f_{\pm 1}^1$ of the polarization moment must appear. According to (5.40) we can write

$$f^1_{\pm 1} = -\sum_{MM'} C^{J'M'}_{J'M1\pm 1} f_{MM'} = -\sum_M C^{J'M\pm 1}_{J'M1\pm 1} f_{MM\pm 1}$$

$$= \pm \sum_M \sqrt{\frac{(J' \pm M \pm 1)(J' \mp M \mp 1 + 1)}{2J'(J' + 1)}} f_{MM\pm 1}. \quad (5.41)$$

Further, one must pay attention to Eq. (5.10) for $f_{MM'}$. In isotropic collisions, when we have $\Gamma_{MM'} = \Gamma_{\Delta M = M - M'}$, and in the absence of an external field, when we have $_{J'}\omega_{MM'} = 0$, linearly polarized excitation yields $f_{MM\pm 1} = -f_{-M\mp 1-M}$. And since $f_{MM\pm 1}$ and $f_{-M\mp 1-M}$ enter into the sum (5.41) with equal coefficients, the aforesaid implies the absence of transversal orientation $f^1_{\pm 1}$. More precisely, the contention concerning the 'antisymmetry' of the respective density matrix elements $f_{MM\pm 1} = -f_{-M\mp 1-M}$ follows from the explicit form of the cyclic components of the polarization vector (see Appendix A), and from the symmetry properties of the Clebsch–Gordan coefficients (see Appendix C).

An analysis of E^q and of the Clebsch–Gordan coefficients shows that for the appearance of non-zero density matrix elements $f_{MM\pm 1}$ at linear polarized excitation it is necessary that the $\hat{\mathbf{E}}(\varphi, \theta)$-vector of light should form with the z quantization axis an angle θ that is different from 0 and $\pi/2$; see Fig. 5.3. Such excitation geometry implies the creation of 'tilted' alignment (see Fig. 2.3(d)) which is described by alignment components $f^2_{\pm 1}$.

Further discussions are obvious. In order to observe *transversal orientation*, either such a perturbation on the part of an external field is required as would lead to

$$_{J'}\omega_{MM\pm 1} \neq {_{J'}\omega_{-M\mp 1-M}}, \quad (5.42)$$

or the occurrence of anisotropic collisions is required, when we have

$$\Gamma_{MM\pm 1} \neq \Gamma_{-M\mp 1-M}. \quad (5.43)$$

The above considerations may also be formulated in a more general way, applying a most fruitful approach proposed by Fano in 1964 [142]. The essence of this approach consists of expanding over multipole moments either the matrix of collision relaxation rates $\Gamma_{MM'}$, or the matrix of energy sublevel energies $\omega_{MM'} = (E_M - E_{M'})/\hbar$, where E_M is the *energy shift* of the M-sublevel energy due to perturbation. Thus, for instance, the multipole moment ω^X_0 of the $\omega_{MM'}$ matrix is

$$\omega^X_0 = \sum_M C^{J'M}_{J'MX0} \omega_{MM}. \quad (5.44)$$

If the component ω^X_0 of *even* rank X differs from zero, we will obtain a transition from alignment to orientation. As can easily be seen, this takes place in the case of quadratic (with respect to the magnetic field

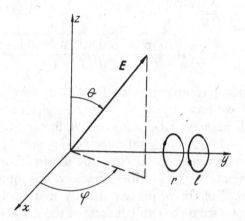

Fig. 5.3. Scheme of excitation and observation for alignment – orientation conversion due to external perturbation.

value) Zeeman or Stark effects [44]. On the other hand, if only the ω_0^X components of *odd* rank X differ from zero, as in the linear Zeeman effect, then such a transition is not possible.

The prerequisite for the creation of orientation in the aligned state can also be formulated in terms of the *time reversal* properties of a Hamiltonian operator which represents the perturbation. As is shown in [276, 277] the alignment–orientation conversion may only take place if the time invariant Hamiltonian is involved. For instance, the Hamiltonian operator of the linear Zeeman effect is odd under time reversal and is thus not able to effect the conversion, whilst the operator of the quadratic Stark effect is even under time reversal and, as a consequence, the quadratic Stark effect can produce alignment–orientation conversion.

It is straightforward to certify the creation of excited state orientation by way of the appearance of fluorescence circularity, namely, the non-zero difference $I_r - I_l$, where I_r, I_l are righthanded and lefthanded circularly polarized light intensities. It is often convenient to measure the normalized quantity, namely the degree of circularity $\mathcal{C} = (I_r - I_l)/(I_r + I_l)$. If light emission is viewed along the z-axis, only $(E')^{-1} = 1$ differs from zero for I_l and $(E')^1 = 1$ for I_r; see Section 2.1 and the discussion on (5.11). As, however, only the transversal orientation $f_{\pm 1}^1$ has been produced in our case, we cannot expect to obtain $\mathcal{C} \neq 0$ if fluorescence light is spreading along the quantization axis. Let us thus choose the y-axis as the observation direction; see Fig. 5.3. This means that the cyclic components $(E')^q$ entering into Eq. (5.11) are to be found by means of Wigner D-matrices; see (A.10), Appendix A. Taking account of this and combining Eqs. (5.10) and (5.11), the following final expressions can be found:

$$I_r - I_l \propto \Gamma_p \frac{\sin 2\theta}{2} \sum_M \frac{\Gamma \sin \varphi + \omega_{MM+1} \cos \varphi}{\Gamma^2 + \omega_{MM+1}{}^2}$$

$$\times \left(C_{J''M+11-1}^{J'M} C_{J''M+110}^{J'M+1} - C_{J''M10}^{J'M} C_{J''M11}^{J'M+1} \right)$$

$$\times \left(C_{J_1''M10}^{J'M} C_{J_1''M11}^{J'M+1} + C_{J_1''M+11-1}^{J'M} C_{J_1''M+110}^{J'M+1} \right), \quad (5.45)$$

$$I_r + I_l \propto \sum_M \frac{\Gamma_p}{\Gamma} \left\{ \frac{\sin^2 \theta}{2} \left[\left(C_{J''M-111}^{J'M} \right)^2 + \left(C_{J''M+11-1}^{J'M} \right)^2 \right] \right.$$

$$\left. + \cos^2 \theta \left(C_{J''M10}^{J'M} \right)^2 \right\}$$

$$\times \left[\frac{1}{2} \left(C_{J_1''M-111}^{J'M} \right)^2 + \left(C_{J_1''M10}^{J'M} \right)^2 + \frac{1}{2} \left(C_{J_1''M+11-1}^{J'M} \right)^2 \right]$$

$$+ \Gamma_p \frac{\sin^2 \theta}{2} \frac{\Gamma \cos 2\varphi - \omega_{M-1M+1} \sin 2\varphi}{\Gamma^2 + \bar{\omega}_{M-1M+1}{}^2}$$

$$\times C_{J''M1-1}^{J'M-1} C_{J''M11}^{J'M+1} C_{J_1''M1-1}^{J'M-1} C_{J_1''M11}^{J'M+1}, \quad (5.46)$$

allowing computation of the signal, $I_r - I_l$ or the degree of circular polarization of fluorescence, C (see Fig. 5.3), under excitation with a linearly polarized light beam of arbitrary direction of $\hat{\mathbf{E}}(\theta, \varphi)$. It is assumed here that $\Gamma_{MM'} = \Gamma$, hence orientation can be effected by an external field only. Note that knowledge of $\omega_{MM'}$, namely of the magnetic sublevel energy set E_M in the case where the external field is applied, or knowledge of $\Gamma_{MM'}$ in the case of collisions, is the only information on perturbation needed to compute the signal.

We will now turn to discussing applications of the given approach to concrete situations. The possibility of conversion from alignment to orientation under the effect of *anisotropic collisions* was first considered by Rebane [323] and Lombardi [275]. They have shown that partial alignment–orientation conversion may be induced in an atomic ensemble by anisotropic collisions when the angle between the collision axis and that of alignment differs from 0 or $\pm \pi/2$. The idea was later confirmed in experimental observations [277, 284]. Thus, in [284] the anisotropic velocity distribution of excited atoms $Ne^*(^3P_2)$ was achieved by frequency shifted linearly polarized laser excitation. The application of a constant magnetic field $\mathbf{B} \perp \hat{\mathbf{E}}$ inclined the axis of optical alignment, thus causing the appearance of 'tilted' alignment ($f_{\pm 1}^2$) components. In this case anisotropic $Ne^* + Ne$ collisions are able to transform $f_{\pm 1}^2$ into transversal orientation $f_{\pm 1}^1$. The effect has been detected as the appearance of fluorescence circularity, in agreement with the prediction given in [285] and with the detailed description given in [310]. A more subtle experiment was performed by Chaika and coworkers [279, 280]. They used 'hidden' alignment in Ne discharge to align Ne $^3P_1$ atoms. The axis of this alignment coincides with the symmetry axis of anisotropic collisions. Again, the alignment axis was turned in a weak magnetic field **B**, thus leading to a collision-induced partial transformation from 'hidden' alignment into orientation. The effect was monitored [279, 280] via intensity changes with B of a linearly polarized probe laser beam passing in the **B** direction through the discharge tube placed between crossed polarizers. The appearance of orientation causes the transmitted laser light due to rotation of its $\hat{\mathbf{E}}$-vector. The collisional alignment–orientation transformation is, in principle, expected for various kinds of collisions including collisions with the surface [325]. Such orientation was observed in direct scattering of N_2 molecules from a single-crystal surface of Ag(111). The orientation appeared along a direction perpendicular to the scattering plane and was interpreted qualitatively by a hard-cube, hard-ellipsoid model as the result of the action of the tangential frictional forces [353, 355]; see also Section 6.2.

The other group of works deals with *electric field* effects. Lombardi [276] describes a circularity signal from He (4^1D_2) in a HF capacitive electrodeless helium discharge. Collisions with electrons served as a source of alignment, whilst the external magnetic field **B** inclined the alignment produced. The electric field $\mathcal{E}$ of the discharge was considered as a perturbing factor able to produce an orientation signal. The authors of [137] observed a circularity signal in fluorescence under linearly polarized laser excitation of Ar^+ in a hollow-cathode discharge, applying both electric and weak magnetic fields. The signal was, however, interpreted rather as an effect on the 'tilted' alignment of collisions with electrons than as an electric field effect.

The **E**-vector action of a sufficiently intensive light, which may be considered as a source of the dynamic Stark effect $(\omega_S \neq 0)$ can also induce orientation of particles. Such alignment–orientation conversion in the depopulated ground state in the presence of a magnetic field will be discussed in Section 5.5

2   An example: quadratic Stark effect

Let us consider the situation [44] where an external stationary homogeneous electric field $\mathcal{E}$ is applied along the z-axis, Fig. 5.3. We will use the quadratic Stark effect energy expression in the form of (5.15), supposing $\theta_0 = \pi/4$. The choice of azimuth angle φ value needs more discussion. First, one must remember that the $I_r - I_l$ signal only appears if the orientation component along the direction of observation possesses non-zero value. In the case where E_M is given by Eq. (5.15), orientation only appears in the direction perpendicular to the $\mathcal{E}$, $\hat{\mathbf{E}}$ plane, hence φ must differ from $\pi/2$ possessing the optimal value $\varphi = 0$. This can be seen from the form of the φ dependence in Eq. (5.45). Indeed, as E_M depends only on M^2 (see Eq. (5.15)), we have $\omega_{MM+1} = -\omega_{-M-1-M}$. The circularity $I_r - I_l$ is zero for $\varphi = \pi/2$ since the terms in Eq. (5.45) contain ω^2_{MM+1} only, whilst the total product of the Clebsch–Gordan coefficients changes sign when we pass from the $M, M+1$ term to the $-M-1, -M$ term. In fact, this might be a sensitive test with which to check the type of Stark effect for a certain molecular state.

In the special case $J'' = J''_1 = 0$, $J' = 1$, the analytic solution of Eqs. (5.15), (5.45) and (5.46) for $\varphi = 0$ may be obtained, leading to a simple formula:

$$C = \frac{(\omega_{01}/\Gamma)}{1 + (\omega_{01}/\Gamma)^2}. \tag{5.47}$$

A result of numeric computation for $J' = 10$ is presented in Fig. 5.4. The degree of circularity C is given as dependent on a dimensionless parameter $k\mathcal{E}^2$, where k is given by (5.16), for all types (P, Q, R) of molecular transitions in excitation and radiation processes. The signal possesses maximal value in the region $\omega_{MM'} \sim \Gamma$, dropping to zero when $\omega_{MM'} \gg \Gamma$, that is, when the M, M' coherence is completely destroyed. As the creation of transversal orientation $f^1_{\pm 1}$ by perturbation leading to E_M in the form of Eq. (5.15) is independent of the type of molecular transition in excitation, the signals caused by the different absorption branches $(P$ or $R)$ have the same signs for the same radiation branches. It is easy to understand that the difference between, say, PR and RR transitions will vanish with $J \to \infty$; see Table 3.6. It is also clear that the degree of circularity for a QQ type of transition must tend to zero with an increase in J.

In considering a concrete electric field scale, let us take the NaK molecule. The following numerical values of parameters characterizing a NaK molecule in a $C^1\Sigma^+$-state were assumed: $B = 0.059$ cm^{-1}, $d_p = 2.0$ D, $\Gamma = 10^8$ s^{-1}. Then, according to (5.16), electric field $\mathcal{E}_{max} \approx 400$ V/cm is needed to obtain the maximal or minimal C value for $J = 1$, and $\mathcal{E}_{max} \approx 8.5$ kV/cm for $J = 10$, Fig. 5.4. The $\mathcal{E}_{max}$ value grows rapidly with J, reaching $\mathcal{E}_{max} \approx 250$ kV/cm for $J = 100$.

3 An example: quadratic Zeeman effect

An interesting possibility of meeting the conditions (5.42) and thus observing alignment–orientation conversion appears when the quadratic contribution to the Zeeman energy (4.53) is taken into account. In particular, the term connected with the non-spherical part of the diamagnetic suscep-

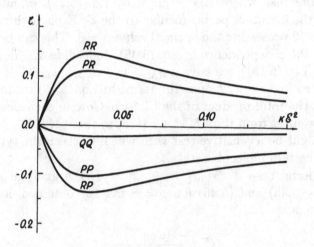

Fig. 5.4. Calculated fluorescence circularity rate under linear polarized excitation as dependent on squared electric field $\mathcal{E}^2$ for different absorption and fluorescence branches.

tibility χ [156, 294] is quadratic with respect to the magnetic field. The M-dependent part of such an interaction for the $^1\Sigma$ state of molecules has the following form in first-order perturbation theory:

$$\Delta E_M^{quadr} = \frac{1}{3}B^2(\chi_\parallel - \chi_\perp)\frac{3M^2 - J(J+1)}{(2J-1)(2J+3)} \qquad (5.48)$$

and is closely connected with the electric quadrupole moment due to electrons.

Another example of the emergence of a quadratic term of Zeeman energy is the magnetic field-induced interaction between levels, between which the rotational quantum number differs by unity, $\Delta J = \pm 1$. Such interaction arises as a result of non-zero off-diagonal (with respect to J) matrix elements of the Zeeman operator. As a concrete illustration we may return to the previously discussed (see Section 4.5) case of the 0_u^+ and $1_u^\pm$ components of the $B^3\Sigma_u^-$ state of the Te$_2$ molecule. In particular, we may write the magnetic energy for the 1_u^- state (which does not interact with the 0_u^+ state in the absence of a magnetic field, owing to the $e \longleftrightarrow\!\!\!/\; f$ selection rule because wave functions with opposite electronic parity cannot interact) [381]:

$$E_M = -\frac{G_z}{J(J+1)}M\mu_B B + G_z^2 A(M,J)\mu_B^2 B^2 + G_\pm^2 B(M,J)\mu_B^2 B^2. \quad (5.49)$$

Here we have applied the Hund's case c basis, in which the Zeeman operator is $-\mu_B\mathbf{B}[g_l\mathbf{J}_a + (g_s - g_l)\mathbf{S}]$ and $G_z = [g_l\Omega + (g_s - g_l)]\langle\Omega|S_z|\Omega\rangle\Omega$ is the J-

independent electronic molecular g-factor; thus, $g_J = G_z/[J(J+1)]$ in the 'pure' state where $\Omega = 1$. The first term in (5.49) describes the *linear Zeeman effect* (4.53). The second term describes the magnetic field-induced $1_u^+ \sim 1_u^-$ *homogeneous* $(\Omega - \Omega') = 0$ interaction of rotational levels with $\Delta J = \pm 1$; see Fig. 5.5. Such interaction mixes the e/f eigenstates. The third term in (5.49) describes the similar *heterogeneous* $|\Omega - \Omega'| = 1$ interaction $1_u^- \sim 0_u^+$; see Fig. 5.5. Here $G_\pm = g_l\langle\Omega|J_{a\pm}|\Omega^*\rangle + (g_s - g_l)\langle\Omega|S_\pm|\Omega^*\rangle$ is the component of the electronic part of the g-factor produced by heterogeneous interaction. Note that $g_l = -1$, $g_s \cong -2.0023$; see Table 4.1. The 'geometric' factor $A(M, J)$ of *homogeneous* $1_u^+ \sim 1_u^-$ interaction may be conveniently expressed through the direction cosines $\alpha_{11}(M, J; M, J \pm 1)$ which emerge (see [198]) in the transition from a molecule-fixed to a space-fixed coordinate system:

$$A(M, J) = \frac{\alpha_{11}^2(M, J; M, J + 1)}{E_{1_u^-}^{v_1 J} - E_{1_u^+}^{v_1 J+1}} + \frac{\alpha_{11}^2(M, J; M, J - 1)}{E_{1_u^-}^{v_1 J} - E_{1_u^+}^{v_1 J-1}}. \tag{5.50}$$

For heterogeneous interaction $0_u^+ \sim 1_u^-$ the 'geometric' factor is of the form

$$B(M, J) = \left[\alpha_{10}^2(M, J; M, J + 1)\sum_{v_0}\frac{\langle v_1^J|v_0^{J+1}\rangle^2}{E_{1_u^-}^{v_1 J} - E_{0_u^+}^{v_0 J+1}}\right.$$

$$\left. + \alpha_{10}^2(M, J; M, J - 1)\sum_{v_0}\frac{\langle v_1^J|v_0^{J-1}\rangle^2}{E_{1_u^-}^{v_1 J} - E_{0_u^+}^{v_1 J-1}}\right]. \tag{5.51}$$

Here v_1, v_0 pertain to the vibrational levels of the 1_u, 0_u^+ states, respectively. It can easily be seen from (5.49), (5.50) and (5.51) that the condition of asymmetric splitting of M-levels (5.42) is satisfied, since

$$\omega_{MM+1} = -\omega_L[1 + (a_J + b_J)(2M + 1)\omega_L], \tag{5.52}$$

where a_J and b_J are M-independent factors arising from $A(J, M)$ and $B(J, M)$, respectively, whilst $\omega_L = -(G_z\mu_B B)/[J(J + 1)\hbar]$.

Substitution of $\omega_{MM'}$ into (5.45) and (5.46) permits the calculation of the expected value of the circularity rate $\mathcal{C}$ emerging at linearly polarized excitation. The results of such a calculation are presented in Fig. 5.6. The parameters employed in the calculation are for the $v_1 = 2$, $J = 96$ level of a $^{130}\text{Te}_2$ molecule in the $B^3\Sigma_u^-(1_u^-)$ state. Fig. 5.6(b) refers to the same geometry as in Fig. 5.4, which corresponds to $\varphi = 0$ in Fig. 5.3, with the difference that here it is not the electric, but the external magnetic, field **B** which is directed along the z-axis. Curve 1 corresponds to the assumption that $B(M, J) = 0$ in (5.49), or that $b_J = 0$ in (5.52). The curve 2 refers to the total Zeeman effect after (5.49). It is assumed in the calculations that $G_z = 1.86$, $G_\pm = 2.9$ [235]. As can be seen, the magnetic field-induced

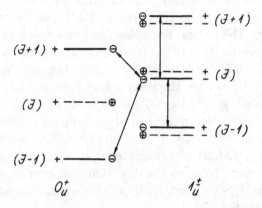

Fig. 5.5.   Magnetic field-induced e/f mixing between 1_u^- and 1_u^+, 0_u^+ state levels with $\Delta J = 1$. For an even isotope J possesses even integer values only, thus the levels shown with a dashed line are not realized; $\oplus$, $\ominus$ denote total parity, $+$, $-$ (or e, f) denote electronic parity.

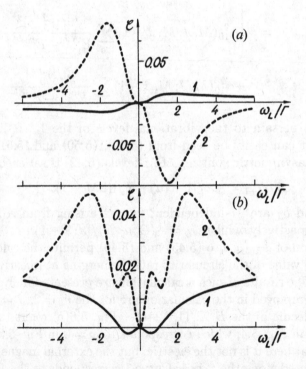

Fig. 5.6.   Calculated circularity rate dependence on ω_L/Γ: (a) when $\hat{\mathbf{E}}$ is directed at $\theta = \pi/4$, $\varphi = \pi/2$; (b) when $\theta = \pi/4$, $\varphi = 0$, see Fig. 5.3. Curve 1 corresponds to $G_\pm = 0$, curve 2 corresponds to $G_\pm = 2.9$.

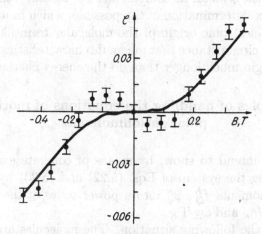

Fig. 5.7. Degree of circular polarization of fluorescence $\mathcal{C}$, as registered [235] at linearly polarized irradiation for the $v_1 = 2(J = 96)$ level of the $B^3\Sigma^-(1_u^-)$ state of $^{130}Te_2$. Solid line obtained by fitting according to (5.45), (5.46), (5.49) at $G_z = 1.86$, $\Gamma = 5.55 \cdot 10^6$ s^{-1}.

$0_u^+ \sim 1_u^+$ interaction is the dominant cause of the appearance of circularity. It may be of interest that the alignment–orientation conversion can also be observed in the case where $\varphi = \pi/2$; see Fig. 5.6(a). This is due to the fact that the Zeeman effect (5.49) possesses both a quadratic and a linear part with respect to the magnetic field. For this reason we have, from the form of $\omega_{MM'}$ following (5.52), and unlike the 'pure' quadratic Stark effect (5.15), that $\omega_{MM+1} \neq -\omega_{-M-1-M}$.

The orientation produced in Fig. 5.6(a) shows a simpler dependence on ω_L, namely of 'dispersion' type, as compared with the dependence in Fig. 5.6(b). The geometry leading to Fig. 5.6(a), where we have $\varphi = \pi/2$ (see Fig. 5.3), was realized in the experiment [235]. Such a scheme is technically more convenient, since it enables observation of fluorescence at right angles to the exciting beam, as well as to the external magnetic field. The results of the experiment are presented for the cycle $(J'' = 95) \rightarrow (J' = 96) \rightarrow (J_1'' = 97)$. As can be seen, the efficiency of alignment–orientation conversion, as may be achieved in the experiment in [235], is sufficiently high. It ought to be noted that, owing to the high sensitivity with which the appearance of circularity can be registered, the above effect of the transformation of alignment into orientation might well be the most sensitive method of testing the linearity of the Zeeman effect. It also permits determination of the magnitudes connected with the non-linear (with respect to the magnetic field) energy terms of the E_M level, thus widening the scope of the magnetic measurements discussed in Section 4.5

in connection with the g-factor measurements. For instance, the data from Fig. 5.7 make direct determination of $G_\pm$ possible, which in its turn, may help to verify the electronic origin of the molecular term [308]. This is connected with the circumstance that magnetic characteristics 'remember' their electronic origin much longer than do the energy characteristics.

5.5   Examples of handling the equations of motion of polarizationmoments

In this section we intend to show, by means of concrete examples, the technique of solving the system of Eqs. (5.22) and (5.23) by expanding the polarization moments f_Q^K, φ_q^κ into a power series of the parameters Γ_p/γ_κ, Γ_p/Γ_K, ω_S/γ_κ and ω_S/Γ_K.

Let us consider the following situation. The molecules are excited in a constant magnetic field $\mathbf{B} \parallel \mathbf{z}$ by linearly polarized light of constant intensity, the polarization vector $\hat{\mathbf{E}}$ being directed at angle θ with respect to the magnetic field $\mathbf{B}$ (see Fig. 5.8). We assume that the absorption rate Γ_p is sufficiently small and does not produce any stimulated transitions, $\Gamma_p \ll \Gamma_K$, yet is strong enough to cause ground state optical polarization, i.e. $\Gamma_p \leq \gamma_\kappa$. We also assume that the centre of the spectral contour of the exciting light does not coincide with the frequency of the molecular transition, so that we have the presence of the *dynamic Stark effect*, with $\omega_S \approx \Gamma_p$, according to (5.24) and (5.25). However, having $\Gamma_p \ll \Gamma_K$, its effect manifests itself only in the ground state. We now proceed directly to the solution of the system of equations (5.22) and (5.23).

In zero-order approximation, pumping light being absent, the answer is obvious:

$$\overset{(0)}{f_Q^K} = 0, \qquad \overset{(0)}{\varphi_q^\kappa} = \delta_{\kappa 0}\delta_{q0}, \tag{5.53}$$

i.e. with excitation switched off, the upper transition level is not populated, whilst on the lower level only that population differs from zero which we, for the sake of convenience and following tradition [303], normalize to unity; see also Appendix D. Here, and subsequently, we will denote the number of the approximation by means of an index over the polarization moment, whilst the answer will be searched for in the form of a series:

$$\left.\begin{array}{l} f_Q^K = \overset{(0)}{f_Q^K} + \overset{(1)}{f_Q^K} + \overset{(2)}{f_Q^K} + \ldots, \\[2mm] \varphi_q^\kappa = \overset{(0)}{\varphi_q^\kappa} + \overset{(1)}{\varphi_q^\kappa} + \overset{(2)}{\varphi_q^\kappa} + \ldots. \end{array}\right\} \tag{5.54}$$

In order to find the next correction of order $n + 1$, it is necessary, in the general case, to substitute the correction of the preceding approximation, of order n (for the first correction it is the zero approximation) into the righthand side of the system of equations (5.22) and (5.23), and solve it in this form. Since we are interested in optical pumping under conditions of excitation that are constant in time, we first make the derivatives in both equations equal to zero, and express the corresponding polarization moments. Omitting the terms responsible for stimulated transitions in both equations, and in the first equation also the term describing the influence of the dynamic Stark effect, we then obtain

$$
\overset{(n+1)}{f_Q^K} = \frac{\Gamma_p}{\Gamma_K - iQ\omega_{J'}} \sum_{X\kappa} {}^K F^{X\kappa} \left\{ \Phi^{(X)} \otimes \overset{(n)}{\varphi^{(\kappa)}} \right\}_Q^K , \tag{5.55}
$$

$$
\overset{(n+1)}{\varphi_q^\kappa} = -\frac{\Gamma_p}{\gamma_\kappa - iq\omega_{J''}} \sum_{X\kappa'} {}^\kappa A_+^{X\kappa'} \left\{ \Phi^{(X)} \otimes \overset{(n)}{\varphi^{(\kappa')}} \right\}_q^\kappa
$$

$$
+ \frac{2i\omega_S}{\gamma_\kappa - iq\omega_{J''}} \sum_{X\kappa'} {}^\kappa A_-^{X\kappa'} \left\{ \Phi^{(X)} \otimes \overset{(n)}{\varphi^{(\kappa')}} \right\}_q^\kappa
$$

$$
+ \Gamma_{J'J''} C_\kappa \delta_{K\kappa} \delta_{Qq} \overset{(n+1)}{f_Q^K} . \tag{5.56}
$$

Here we have digressed, to a certain extent, from the general scheme of building up approximations. Since, after having solved the first equation, we already have $\overset{(n+1)}{f_Q^K}$, we have substituted not $\overset{(n)}{f_Q^K}$ but $\overset{(n+1)}{f_Q^K}$ into the righthand side of (5.56) in order to accelerate the convergence of the series (5.54).

1 Linear approximation: excited state Hanle effect

Let us consider the transition $(J'' = 1 \leftrightarrow J' = 1)$ which is excited by linearly polarized light, the polarization vector $\hat{\mathbf{E}}$ being directed at angles θ, φ; Fig. 5.8. Substituting the values of the Dyakonov tensor Φ_ξ^X from Table 2.1, and the coefficients from Table 5.2 into the system of equations (5.55) and (5.56), we obtain for $\overset{(1)}{f_Q^K}$:

$$
\overset{(1)}{f_0^0} = \frac{\Gamma_p}{3\Gamma_0}, \tag{5.57}
$$

$$
\overset{(1)}{f_0^2} = \frac{\Gamma_p}{6\sqrt{10}\Gamma_2} \left(3\cos^2\theta - 1 \right), \tag{5.58}
$$

$$
\overset{(1)}{f_{\pm 1}^2} = \pm \frac{\Gamma_p}{4\sqrt{15}(\Gamma_2 \mp i\omega_{J'})} \sin 2\theta e^{\pm i\varphi}, \tag{5.59}
$$

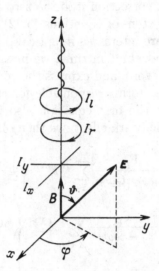

Fig. 5.8. Geometry for calculations.

$$\overset{(1)}{f^2_{\pm 2}} = \frac{\Gamma_p}{4\sqrt{15}(\Gamma_2 \mp 2i\omega_{J'})} \sin^2 \theta e^{\pm 2i\varphi}. \tag{5.60}$$

The result obtained reflects *linear approximation*, i.e. manifestation of the usual linear (observable signal is proportional to Γ_p) *Hanle effect* (Section 4.2). This contention becomes obvious if we substitute the polarization moments obtained into the expression for fluorescence intensity (5.34) and find the form of the degree of polarization $\mathcal{P} = (I_y - I_x)/(I_y + I_x)$ (Fig. 5.8). For instance, for a radiative transition $(J' = 1) \leftrightarrow (J''_1 = 1)$, and using Table C.3 for numerical values of $6j$-symbols and for values of Φ^X_ξ from Table 2.1, we obtain

$$\mathcal{P} = -\frac{\sqrt{30}\mathrm{Re}f^2_2}{2\sqrt{2}f^0_0 - \sqrt{5}f^0_0}. \tag{5.61}$$

Further, substituting (5.57), (5.58) and (5.60) into (5.61), we have

$$\mathcal{P} = -\frac{3\sin^2\theta(\Gamma_2\cos 2\varphi - 2\omega_{J'}\sin 2\varphi)}{(\Gamma_2^2 + 4\omega_{J'}^2)\left(\frac{8}{\Gamma_0} - \frac{3\cos^2\theta - 1}{\Gamma_2}\right)}, \tag{5.62}$$

which, at $\theta, \varphi = \pi/2$ (see Fig. 5.8), corresponds to the Lorentzian form, whilst at $\theta = \pi/2$, $\varphi = \pi/4$ it corresponds to the dispersion form of the excited state Hanle effect. The classical $(J \to \infty)$ analogue of Eq. (5.62) for the $\theta = \pi/2$ is Eq. (4.26) from Section 4.2.

2   Ground state Hanle effect

The first-order approximation for the ground state polarization moments formed in the transition $J'' = 1 \leftrightarrow J' = 1$ is of the form:

$$\overset{(1)}{\varphi_0^0} = -\frac{\Gamma_p}{3\gamma_0}\left(1 - \frac{\Gamma_{J'J''}}{\Gamma_0}\right), \tag{5.63}$$

$$\overset{(1)}{\varphi_0^2} = -\frac{\Gamma_p}{6\sqrt{10}\gamma_2}\left(3\cos^2\theta - 1\right)\left(1 + \frac{\Gamma_{J'J''}}{2\Gamma_2}\right), \tag{5.64}$$

$$\overset{(1)}{\varphi_{\pm1}^2} = \mp\frac{\Gamma_p\sin 2\theta}{4\sqrt{15}(\gamma_2 \mp i\omega_{J''})}e^{\pm i\varphi} \times \left(1 + \frac{\Gamma_{J'J''}}{2(\Gamma_2 \mp i\omega_{J'})}\right), \tag{5.65}$$

$$\overset{(1)}{\varphi_{\pm2}^2} = -\frac{\Gamma_p\sin^2\theta}{4\sqrt{15}(\gamma_2 \mp 2i\omega_{J''})}e^{\pm 2i\varphi} \times \left(1 + \frac{\Gamma_{J'J''}}{2(\Gamma_2 \mp 2i\omega_{J'})}\right). \tag{5.66}$$

This leads to the manifestation of the *non-linear Hanle effect* in the ground state (the observed signal is proportional to Γ_p^2). The non-linearity of the effect is not obvious in this case and consists of the following fact. Despite the fact that the approximation (5.63), (5.64), (5.65) and (5.66) is linear with respect to the light field (proportional to Γ_p), it is necessary for observation of the Hanle signal (5.61) to calculate the second approximation $\overset{(2)}{f_Q^K}$ which emerges, when $\overset{(1)}{\varphi_q^k}$ is transferred to the excited level and, according to (5.61), appears in fluorescence. In order to avoid cumbersome formulae, we present the explicit form of $\overset{(2)}{f_Q^K}$ for the special case of 'traditional' geometry $\theta, \varphi = \pi/2$, assuming that $\Gamma_{J'J''} = 0$. We then have

$$\overset{(2)}{f_0^0} = \frac{\Gamma_p}{\Gamma_0}\left(-\frac{\Gamma_p}{9\gamma_0} - \frac{\Gamma_p}{72\gamma_2} - \frac{\Gamma_p\gamma_2}{24(\gamma_2^2 + 4\omega_{J''}^2)}\right), \tag{5.67}$$

$$\overset{(2)}{f_0^2} = \frac{\Gamma_p}{6\sqrt{10}\Gamma_2}\left(-\frac{\Gamma_p}{6\gamma_2} + \frac{\Gamma_p}{3\gamma_0} + \frac{\Gamma_p\gamma_2}{2(\gamma_2^2 + 4\omega_{J''}^2)}\right), \tag{5.68}$$

$$\overset{(2)}{f_{\pm2}^2} = \frac{\Gamma_p}{24\sqrt{15}(\Gamma_2 \mp 2i\omega_{J'})}\left(\frac{\Gamma_p}{\gamma_2} + \frac{2\Gamma_p}{\gamma_0}\right). \tag{5.69}$$

As can be seen, the signal (5.61) also reflects the ground state Hanle effect signal which is proportional to Γ_p^2 as it must be in the second-order approximation. It may be of interest that we do not find any terms containing $\omega_{J''}$ in $f_{\pm2}^2$. This peculiarity is due to the zero value of the coefficient at $\overset{(1)}{\varphi_{\pm2}^2}$ in the calculation of $\overset{(2)}{f_{\pm2}^2}$. This is specific for the transition $(J'' = 1) \leftrightarrow (J' = 1)$ and does not take place in transitions between

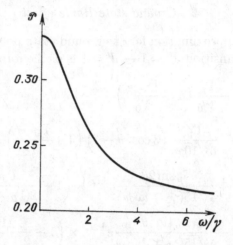

Fig. 5.9.   Ground state Hanle effect in molecules for the transition $(J'' = 1) \leftrightarrow (J' = 1) \leftrightarrow (J_1' = 1)$ at $\gamma_\kappa = \gamma$, $\Gamma_p \ll \Gamma_K = \Gamma$, $\chi = \Gamma_p/\gamma = 20$, $\omega_{J''}/\gamma \gg \omega_{J'}/\Gamma$ [29].

states with different angular momentum values, as pointed out in [29]. Let us substitute into (5.61) the polarization moments f_Q^K, expressed in the form of the sum $\overset{(1)}{f_Q^K} + \overset{(2)}{f_Q^K}$, where $\overset{(1)}{f_Q^K}$ is obtained from (5.57) (5.58), (5.59) and (5.60) for $\theta = \varphi = \pi/2$, and $\overset{(2)}{f_Q^K}$ from (5.67), (5.68) and (5.69). We then come to a non-typical form of the ground state Hanle signal. In this case the ground state Hanle effect does not cause an increase in the degree of polarization $\mathcal{P}$ with increase in magnetic field strength, as is usual (Section 4.3, Figs. 4.11 and 4.12), but a decrease. This may be more clearly illustrated by the precise computer solution of Eqs. (5.22) and (5.23), the result of which is demonstrated in Fig. 5.9.

3 Dynamic Stark effect

In the case of strong excitation, in addition to the effects considered in Section 5.4, under the action of the dynamic Stark effect, transition from *alignment* to *orientation* may also take place [36, 243], which means that polarization moments of odd rank may emerge. This manifests itself in the second order of the expansion. Thus, we have

$$\overset{(2)}{\varphi_q^1} = \frac{2i\omega_S}{\gamma_1 - iq\omega_{J''}} {}^1A_-^{22} \left\{ \Phi^{(2)} \otimes \overset{(1)}{\varphi^{(2)}} \right\}_q^1 . \tag{5.70}$$

Expanding Eq. (5.70), we obtain, in the case of our example,

$$
\overset{(2)}{\varphi_0^1} = \frac{\sqrt{2}}{12} \frac{\omega_S \Gamma_p}{\gamma_1} \sin^2 \theta \left(\frac{\omega' \sin^2 \theta}{\gamma_2^2 + 4\omega_{J''}^2} + \frac{\omega'' \cos^2 \theta}{\gamma_2^2 + \omega_{J''}^2} \right), \tag{5.71}
$$

$$
\overset{(2)}{\varphi_{\pm 1}^1} = \pm i e^{\pm i \varphi} \frac{\omega_S \Gamma_p \sin 2\theta}{48(\gamma_1 \mp i\omega_{J''})} \left[\sin^2 \theta \left(\frac{o_1}{\gamma_2 \mp 2i\omega_{J''}} - \frac{o_2^*}{\gamma_2 \pm i\omega_{J''}} \right) \right.
$$
$$
\left. + (3\cos^2 \theta - 1) \left(\frac{o_2}{\gamma_2 \mp i\omega_{J''}} - \frac{o_3}{\gamma_2} \right) \right], \tag{5.72}
$$

where

$$
\omega' = \omega_{J''} + \frac{1}{2} \frac{\Gamma_{J'J''}(\omega_{J'}\gamma_2 + \omega_{J''}\Gamma_2)}{\Gamma_2^2 + 4\omega_{J'}^2},
$$

$$
\omega'' = \omega_{J''} + \frac{1}{2} \frac{\Gamma_{J'J''}(\omega_{J'}\gamma_2 + \omega_{J''}\Gamma_2)}{\Gamma_2^2 + \omega_{J'}^2},
$$

$$
o_1 = 1 + \frac{1}{2} \frac{\Gamma_{J'J''}}{\Gamma_2 \mp 2i\omega_{J'}}, \qquad o_2 = 1 + \frac{1}{2} \frac{\Gamma_{J'J''}}{\Gamma_2 \mp i\omega_{J'}},
$$

$$
o_3 = 1 + \frac{1}{2} \frac{\Gamma_{J'J''}}{\Gamma_2}.
$$

An analysis of Eqs. (5.71) and (5.72) shows that orientation of the ground (initial) level J'' emerges only under conditions of the dynamic Stark effect, when $\omega_S \neq 0$, and at non-zero angle between the magnetic field **B** and the light vector $\hat{\mathbf{E}}$ (see Fig. 5.8). The shape of the dependence of $\overset{(2)}{\varphi_q^1}$ on the frequency of Zeeman splitting $\omega_{J''}$ shows that at zero, as well as at very large field strengths, orientation equals zero, reaching maximum value at intermediate magnetic field strengths. Thus, e.g., in the geometry $\theta = \varphi = \pi/2$, at $\omega_{J'} = \Gamma_{J'J''} = 0$ we have

$$
\overset{(2)}{\varphi_0^1} = \frac{\sqrt{2}}{12} \frac{\omega_S \Gamma_p \omega_{J''}}{\gamma_1 (\gamma_2^2 + 4\omega_{J''}^2)}, \tag{5.73}
$$

i.e. the $\omega_{J''}$ dependence of longitudinal orientation $\overset{(2)}{\varphi_0^1}$ is of dispersion type, showing extreme values at $\omega_{J''}/\gamma_2 = \pm 0.5$. Here we may observe another peculiarity in the effect of the transition from alignment to orientation: at 'traditional' geometry ($\theta = \pi/2$), as can be seen from (5.72), no components of transversal orientation $\overset{(2)}{\varphi_{\pm 1}^1}$ are formed. Finally, the general formula (5.70) permits us to follow up how the effect depends on the value of the angular momenta of the optical transition levels. Orientation is proportional to the coefficient $^1A_{-}^{22}$, which may be found for various types of molecular transitions [36], by applying the explicit expressions

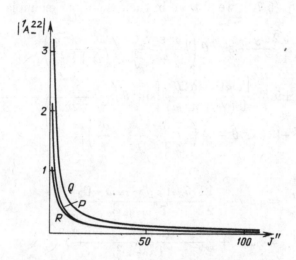

Fig. 5.10. Angular momentum J'' dependence of $^1A_-^{22}$ for P-, Q- and R-type molecular transitions.

for the $6j$-symbols [379] and substituting them into (5.30):

$$^1A_-^{22}(P) = \frac{5}{\sqrt{3}}\sqrt{\frac{2J''-1}{2J''+3}}\frac{2J''^2+5J''+3}{J''(J''+1)(2J''+1)}, \qquad (5.74)$$

$$^1A_-^{22}(Q) = \frac{5[3-4J''(J''+1)]}{J''(J''+1)\sqrt{3(2J''-1)(2J''+3)}}, \qquad (5.75)$$

$$^1A_-^{22}(R) = \frac{5}{\sqrt{3}}\frac{\sqrt{(2J''+3)(2J''-1)}}{(2J''+1)(J''+1)}. \qquad (5.76)$$

In all three cases, we obtain $^1A_-^{22} \sim J''^{-1}$ at large J'' values. Thus, the effect of transition from alignment to orientation under the action of the dynamic Stark effect is a *purely quantum mechanical* phenomenon and disappears at $J \to \infty$.

Fig. 5.10 shows the angular momentum J'' dependence of $^1A_-^{22}$. However, despite the fast decrease in $|^1A_-^{22}|$ with increasing J'', the emerging orientation may still be observable at values of $J'' \approx 20$ and more. Thus, Fig. 5.11 shows the result of the computer solution of the system (5.22) and (5.23) [38], demonstrating the $\omega_{J''}/\gamma$ dependence of the degree of circular polarization C for the transition $(J'' = 18) \leftrightarrow (J' = 17) \leftrightarrow (J_1'' = 16)$ in a Na_2 molecule. Values assumed for calculations are $\gamma_\kappa = \gamma = 0.3$ μs^{-1}, $\Gamma_K = \Gamma = 83.3\ \mu s^{-1}$, $\Gamma_{J'J''} = 0.83\ \mu s^{-1}$ and $\omega_{J'}/\omega_{J''} = -1$. As may be seen, in the case where $\Gamma_p = \omega_S$ exceeds the ground state relaxation rate γ by only a few, we obtain a degree of circularity of radiation above 0.01, which may be observed experimentally.

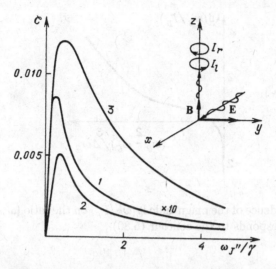

Fig. 5.11. Appearance of circularity degree C in fluorescence at transition from alignment to orientation. It is assumed, in the calculation, that: 1: $\Gamma_p = \omega_S = 0.3\ \mu s^{-1}$; 2: $\Gamma_p = \omega_S = 1.0\ \mu s^{-1}$; 3: $\Gamma_p = \omega_S = 3.0\ \mu s^{-1}$.

The following qualitative interpretation of the effect of transition from alignment to orientation may be proposed [41]. In the excitation process of molecules by strong linearly polarized light and in a weak magnetic field, coherence of nearly degenerate magnetic sublevels of states J'' and J' is created. An increase in magnetic field strength causes shifting of the Zeeman components, and destruction of coherence. This destruction takes place with equal efficiency for all magnetic sublevels in the absence of the dynamic Stark effect, independently of the value of the magnetic quantum number, owing to equidistance of the Zeeman components. In the presence of the dynamic Stark effect the equidistance of the magnetic sublevels is disturbed [74], hence coherence for positive and negative sublevels is destroyed with varying efficiency. This produces various conditions for the absorption of light from these sublevels, which, as known, in its turn implies the presence of orientation.

Finally, it is necessary to discuss whether it is possible to implement experimentally a noticeable (in comparison with Γ_p) value of the Stark shift ω_S. From the definition of Γ_p (5.24) and ω_S (5.25) we have

$$\frac{\omega_S}{\Gamma_p} = \frac{1}{2\pi i(\omega_0)} \text{v.p.} \int \frac{i(\omega_l)}{\omega_l - \omega_0} d\omega_l. \tag{5.77}$$

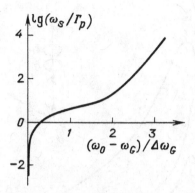

Fig. 5.12. Dependence of the magnitude $\lg(\omega_S/\Gamma_p)$ on the ratio $(\omega_0-\omega_G)/\Delta\omega_G$. The notation corresponds to the contour (5.80).

Assuming the Lorentzian shape of the excitation line spectral profile

$$i(\omega_l) = \frac{i_o}{1 + \left(\frac{\omega_l - \omega_L}{\Delta\omega_L}\right)^2} \tag{5.78}$$

with centrum at frequency ω_L and with halfwidth $\Delta\omega_L$, we obtain

$$\frac{\omega_S}{\Gamma_p} = \frac{\omega_0 - \omega_L}{\Delta\omega_L}. \tag{5.79}$$

For a Gaussian contour with centrum at frequency ω_G and with width parameter $\Delta\omega_G$:

$$i(\omega_l) = i_0 \exp\left(-\frac{(\omega_l - \omega_G)^2}{\Delta\omega_G^2}\right). \tag{5.80}$$

The results of the numerical integration of (5.77) for a Gaussian contour [38] are presented in Fig. 5.12. We may thus conclude that a situation where ω_S and Γ_p are of the same order of magnitude is perfectly feasible.

This concludes our discussion of examples of solving equations of motion systems. We see that such solutions require a large amount of calculations. This amount can be reduced in the case of molecular states with sufficiently large J', J'' values. With this aim we must pass, in the system of Eqs. (5.22) and (5.23), to the limit $J \to \infty$ in the summation coefficients of angular momenta. The following section will be devoted to this task.

5.6 Asymptotic equations for angular momenta

As mentioned in the preceding section, the law of conservation of angular momentum in the process of optical pumping of molecules is reflected in the coefficients $^K F^{X\kappa}$, $^\kappa F_1^{XK}$, $^\kappa A_\pm^{X\kappa'}$, $^K A_{1\pm}^{XK'}$ and C_κ in Eqs. (5.28), (5.29), (5.30), (5.31) and (5.32). For optical transitions between states with large angular momentum quantum numbers J', J'' the explicit form of these coefficients is considerably simplified [15, 39]. These simplifications are due to the possibility of employing asymptotic expressions of the $6j$-symbols presented in Appendix C. Applying Eq. (C.12) for the coefficients $^\kappa A_\pm^{X\kappa'}$ and $^K A_{1\pm}^{XK'}$ we obtain

$$^{L_1}A_+^{L_2L_3} = (-1)^\Delta \sqrt{\frac{(2L_2+1)(2L_3+1)}{2L_1+1}} C_{1\Delta1-\Delta}^{L_20} C_{L_20L_10}^{L_30}, \quad (5.81)$$

$$^{L_1}A_{1+}^{L_2L_3} = (-1)^\Delta \sqrt{\frac{(2L_2+1)(2L_3+1)}{2L_1+1}} C_{1-\Delta1\Delta}^{L_20} C_{L_20L_10}^{L_30}, \quad (5.82)$$

$$^{L_1}A_-^{L_2L_3} = {}^{L_1}A_{1-}^{L_2L_3} = 0. \quad (5.83)$$

In order to be able to apply Eq. (C.12) to the coefficients $^K F^{X\kappa}$ and $^\kappa F_1^{XK}$, the $9j$-symbols entering into Eqs. (5.28) and (5.29) must first be expanded over $6j$-symbols (C.14). As a result we obtain the formulae

$$^{L_1}F^{L_2L_3} = (-1)^\Delta \sqrt{\frac{(2L_2+1)(2L_3+1)}{2L_1+1}} C_{1\Delta1-\Delta}^{L_20} C_{L_20L_10}^{L_30}, \quad (5.84)$$

$$^{L_1}F_1^{L_2L_3} = (-1)^\Delta \sqrt{\frac{(2L_2+1)(2L_3+1)}{2L_1+1}} C_{1-\Delta1\Delta}^{L_20} C_{L_20L_10}^{L_30}, \quad (5.85)$$

completely coinciding with Eqs. (5.81) and (5.82) respectively.

The asymptotic expressions for the coefficients C_κ may be found with the aid of (C.13) and are of simple form, namely

$$C_\kappa = 1. \quad (5.86)$$

Thus, introducing unified notation for coinciding coefficients:

$$^{L_1}A_+^{L_2L_3} = {}^{L_1}F^{L_2L_3} = {}^{L_1}S^{L_2L_3}, \qquad ^{L_1}A_{1+}^{L_2L_3} = {}^{L_1}F_1^{L_2L_3} = {}^{L_1}\tilde{S}^{L_2L_3},$$

we obtain the desired system of *asymptotic equations* of motion for polarization moments

$$\dot{f}_Q^K = \Gamma_p \sum_{X\kappa} {}^K S^{X\kappa} \left\{ \Phi^{(X)} \otimes \varphi^{(\kappa)} \right\}_Q^K$$

$$-\Gamma_p \sum_{XK'} {}^K \tilde{S}^{XK'} \left\{ \Phi^{(X)} \otimes f^{(K')} \right\}_Q^K$$

$$-(\Gamma_K - iQ\omega_{J'}) f_Q^K, \tag{5.87}$$

$$\dot{\varphi}_q^\kappa = -\Gamma_p \sum_{X\kappa'} {}^\kappa S^{X\kappa'} \left\{ \Phi^{(X)} \otimes \varphi^{\kappa'} \right\}_q^\kappa$$

$$+\Gamma_p \sum_{XK} {}^\kappa \tilde{S}^{XK} \left\{ \Phi^{(X)} \otimes f^{(K)} \right\}_q^\kappa$$

$$-(\gamma_\kappa - iq\omega_{J''}) \varphi_q^\kappa + \Gamma_{J'J''} \delta_{K\kappa} \delta_{Qq} f_Q^K + \lambda_q^\kappa \delta_{\kappa 0} \delta_{q0}. \tag{5.88}$$

Owing to the coincidence between a number of coefficients, the symmetry of the equations obtained is considerably higher, as compared to (5.22) and (5.23). In addition, the terms responsible for the dynamic Stark effect disappear, which agrees perfectly with the analysis performed in the preceding section concerning the influence of the dynamic Stark effect on optical polarization of molecules.

The rate of convergence of the equations of optical pumping of molecules to their asymptotic limit characterizes the convergence of the coefficients $^{L_1} S^{L_2 L_3}$, $^{L_1} \tilde{S}^{L_2 L_3}$ and C_κ. Putting $J', J'' \to \infty$, we admit an error proportional to J^{-1} in these magnitudes for given J values [15, 39].

A new circumstance, in comparison with Eqs. (5.22) and (5.23), is the fact that the asymptotic system (5.87), (5.88) contains an infinite number of equations ($0 \leq K \leq \infty$, $0 \leq \kappa \leq \infty$). If the solution for polarization moments is found by way of expansion into a series over a small parameter, then the argument produced in Section 5.4 can be applied here without change. However, if the system of equations (5.87) and (5.88) is solved by computer, then we are deprived, in principle, of the possibility of accounting for all emerging polarization moments. What could be done in this case?

We know (see Figs. 3.3, 3.4 and 3.5) that in the initial (ground) state the magnitude of the polarization moment created by absorption decreases in modulus with increase in its rank κ. In the excited state the polarization moments reveal the same property. In addition, the fluorescence signal is directly influenced only by the moments of the excited state possessing $K \leq 2$ (2.24) and in the ground state possessing $\kappa \leq 4$, in accordance with (3.21) and its discussion. One may therefore obtain a sufficiently correct result if we consider only a limited number of polarization moments in the system of equations (5.87) and (5.88).

This contention is illustrated by Fig. 5.13. The diagram represents the transition $Q \uparrow Q \downarrow$ in the geometry of excitation and observation,

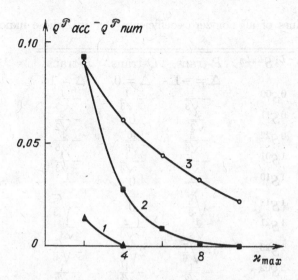

Fig. 5.13. Accuracy of numerical calculation of degree of polarization, as dependent on the number of ground state polarization moments taken into account: 1: $\chi = 1$; 2: $\chi = 10$; 3: $\chi = 100$.

as in Fig. 2.5. The chosen parameters are $\Gamma_p \ll \Gamma_K = \Gamma$, $\gamma_\kappa = \gamma$, $\chi = \Gamma_p/\gamma = 100, 10$ and 1, and $\omega_{J'} = \omega_{J''} = \Gamma_{J'J''} = 0$. The difference is shown between their accurate polarization value $\mathcal{P}_{acc}$, as calculated from Table 3.6, and the value $\mathcal{P}_{num}$ obtained by computer solution of the system of equations (5.87) and (5.88), for f_Q^K of rank $K \leq 2$ and φ_q^κ of rank $\kappa \leq \kappa_{max}$. As can be seen, for sufficiently accurate results at $\chi = 1$, one may limit oneself to accounting for moments of rank $\kappa \leq 4$; for $\chi = 10$ it is necessary to consider 8–10 polarization moments, and for $\chi = 100$ even accounting for the first ten moments we do not obtain satisfactory results. In our example, in which stimulated transitions are absent, i.e. at $\Gamma \gg \Gamma_p$, it is sufficient to account for the excited state moments f_Q^K of rank $K \leq 2$ only. In the presence of stimulated transitions, however, a larger number of moments also has to be taken into consideration for the excited level.

In cases where there is no possibility of comparing the results of numerical calculations of the observed signal with an accurate solution, for instance in the presence of a magnetic field, the number of polarization moments that have to be accounted for can be determined from the convergence of the solution with increasing K_{max}, κ_{max} values.

In order to simplify the analytical solution we present the numerical values of all non-zero coefficients $^{L_1}S^{L_2 L_3}$ in which the values of the indices L_1, L_2 and L_3 do not exceed 4. It follows from the properties of the

Table 5.3. Values of all non-zero coefficients $^{L_1}S^{L_2 L_3}$ whose indices do not exceed 4

$^{L_1}S^{L_2 L_3}$	P-trans. $\Delta = -1$	Q-trans. $\Delta = 0$	R-trans. $\Delta = 1$
$_0S^{00}$	$-\frac{1}{\sqrt{3}}$	$-\frac{1}{\sqrt{3}}$	$-\frac{1}{\sqrt{3}}$
$_0S^{11}$	$\frac{3}{\sqrt{2}}$	0	$-\frac{3}{\sqrt{2}}$
$_0S^{22}$	$-\frac{5}{\sqrt{6}}$	$\frac{10}{\sqrt{6}}$	$-\frac{5}{\sqrt{6}}$
$_1S^{01}$	$-\frac{1}{\sqrt{3}}$	$-\frac{1}{\sqrt{3}}$	$-\frac{1}{\sqrt{3}}$
$_1S^{10}$	$-\frac{1}{\sqrt{6}}$	0	$\frac{1}{\sqrt{6}}$
$_1S^{12}$	$\sqrt{\frac{5}{3}}$	0	$-\sqrt{\frac{5}{3}}$
$_1S^{21}$	$\frac{1}{\sqrt{3}}$	$-\frac{2}{\sqrt{3}}$	$\frac{1}{\sqrt{3}}$
$_1S^{23}$	$-\sqrt{\frac{7}{6}}$	$2\sqrt{\frac{7}{6}}$	$-\sqrt{\frac{7}{6}}$
$_2S^{02}$	$-\frac{1}{\sqrt{3}}$	$-\frac{1}{\sqrt{3}}$	$\frac{1}{\sqrt{3}}$
$_2S^{11}$	$-\frac{3}{5}$	0	$\frac{3}{5}$
$_2S^{13}$	$\frac{3\sqrt{7}}{5\sqrt{2}}$	0	$-\frac{3\sqrt{7}}{5\sqrt{2}}$
$_2S^{20}$	$-\frac{1}{\sqrt{30}}$	$\frac{2}{\sqrt{30}}$	$-\frac{1}{\sqrt{30}}$
$_2S^{22}$	$\frac{5}{\sqrt{105}}$	$-\frac{10}{\sqrt{105}}$	$\frac{5}{\sqrt{105}}$
$_2S^{24}$	$-\frac{3\sqrt{3}}{\sqrt{35}}$	$\frac{6\sqrt{3}}{\sqrt{35}}$	$-\frac{3\sqrt{3}}{\sqrt{35}}$
$_3S^{03}$	$-\frac{1}{\sqrt{3}}$	$-\frac{1}{\sqrt{3}}$	$-\frac{1}{\sqrt{3}}$
$_3S^{12}$	$-\frac{3\sqrt{5}}{7\sqrt{2}}$	0	$\frac{3\sqrt{5}}{7\sqrt{2}}$
$_3S^{14}$	$\frac{3\sqrt{6}}{7}$	0	$-\frac{3\sqrt{6}}{7}$
$_3S^{21}$	$-\frac{3}{7\sqrt{2}}$	$\frac{6}{7\sqrt{2}}$	$-\frac{3}{7\sqrt{2}}$
$_3S^{23}$	$\frac{\sqrt{2}}{3}$	$-\frac{2\sqrt{2}}{3}$	$\frac{\sqrt{2}}{3}$
$_4S^{04}$	$-\frac{1}{\sqrt{3}}$	$-\frac{1}{\sqrt{3}}$	$-\frac{1}{\sqrt{3}}$
$_4S^{13}$	$-\frac{\sqrt{14}}{3\sqrt{3}}$	0	$\frac{\sqrt{14}}{3\sqrt{3}}$
$_4S^{22}$	$-\frac{5}{3\sqrt{21}}$	$\frac{10}{3\sqrt{21}}$	$-\frac{5}{3\sqrt{21}}$
$_4S^{24}$	$\frac{5\sqrt{2}}{\sqrt{231}}$	$-\frac{10\sqrt{2}}{\sqrt{231}}$	$\frac{5\sqrt{2}}{\sqrt{231}}$

Clebsch–Gordan coefficients (C.3), that the magnitudes $^{L_1}S^{L_2 L_3}$ possess certain symmetry properties:

$$^{L_1}S^{L_2 L_3} = (-1)^{L_2} \left(\frac{2L_3 + 1}{2L_1 + 1} \right)^{3/2} {}^{L_3}S^{L_2 L_1}. \tag{5.89}$$

In addition to (5.89), the following statements can be proved. For (P, R)-, and Q-transitions, at $L_2 = 0$, the coefficients $^{L_1}S^{0 L_3}$ coincide. At $L_2 = 1$

we have $^{L_1}S^{1L_3} = 0$ for Q-transitions, whilst in P- and R-transitions $^{L_1}S^{1L_3}$ differ in sign. In cases where $L_2 = 2$, the coefficients $^{L_1}S^{2L_3}$ coincide for (P, R)-transitions, whilst the coefficient for Q-transition is twice as large in absolute value, and of opposite sign. Table 5.3 illustrates these symmetry properties. All these properties also remain in force for the magnitudes $^{L_1}\tilde{S}^{L_2 L_3}$ however, Table 5.3 may be applied to them only after changing the sign of the numerical coefficient for $L_2 = 1$.

After determining the solution for the polarization moments φ_q^κ, f_Q^K, it is necessary to calculate the observed signal. At this stage, too, one may obtain certain simplifications for molecular states with large angular momentum. Turning to the asymptotic limit $J', J_1'' \to \infty$ we have for fluorescence intensity $I(\hat{\mathbf{E}}')$ in Eq. (5.34), and applying Eq. (C.12) from Appendix C [15]:

$$I(\hat{\mathbf{E}}') = I_0|(J'\|d\|J_1'')|^2(-1)^{\Delta'} \sum_K \sqrt{2K+1}C_{1-\Delta'1\Delta'}^{K0}$$
$$\times \sum_Q (-1)^Q f_Q^K \Phi_{-Q}^K(\hat{\mathbf{E}}'). \tag{5.90}$$

This expression coincides with (2.24), obtained on the basis of *classical* concepts. Such a coincidence is understandable from the point of view of the *correspondence principle*. One must bear in mind that the limit $J \to \infty$ means nothing other than that the number of projections of the angular momentum upon the z-axis, permitted by the rule of space quantization and equalling $2J + 1$, becomes infinitely large, and the angular momentum becomes classical.

In just the same way as in the case of fluorescence intensity, the asymptotic equations of motion of polarization moments (5.54), (5.55) and (5.87), (5.88) must coincide with the corresponding equation of motion of classical multiple moments, as introduced by Eq. (2.16). We will show that this is indeed so in the following section.

5.7 Equations of motion of probability density and of classical polarization moments

In order to obtain the equations of motion of *classical polarization moments*, we must base our methods on the system of equations of motion of the probability density $\rho_a(\theta, \varphi)$ and $\rho_b(\theta, \varphi)$ for a molecule with angular momentum vector $\mathbf{J}(\theta, \varphi)$, under conditions of optical pumping. For a number of maximally simplified situations, where the probability density in the ground state $\rho_a(\theta, \varphi)$ does not depend on that of the excited state $\rho_b(\theta, \varphi)$, we have already encountered such equations in preceding chapters; see e.g., (3.4), or (4.5) and (4.6).

The system of equations of motion of $\rho_a(\theta, \varphi)$ and $\rho_b(\theta, \varphi)$ which correlates with the system of equations for the *density matrix* (5.1) and (5.2) in completeness of accounting for various factors affecting the process of optical pumping, may be written as follows [19]:

$$\dot{\rho}_b(\theta, \varphi, t) = \Gamma_p \int G_2'(\theta, \varphi; \theta' \varphi') \rho_a(\theta', \varphi', t) \sin \theta' d\theta' d\varphi'$$

$$-\Gamma_p G_1(\theta, \varphi) \rho_b(\theta, \varphi)$$

$$-\int \Gamma(\theta, \varphi; \theta', \varphi') \rho_b(\theta', \varphi', t) \sin \theta' d\theta' d\varphi'$$

$$+\omega_{J'} \frac{\partial}{\partial \varphi} \rho_b(\theta, \varphi, t), \qquad (5.91)$$

$$\dot{\rho}_a(\theta, \varphi, t) = -\Gamma_p G_2(\theta, \varphi) \rho_a(\theta, \varphi, t)$$

$$+\Gamma_p \int G_1'(\theta, \varphi; \theta', \varphi') \rho_b(\theta', \varphi', t) \sin \theta' d\theta' d\varphi'$$

$$-\int \gamma(\theta, \varphi; \theta', \varphi') \rho_a(\theta', \varphi', t) \sin \theta' d\theta' d\varphi'$$

$$+\omega_{J''} \frac{\partial}{\partial \varphi} \rho_a(\theta, \varphi, t)$$

$$+\int \Gamma_{J'J''}(\theta, \varphi; \theta', \varphi') \rho_b(\theta', \varphi', t) \sin \theta' d\theta' d\varphi' + \lambda_a.$$

$$(5.92)$$

The individual terms require some explanation. In both equations the first term on the righthand side describes absorption of light which raises the molecule with orientation $\mathbf{J}_a(\theta', \varphi')$ of the angular momentum in the ground state to an excited state with orientation $\mathbf{J}_b(\theta, \varphi)$. At this stage as the result of absorption of light, the angular momentum of the molecule may change its spatial orientation. The different form of the terms responsible for the effect of light absorption in the ground and in the excited state may be explained by the following fact. For the excited state b (see Fig. 1.2) it is necessary to know the probability $G_2'(\theta, \varphi; \theta', \varphi')$ for the molecule with orientation of angular momentum $\mathbf{J}_a(\theta', \varphi')$ in the ground state a, in order to assume an angular momentum $\mathbf{J}_b(\theta, \varphi)$ in the excited state after absorption of light. For the ground state it is only important to know the probability $G_2(\theta, \varphi)$ for the molecule with angular momentum orientation θ, φ to absorb light. From the point of view of the ground state, what is going to happen to the molecule after it has left level a is not important.

The second term in both equations describes stimulated light emission in a similar way. Keeping in mind that stimulated emission may be considered, from the point of view of conservation of angular momentum, as absorption on $b \to a$ transition, the physical meaning of the functions $G_1'(\theta, \varphi; \theta', \varphi')$ and $G_1(\theta, \varphi)$ becomes clear.

The third term in the equations describes relaxation processes. The functions $\Gamma(\theta, \varphi; \theta', \varphi')$ and $\gamma(\theta, \varphi; \theta', \varphi')$ characterize the efficiency of this relaxation. They take account of the fact that, for example, the molecule which had an orientation $\mathbf{J}(\theta', \varphi')$ before the collision may display a turned angular momentum with orientation $\mathbf{J}(\theta, \varphi)$ after the collision.

The fourth term in both equations considers precession of the angular momentum in an external magnetic field, in agreement with Eq. (4.4) and its discussion.

The penultimate term in (5.92) deals with reverse spontaneous transitions. The function $\Gamma_{J'J''}(\theta, \varphi; \theta', \varphi')$ represents the probability of the molecule with angular momentum orientation $\mathbf{J}_b(\theta', \varphi')$ in the excited state arriving with orientation $\mathbf{J}_a(\theta, \varphi)$ at the ground state.

Finally, the last term in Eq. (5.92) describes isotropic relaxation processes in the ground state.

In order to obtain equations of motion of multipole moments we must now expand the system of equations (5.91), (5.92) over spherical functions (2.14). Let us consider the result of such an expansion for each term of the system.

It was found in Section 2.1 that in absorption of light the modulus of the angular momentum of the molecule may change by $\Delta = 0, \pm 1$. This result was interpreted as a manifestation of angular momentum conservation in the absorption of a photon possessing spin equal to unity. This means that, in the case where $J \gg 1$, the angular momentum of the molecule in absorption, as well as in emission of light, does not turn, i.e. $G_1(\theta, \varphi) = G'_1(\theta, \varphi; \theta', \varphi')$ and $G_2(\theta, \varphi) = G'_2(\theta, \varphi; \theta', \varphi')$. The magnitude $G_2(\theta, \varphi)$, which describes absorption of light, may be calculated according to (2.8). A similar value $G_1(\theta, \varphi)$ for stimulated emission of light may be obtained if we change the sign before $\Delta = J' - J''$ to the opposite one in (2.8). Further, assuming the same procedure as the one used to obtain Eqs. (3.19), (3.20) and (3.21) we have:

$$_b\dot{\rho}_Q^K = \Gamma_p \sum_{X\kappa} {}^K S^{X\kappa} \left\{ \Phi^{(X)} \otimes {}_a\rho^{(\kappa)} \right\}_Q^K$$
$$- \Gamma_p \sum_{XK'} {}^K \tilde{S}^{XK'} \left\{ \Phi^{(X)} \otimes {}_b\rho^{(K')} \right\}_Q^K, \qquad (5.93)$$

$$_a\dot{\rho}_q^\kappa = -\Gamma_p \sum_{X\kappa'} {}^\kappa S^{X\kappa'} \left\{ \Phi^{(X)} \otimes {}_a\rho^{(\kappa')} \right\}_q^\kappa$$
$$+ \Gamma_p \sum_{XK} {}^\kappa \tilde{S}^{XK} \left\{ \Phi^{(X)} \otimes {}_b\rho^{(K)} \right\}_q^\kappa, \qquad (5.94)$$

where $^{L_1}S^{L_2L_3}$ and $^{L_1}\tilde{S}^{L_2L_3}$ can be calculated following (5.84) and (5.85), taking the corresponding discussion into consideration as well.

We will now dwell on relaxation processes. Since we intend to discuss only relaxation which is isotropic with respect to orientations of $\mathbf{J}$, its probability depends only on the angle Θ between the initial θ', φ' and the final θ, φ position, and does not depend on the concrete values of θ', φ' and θ, φ. Hence, the functions $\Gamma(\Theta)$ and $\gamma(\Theta)$ may be expanded over bipolar harmonics [379] in the form

$$\Gamma(\theta, \varphi; \theta', \varphi') = \Gamma(\Theta) = \sum_{K=0}^{\infty} \Gamma_K \sum_{Q=-K}^{K} Y_{KQ}(\theta', \varphi') Y_{KQ}^*(\theta, \varphi). \quad (5.95)$$

Substituting expressions of type (5.95) for $\Gamma(\Theta)$ and $\gamma(\Theta)$ into the system of equations (5.91) and (5.92), expanding over multipole moments and accounting for the orthogonality of spherical functions (B.3), we obtain the terms $-\gamma_{\kappa a} \rho_q^\kappa$ and $-\Gamma_{Kb} \rho_Q^K$ which have to be added to the righthand sides of (5.94) and (5.93).

Reverse spontaneous transitions cannot change the orientation of the angular momentum of the molecule, hence we have $\Gamma_{J'J''}(\theta, \varphi; \theta', \varphi') = \Gamma_{J'J''}(\theta, \varphi)$. Since the probability of spontaneous transition is invariant with respect to turn of coordinates, we may make an even stronger assertion: $\Gamma_{J'J''}(\theta, \varphi) = \Gamma_{J'J''} = \text{const}$. According to the aforesaid we obtain a term that is responsible for reverse spontaneous transitions in the form $\Gamma_{J'J''b} \rho_Q^K \delta_{K\kappa} \delta_{Qq}$, which has to be added to the righthand side of Eq. (5.94).

Finally, performing multipole expansion of the terms representing the magnetic field effect, as shown in the deduction of Eq. (4.9), we obtain the terms $iq\omega_{J''a} \rho_q^\kappa$ and $iQ\omega_{J'b} \rho_Q^K$, which also have to be added to the righthand sides of (5.94) and (5.93).

Summing up the above, we may conclude that the 'classical' system of equations (5.93) and (5.94), together with the above given additional terms, coincides perfectly with the asymptotic system of equations of motion of quantum mechanical polarization moments (5.87) and (5.88). This result was actually to be expected from correspondence principle considerations.

5.8 General restrictions for the relaxation rates of polarization moments

As can be seen, for any isotropic relaxation process different rank polarization moments change independently. The question arises: can the relaxation constants γ_κ and Γ_K be arbitrary? The following simple example [135] shows that this is not the case. Let us assume that the angular momentum $J = 1$ and the initial state are such that only level $M = 1$

is populated, so that $f_{MM} = 1$, while the other elements of the density matrix are zero at $t = 0$. Then the following quantum mechanical components of orientation and alignment exist (see Appendix (D.13))

$$f_0^1 = \frac{1}{\sqrt{2}} (f_{11} - f_{-1-1}), \tag{5.96}$$

$$f_0^2 = \frac{1}{\sqrt{10}} (f_{11} + f_{-1-1} - 2f_{00}), \tag{5.97}$$

and according to our assumption, $f_0^1 = 1/\sqrt{2}$, $f_0^2 = 1/\sqrt{10}$ at $t = 0$. It is easy to see that the alignment f_0^2 cannot be destroyed as quickly as one may wish compared to the orientation f_0^1. Indeed, if one assumes that $\Gamma_2 \gg \Gamma_1$ then after an interval of time t, such that $\Gamma_2^{-1} \ll t \ll \Gamma_1^{-1}$, the quantity f_0^2 will practically disappear, whilst f_0^1 will remain very close to $1/\sqrt{2}$. This will contradict, however, the normalization condition $f_{11} + f_{00} + f_{-1-1} = 1$ and the requirement that the population of the magnetic sublevels should be positive. Thus, it is clear that the relaxation rates Γ_1 and Γ_2 cannot be absolutely arbitrary and must satisfy some inequality.

Thus, in [135] the system of inequalities for Γ_K

$$\sum_K (2K + 1)(-1)^{M-M_1} \begin{pmatrix} J & J & K \\ M & -M & 0 \end{pmatrix} \begin{pmatrix} J & J & K \\ M_1 & -M_1 & 0 \end{pmatrix} \tilde{\Gamma}_K \leq 0,$$

$$\text{for } M \neq M_1, \tag{5.98}$$

has been obtained, where $\tilde{\Gamma}_K$ is $\Gamma_K - \Gamma_0$ and the quantity in brackets is a $3j$-symbol; see Appendix C. Another system of inequalities

$$(-1)^{\Upsilon+2J+1} \sum_K (-1)^K (2K+1) \begin{Bmatrix} J & J & K \\ J & J & \Upsilon \end{Bmatrix} \tilde{\Gamma}_K \geq 0, \quad 1 \leq \Upsilon \leq 2J, \tag{5.99}$$

different from the preceding one, is given in [304]; the quantity in curly brackets is a $6j$-symbol; see Appendix C.

Apart from the general significance, the given inequalities should be taken into account when relaxation processes are described phenomenologically as well as when fitting the experimental data. Systems (5.98) and (5.99) contain different numbers of independent inequalities. Taking into consideration the symmetry properties of $3j$-symbols (C.3), it is quite simple to make sure that the system (5.98) contains $J(J+1)$ independent inequalities when J is an integer, and $J(J+1)+1/4$ inequalities when J is a half integer. This means that for $J \geq 3/2$ the number of inequalities exceeds (and for large J values by a considerable amount) the number of relaxation constants $\tilde{\Gamma}_K$. In the case of (5.99) the number of inequalities

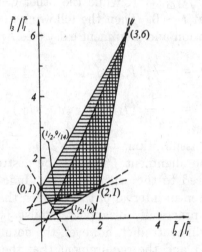

Fig. 5.14. Allowed region of the parameters $\tilde{\Gamma}_K/\tilde{\Gamma}_1$ in the case where $J = 3/2$, determined by the systems of inequalities (5.98) and (5.99).

coincides with the number of constants $\tilde{\Gamma}_K$. Let us assume that all relaxation constants $\tilde{\Gamma}_K$ are measured in units of one of them, say $\tilde{\Gamma}_1$. In this case inequalities (5.98) or (5.99) in $(2J-1)$-dimensional space define the region of allowed values of $\tilde{\Gamma}_K/\tilde{\Gamma}_1$. As an illustration, this region is shown in Fig. 5.14 for $J = 3/2$. Each straight line in the diagram represents one inequality. The region of $\tilde{\Gamma}_K/\tilde{\Gamma}_1$ allowed by inequalities (5.98) is denoted by horizontal strokes, and the region allowed by inequalities (5.99) is denoted by vertical ones. As can be seen, in this case Eq. (5.99) imposes more strict limitations on the constants $\tilde{\Gamma}_K$. It follows from the results of computer analysis that a similar situation occurs for arbitrary J values (at least for $J \leq 70$).

In the general case of arbitrary J values it is not easy to present, in $(2J-1)$-dimensional space, the area of allowed $\tilde{\Gamma}_K/\tilde{\Gamma}_1$ values. In order to characterize this area, it is possible to propose the following approach. Of all permitted $\tilde{\Gamma}_K/\tilde{\Gamma}_1$ values it is necessary to determine the minimal and the maximal value for each rank K. Here it is necessary to take into consideration the fact that all maximum or all minimum values of $\tilde{\Gamma}_K/\tilde{\Gamma}_1$ cannot always be reached simultaneously. This assertion may be illustrated by Fig. 5.14 in the case of the system of inequalities (5.98). The minimum value $\tilde{\Gamma}_2/\tilde{\Gamma}_1 = 0$ is attained when $\tilde{\Gamma}_3/\tilde{\Gamma}_1 = 1$, but the minimum value $\tilde{\Gamma}_3/\tilde{\Gamma}_1 = 1/6$ is attained when $\tilde{\Gamma}_2/\tilde{\Gamma}_1 = 1/2$.

An analysis of (5.98), in the case where the number of inequalities considerably exceeds the number of constants $\tilde{\Gamma}_K/\tilde{\Gamma}_1$, shows [42] that it is possible to make use of the Monte-Carlo method, i.e. one must generate by the computer random points in the $(2J-1)$-dimensional space of $\tilde{\Gamma}_K/\tilde{\Gamma}_1$, and from all points hit into the allowed region (simultaneously satisfying all inequalities (5.98)) one must choose the minimal and the maximal values of constants $\tilde{\Gamma}_K/\tilde{\Gamma}_1$ for each rank K.

In the analysis of (5.99), when the number of inequalities is equal to $2J$ it is possible to use another method of examination of the permitted region for $\tilde{\Gamma}_K/\tilde{\Gamma}_1$. One can turn from the system of inequalities to a system of equalities and solve it by alternately 'switching off' one equality. In this way we will find in the $(2J-1)$-space 'corner point' coordinates (see Fig. 5.14 in the case of inequalities (5.99)), it means the coordinates for the region of allowed values of $\tilde{\Gamma}_K/\tilde{\Gamma}_1$. From all coordinates of these corner points we must select those which correspond to the minimal and the maximal values of $\tilde{\Gamma}_K/\tilde{\Gamma}_1$ for each rank K.

Using these methods an analysis of the system of inequalities (5.98) and (5.99) for $J \le 70$ was carried out in [42]. It is shown there that for these J values the system (5.99) introduces more strict limitations on the minimal values of $\tilde{\Gamma}_K/\tilde{\Gamma}_1$ and both systems (5.98) and (5.99) introduce the same limitations on the maximal values of $\tilde{\Gamma}_K/\tilde{\Gamma}_1$.

The results of absolute *minimal* values of $\tilde{\Gamma}_K/\tilde{\Gamma}_1$ given by (5.99) are presented in Table 5.4 for $J \le 10$. The simultaneously existing minimal values of $\tilde{\Gamma}_K/\tilde{\Gamma}_1$ obtained from (5.99) are presented in Table 5.5 for the same J values. It can easily be seen that in both cases all these values are positive and do not exceed 0.6. A similar situation also holds for larger J values, at least for $J \le 70$.

Absolute *maximal* values of $\tilde{\Gamma}_K/\tilde{\Gamma}_1$ can all be approached simultaneously and equal $K(K+1)/2$. These values are independent of J.

These restrictions lead to one very general consequence to which we want to draw attention. *If it is known that the relaxation rate of population Γ_0 equals one particular rate Γ_K with $K \ne 0$, then all relaxation rates Γ_K must be equal.* This statement may be important in the analysis of some relaxation processes in molecules, see, for example, Sections 2.5 and 3.6.

In some special cases, analytical expressions can be derived for minimal and maximal values of $\tilde{\Gamma}_K/\tilde{\Gamma}_1$. The following expansion of $\tilde{\Gamma}_K$ is given in [304]:

$$\tilde{\Gamma}_K = \sum_{\Upsilon \ge 1} a_{K\Upsilon} A_\Upsilon, \tag{5.100}$$

with

$$a_{K\Upsilon} = (2J+1)^{-1} - (-1)^{K+\Upsilon+2J} \left\{ \begin{matrix} J & J & K \\ J & J & \Upsilon \end{matrix} \right\}, \tag{5.101}$$

Table 5.4.   Absolute minimum values for $\tilde{\Gamma}_K/\tilde{\Gamma}_1$

J	K 2	3	4	5	6	7	8	9	10
1	0.6000								
3/2	0.5000	0.6429							
2	0.4286	0.6429	0.5952						
5/2	0.3750	0.4792	0.5694	·0.5846					
3	0.3333	0.4000	0.5455	0.5758	0.5711				
7/2	0.3000	0.3636	0.5227	0.5717	0.5612	0.5626			
4	0.2727	0.3506	0.5012	0.5711	0.5524	0.5559	0.5555		
9/2	0.2500	0.3510	0.4808	0.5240	0.5442	0.5510	0.5499	0.5500	
5	0.2308	0.3590	0.4615	0.4744	0.5362	0.5475	0.5451	0.5455	0.5455
11/2	0.2143	0.3714	0.4435	0.4338	0.4864	0.5452	0.5410	0.5418	0.5417
6	0.2000	0.3864	0.4265	0.4011	0.4522	0.5440	0.5372	0.5387	0.5384
13/2	0.1875	0.4026	0.4105	0.3749	0.4350	0.5436	0.5336	0.5362	0.5356
7	0.1765	0.3866	0.3956	0.3543	0.4307	0.5287	0.5301	0.5341	0.5332
15/2	0.1667	0.3754	0.3816	0.3383	0.4355	0.5041	0.5238	0.5326	0.5309
8	0.1579	0.3699	0.3684	0.3263	0.4469	0.4810	0.4878	0.5292	0.5289
17/2	0.1500	0.3688	0.3561	0.3176	0.4627	0.4595	0.4580	0.4985	0.5270
9	0.1429	0.3707	0.3444	0.3116	0.4711	0.4396	0.4339	0.4804	0.5252
19/2	0.1364	0.3750	0.3335	0.3078	0.4631	0.4213	0.4151	0.4732	0.5234
10	0.1304	0.3809	0.3232	0.3060	0.4552	0.4046	0.4012	0.4747	0.5216

J	K 11	12	13	14	15	16	17	18	19	20
11/2	0.5417									
6	0.5385	0.5385								
13/2	0.5357	0.5357	0.5357							
7	0.5334	0.5333	0.5333	0.5333						
15/2	0.5313	0.5312	0.5313	0.5312	0.5313					
8	0.5295	0.5294	0.5294	0.5294	0.5294	0.5294				
17/2	0.5280	0.5277	0.5278	0.5278	0.5278	0.5278	0.5278			
9	0.5266	0.5262	0.5263	0.5263	0.5263	0.5263	0.5263	0.5263		
19/2	0.5255	0.5249	0.5250	0.5250	0.5250	0.5250	0.5250	0.5250	0.5250	
10	0.5245	0.5236	0.5239	0.5238	0.5238	0.5238	0.5238	0.5238	0.5238	0.5238

Table 5.5.   Simultaneously existing minimum values for $\tilde{\Gamma}_K/\tilde{\Gamma}_1$

J	K 2	3	4	5	6	7	8	9	10
1	0.6000								
3/2	0.5000	0.6429							
2	0.4286	0.6429	0.5952						
5/2	0.3750	0.6528	0.5694	0.5846					
3	0.3333	0.6667	0.5455	0.5758	0.5711				
7/2	0.3000	0.6818	0.5227	0.5717	0.5612	0.5626			
4	0.2727	0.6970	0.5012	0.5711	0.5524	0.5559	0.5555		
9/2	0.2500	0.7115	0.4808	0.5731	0.5442	0.5510	0.5499	0.5500	
5	0.2308	0.7253	0.4615	0.5769	0.5362	0.5475	0.5451	0.5455	0.5455
11/2	0.2143	0.7381	0.4435	0.5821	0.5282	0.5452	0.5410	0.5418	0.5417
6	0.2000	0.7500	0.4265	0.5882	0.5201	0.5440	0.5372	0.5387	0.5384
13/2	0.1875	0.7610	0.4105	0.5950	0.5120	0.5436	0.5336	0.5362	0.5356
7	0.1765	0.7712	0.3956	0.6022	0.5038	0.5441	0.5301	0.5341	0.5332
15/2	0.1667	0.7807	0.3816	0.6096	0.4956	0.5452	0.5266	0.5326	0.5309
8	0.1579	0.7895	0.3684	0.6172	0.4874	0.5469	0.5231	0.5314	0.5289
17/2	0.1500	0.7976	0.3561	0.6248	0.4792	0.5491	0.5196	0.5305	0.5270
9	0.1429	0.8052	0.3444	0.6324	0.4711	0.5518	0.5159	0.5300	0.5252
19/2	0.1364	0.8123	0.3335	0.6399	0.4631	0.5548	0.5122	0.5298	0.5234
10	0.1304	0.8188	0.3232	0.6473	0.4552	0.5581	0.5084	0.5300	0.5216

J	K 11	12	13	14	15	16	17	18	19	20
11/2	0.5417									
6	0.5385	0.5385								
13/2	0.5357	0.5357	0.5357							
7	0.5334	0.5333	0.5333	0.5333						
15/2	0.5313	0.5312	0.5313	0.5312	0.5313					
8	0.5295	0.5294	0.5294	0.5294	0.5294	0.5294				
17/2	0.5280	0.5277	0.5278	0.5278	0.5278	0.5278	0.5278			
9	0.5266	0.5262	0.5263	0.5263	0.5263	0.5263	0.5263	0.5263		
19/2	0.5255	0.5249	0.5250	0.5250	0.5250	0.5250	0.5250	0.5250	0.5250	
10	0.5245	0.5236	0.5239	0.5238	0.5238	0.5238	0.5238	0.5238	0.5238	0.5238

but $A_\Upsilon > 0$. It is easy to see that

$$\left(\frac{a_{K\Upsilon}}{a_{1\Upsilon}}\right)_{min} \leq \frac{\tilde{\Gamma}_K}{\tilde{\Gamma}_1} \leq \left(\frac{a_{K\Upsilon}}{a_{1\Upsilon}}\right)_{max}. \tag{5.102}$$

In the last expression the minimum and the maximum are assumed from the parameter Υ. As demonstrated in [304],

$$\left(\frac{a_{2\Upsilon}}{a_{1\Upsilon}}\right)_{min} = \frac{a_{2(2J)}}{a_{1(2J)}} \tag{5.103}$$

and

$$\left(\frac{a_{2\Upsilon}}{a_{1\Upsilon}}\right)_{max} = \frac{a_{21}}{a_{11}}. \tag{5.104}$$

Unfortunately, it is not possible to verify the generalization of (5.103) and (5.104) for arbitrary K values. Yet numerical analysis of (5.98) and (5.99) demonstrates [42] that the generalization of (5.104):

$$\left(\frac{a_{K\Upsilon}}{a_{1\Upsilon}}\right)_{max} = \frac{a_{K1}}{a_{11}} = \frac{K(K+1)}{2}, \tag{5.105}$$

is valid for arbitrary K values. In order to derive the analytical formula for a_{K1}/a_{11} we make use of analytical expressions of $6j$-symbols from Appendix C.

With the aid of analytical formulae for $6j$-symbols it is also possible to obtain

$$\frac{a_{K(2J)}}{a_{1(2J)}} = \frac{1}{1 + J/(J+1)}$$
$$\times \left(1 - (-1)^K \frac{(2J - K + 1)(2J - K + 2)\dots(2J)}{(2J + 2)(2J + 3)\dots(2J + K + 1)}\right), \tag{5.106}$$

but Eq. (5.106) does not lead to the results of Table 5.4. This means that for arbitrary K values this expression does not give the values of the absolute minimum of $\tilde{\Gamma}_K/\tilde{\Gamma}_1$. Nevertheless, it does give simultaneously existing minimal values of $\tilde{\Gamma}_K/\tilde{\Gamma}_1$, presented in Table 5.5, and also for larger J values, which are not contained in the table.

In this connection it is useful to derive expressions for $a_{K(2J)}/a_{1(2J)}$ in some special cases. Thus

$$\left(\frac{\tilde{\Gamma}_{2J}}{\tilde{\Gamma}_1}\right)_{min(simult)} = \frac{a_{2J(2J)}}{a_{1(2J)}} = \frac{J+1}{2J+1} - (-1)^{2J} \frac{\sqrt{\pi}}{2^{4J-1}} \frac{\Gamma(2J)}{\Gamma(2J+1/2)}, \tag{5.107}$$

$\Gamma(x)$ being a gamma function. These formulae in the high J limit give

$$\left(\frac{\tilde{\Gamma}_{2J}}{\tilde{\Gamma}_1}\right)_{min(simult)} \approx \frac{J+1}{2J+1}. \tag{5.108}$$

One must understand that the limitations obtained are less strict than inequalities (5.98) and (5.99). They do not characterize the shape of the allowed region of relaxation constants on the full scale. The restrictions define the multidimensional parallelepiped in $(2J-1)$-dimensional space which is encircled around the allowed region of $\tilde{\Gamma}_K/\tilde{\Gamma}_1$, satisfying the restrictions obtained. Nevertheless, the restrictions obtained are much more obvious and they give a good idea of the limitations imposed by inequalities (5.98) and (5.99).

Tables 5.4 and 5.5 and formulae for the minimal values and the maximal values of $\tilde{\Gamma}_K/\tilde{\Gamma}_1$ seem to be very useful in the case where not all the relaxation constants with $0 \leq K \leq 2J$ are taken into consideration. In this situation it is impossible to use inequalities (5.98) and (5.99) directly, but limitations obtained in [42] for $\tilde{\Gamma}_K/\tilde{\Gamma}_1$ are still valid. This is especially important in the case of molecules where the angular momentum values are, as a rule, large and not all the possible polarization moments are taken into account in the description of the physical process.

6
Other methods of alignment and orientation of molecules

The material presented so far was based on the creation of anisotropic distribution of molecular angular momenta under the direct effect of light absorption. We are now going to discuss briefly some ideas and examples of experimental realization of other methods leading to the production of polarized molecules, including those which are not directly connected with light effects, such as polarization caused by collisions and external electric or magnetic fields.

6.1 Photodissociation and photoionization

It was assumed in the preceding chapters that optical transitions between bonded states of molecules take place with no change in their chemical composition. *Photodissociation* and *photoionization* form a general class of *photofragmentation* processes in collisions between a molecule and a photon which lead to simple chemical reactions of disintegration into atomic or molecular fragments, or into ions and electrons.

1   The photodissociation process

The *photodissociation* process takes place most frequently at excitation of the molecule to a non-bonded state, with subsequent dissociation into products. Since the angular part of the transition probability, according to Chapter 2, is still dependent on the mutual orientation of the $\hat{E}$-vector of the initiating light beam and on the transition dipole moment **d**, one may expect *spatial anisotropy* of angular momenta distribution both in the dissociation products and in the set of molecules which remains undestroyed.

Let us start with the second effect which is most close to the ideas discussed in Chapter 3 in connection with optical polarization by 'depopulation'. Indeed, if the photodissociation process proceeds via one fixed (upper) electronic state, then, firstly, the expressions for angular transition coefficients $G(\theta, \varphi)$, as considered in Chapter 2 (see formula (2.8)), remain valid. Secondly, the model of creating anisotropy of momenta $\mathbf{J}(\theta, \varphi)$ distribution through angularly selective depopulation of those molecular states from which photofragmentation takes place also remains valid. The 'open cycle' model (see Section 3.1) may be viable here even more than for transitions between bonded states (see Fig. 1.2), since here the role of reverse spontaneous and stimulated radiative transitions practically disappears. Non-radiative angular momenta randomization in the ground (initial) state continues to take place as the result of collisional processes and fly-through of molecules moving through the light beam. The slightly different relaxation mechanism is due to the fact that, instead of one rovibrational level, as assumed in the case shown in Fig. 1.2 and Section 3.1, photodissociation usually 'pumps out' a large group of rovibronic levels connected with the continuous upper state by a sufficiently high Franck–Condon factor.

A matter of principal importance here consists of the fact that, simultaneously, P-type and R-type transitions take place for a parallel transition, or P-, R-, and Q-type transitions for a perpendicular transition from the fixed ground state rovibronic level (see Section 1.4). For this reason at excitation by linearly polarized light a parallel transition produces anisotropy of the remaining molecules with much greater efficiency. Indeed, as may be seen from Fig. 3.2 (lower picture), simultaneous realization of both Q-transitions (left) as well as of (P, R)-transitions (right) leads to the simultaneous creation of negative and positive alignment, which in sum produces a less anisotropic distribution of ground state angular momenta than is the case in parallel transition where Q-absorption does not occur. We also wish to point out that even in the latter case one cannot achieve zero probability of photodissociation in this situation for magnetic sublevels with $M = \pm J''$, as was the case at 'total P-polarization' in bonded–bonded transition (see Section 3.3), since it is not possible to avoid the presence of an R-branch.

If photodissociation takes place, as is frequently the case, under the action of a laser pulse, short with respect to relaxation time, then the ratio $\Delta N / N$ for the remaining undecayed molecules in the $|JM\rangle$ state directly after the pulse is simply

$$\frac{\Delta N}{N} = \exp\left[-3\sigma_{ph}F\left(_PC^2 + _QC^2 + _RC^2\right)\right], \qquad (6.1)$$

where σ_{ph} is the cross-section of photodissociation, F is the photon flux in the pulse and $_iC^2$ are the angular coefficients of probability of photodissociation. Their values, for arbitrary J, coincide with the coefficients at χ in the formulae given in Table 3.6, where one has to find the respective branch of absorption, $i = P, Q$ or R (not depending on emission branch), and use the $\mathcal{P}$ or $\mathcal{R}$ expression for linearly polarized excitation and the $\mathcal{C}$ expression for circularly polarized excitation. The $_iC^2$ coefficients converge to $G(\theta, \varphi)$ according to (2.8) for $J \to \infty$.

It may be of interest that it is just this method that was first realized for obtaining an ensemble of optically polarized ground state molecules, even before the 'laser' era, by Dehmelt and coworkers [112, 211, 212, 326] on a simple molecular object, namely on a molecular hydrogen ion H_2^+. The ions were created by short electron bombardment pulses and were kept in radiofrequency quadrupole traps (for their evolvement Dehmelt was awarded the Nobel prize in physics in 1989). *M-selective photodissociation* (the term was introduced by Dehmelt) was produced by irradiation with linear polarized light from a high intensity Hg arc. The effect of alignment in H_2^+ remaining in the electronic ground state with rotational quantum numbers not exceeding two was detected by the method of magnetic resonance between Zeeman sublevels in a magnetic field (50–115)$\cdot 10^{-7}$ T. The experiments yielded information on Landé factors and hyperfine interaction constants for H_2^+.

The idea and the theoretical description of creating alignment in neutral molecules which remained undestroyed after photodissociation may be found in the works by Bersohn and Lin [68], Zare [400] and Ling and Wilson [272]. The experimental proof of alignment of non-decayed molecules, as a rule, was of an indirect nature.

As a specific example, let us consider photodissociation of a Ca_2 molecular beam into Ca atoms. According to estimates [383], under the action of 406 nm and 413 nm lines of a Kr^+-laser the energy width of noticeable Franck–Condon factor density in the transition $^1\Sigma_g^+ \to {}^1\Pi_u$ amounts to 10^2 cm^{-1}, the Franck–Condon factor density itself not exceeding 10^{-2} cm. If, following the authors of [383], we assume that the rate of photodissociation at maximum probability is about $\Gamma_p \approx 10^9$ s^{-1}, the fly-through time of the molecular beam through the light beam equalling $T_0 \approx 10^{-7}$ s, then we may expect noticeable depopulation of a sufficiently large group of levels. Thus, we may observe the whole spectral range of the laser radiation working here with sufficient efficiency. It proved possible to achieve disintegration of up to 5% of the Ca_2 molecules. Clearly, the process is, as before, governed by the parameter Γ_p/γ_Σ (see Section 3.1). The nonlinearity of the $Ca(^1P)$ fluorescence dependence on laser power indicates the effective depopulation of the Ca_2 molecule ground state in the process of $^1\Sigma_g^+ - {}^1\Pi_u$ dissociation, and one may therefore expect molecular ground state optical polarization via photodissociation in this situation.

Another indirect confirmation of the effect follows from the experiment of De Vries, Martin *et al.* [117] who produced polarization in an effusion beam of IBr molecules through their dissociation by a 532.4 nm laser line (second harmonic of a Nd-YAG laser). A high degree of polarization of the remaining undestroyed molecules was provided by the mainly parallel nature of the transition, at photodissociation cross-section of $8\cdot10^{-19}$ cm^2. Applying a photon flux F of $2\cdot10^{14}$ cm^{-2} in a pulse it is possible, according to (6.1), to achieve a considerable ratio $\Delta N/N$ and, thus, a high degree of polarization of the remaining non-dissociated molecules ($\approx$ 20 p.c.). The beam of IBr molecules was then crossed with a supersonic atomic beam of metastable Xe*($^3P_2, {}^3P_0$) atoms. The luminescence of the excimeric XeBr and XeI molecules was registered in the reaction

$$\text{Xe}^* + \text{IBr} \rightarrow \left\{ \begin{array}{l} (\text{XeBr})^* + \text{I} \\ (\text{XeI})^* + \text{Br} \end{array} \right. \tag{6.2}$$

as dependent on the angle between the light vector of excitation $\hat{\mathbf{E}}$ and the direction of the beam. The results showed that the reaction probability is highest in the situation where the $\hat{\mathbf{E}}$-vector is orthogonal with respect to the relative beam velocity $\mathbf{v}_{rel}$, i.e. when $\mathbf{v}_{rel}$ lies in the plane of rotation of the IBr molecule. The quantitative information would require control of the degree of alignment of undecayed ground state IBr molecules, to which purpose one may apply optical methods, say, laser fluorescence with determination of polarization degree in the fluorescence (see Section 3.3).

So far we have been considering the remaining non-decayed reagent molecules. Let us now turn to anisotropic effects on the *products*, as has been the subject of most investigations. The most general and fruitful approach is the concept of correlation between the vector quantities of the initial molecule and the photofragments. As sources of information on these topics one might recommend, e.g., [66, 69, 164, 199, 262, 351, 402] and the bibliographies contained therein. The interest is mostly focussed on the connection between the bonded–non-bonded transition dipole moments $\mathbf{d}$ of the parent molecules and the relative velocity $\mathbf{v}_f$ of the separating fragments or the angular momenta $\mathbf{J}_f$ of the fragments (Fig. 6.1). Indeed, since $\mathbf{d}$, $\mathbf{v}_f$ and $\mathbf{J}_f$ refer to different objects, we are only entitled to judge how far $\mathbf{v}_f$ and $\mathbf{J}_f$ retain the 'memory' on their connection with vector $\mathbf{d}$ (and through it with $\hat{\mathbf{E}}$), in other words, to what extent these vectors *correlate*. Due to this correlation the angular dependence of the action of $\hat{\mathbf{E}}$ on the transition dipole moment $\mathbf{d}$ (see Section 2.1) creates two types of anisotropy: (a) the anisotropy of *angular distribution* of the photodissociation products and (b) the angular momenta *alignment* (orientation) of individual fragments.

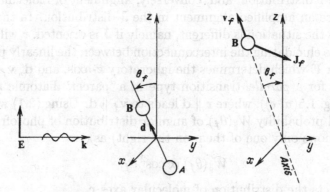

Fig. 6.1. Schemes of $\hat{\mathbf{E}} - \mathbf{d} - \mathbf{v}_f$ and of $\hat{\mathbf{E}} - \mathbf{d} - \mathbf{J}_f$ correlation at photodissociation of a non-rotating (during dissociation) molecule AB through a parallel transition. The $\hat{\mathbf{E}}$-vector fixes the laboratory coordinate system, the absorption probability $(\hat{\mathbf{E}}^*\mathbf{d})^2$ selects the direction of $\mathbf{d}$ and thus also that of the internuclear axes, with which the velocity $\mathbf{v}_f$ and the angular momentum $\mathbf{J}_f$ of the photofragment are connected.

Let us consider the first type which is a direct consequence of the connection between the transition dipole moment $\mathbf{d}$ with the molecular axis $\mathbf{r}$; see Section 1.3, Fig. 1.5. Indeed if the time of photodissociation is much shorter than the period of rotation of the molecule, which is a fair approximation for direct photodissociation, one might expect that the departure of the photofragments will take place in the direction of the vibrational motion (the axial recoil approximation). Thus, the fragments will have the same final direction as their mutual orientation at the moment of photodissociation, as shown in Fig. 6.1. In other words, angular anisotropy of fragment distribution reflects the anisotropy of $\mathbf{r}$ distribution in the 'parent' molecule created by light action. It is therefore useful to trace the connection between the distribution of molecular angular momenta $\mathbf{J}$ (which was considered in the preceding discussion) and that of molecular axis $\mathbf{r}$. If we are describing the anisotropic $\mathbf{J}$ distribution over the spherical angle θ_J by means of Legendre polynomials $P_K(\cos\theta_J)$ (see (D.50)), the corresponding angular distribution θ_r of molecular axes $\mathbf{r}$ can be written for a linear molecule as [158]

$$\langle P_K(\cos\theta_r)\rangle = P_K(0)\langle P_K(\cos\theta_J)\rangle = P_K(0)\frac{2a_K n_J}{2K+1}, \qquad (6.3)$$

where a_K is defined according to (D.53), and the averaging is performed over the molecular ensemble. The polynomial $P_K(0)$ emerges because of $\mathbf{J} \perp \mathbf{r}$; see Fig. 1.5. Here $P_K(0) = 0$ for odd K, and $P_K(0) = -1/2, 3/8, \ldots$ for $K = 2, 4, \ldots$. Thus, any alignment of angular momenta $\mathbf{J}$ is attenuated in the **r** distribution, and, conversely, alignment of molecular axes **r** will produce an amplified alignment in the **J** distribution. In the case of orientation the situation is different, namely if **J** is oriented, **r** will not be.

A scheme elucidating the interconnection between the linearly polarized light vector $\hat{\mathbf{E}}$ which determines the laboratory z-axis, and **d**, $\mathbf{v}_f$ is given in Fig. 6.1 for a *parallel* transition type in a 'parent' diatomic molecule AB (see Fig. 1.5(a, c)), where **r** $\parallel$ **d** leads to $\mathbf{v}_f \parallel$ **d**. Using (2.1) we obtain the desired probability $W_\parallel(\theta_f)$ of angular distribution of photofragments (Fig. 6.1 shows only one of them on the right) as

$$W_\parallel(\theta_f) \propto \cos^2 \theta_f, \tag{6.4}$$

which reflects the distribution of molecular axes **r**.

For a *perpendicular* type of molecular transition (see Fig. 1.5(b, d)), when **d** is orthogonal with respect to the molecular axis, the separation of the fragments proceeds mainly orthogonally with respect to **d**, and the corresponding probability is

$$W_\perp(\theta_f) \propto \sin^2 \theta_f. \tag{6.5}$$

The probability of angular distribution of photofragments has the same origin as the probability $G(\theta, \varphi)$ in (2.8), but now it is a transition from a bonded state into a non-bonded one.

Eqs. (6.4) and (6.5) lead to the *cylindrical* symmetry of the final photofragment angular distribution $W(\theta_f, \varphi_f)$ in the form of a 'dumbbell' and a toroid, which are symmetrical with respect to the $\hat{\mathbf{E}}$-vector (the z-axis). The distribution $W(\theta_f, \varphi_f)$ is proportional to a differential photodissociation cross-section in the laboratory frame, $f(\theta_f, \varphi_f) = d\sigma_{ph}/d\Omega$. For a proper description of its symmetry properties it is usually [376, 402] expanded in a set of spherical harmonics Y_{KQ}. The cylindric symmetry in this case means that only spherical functions Y_{00} and Y_{20} appear with non-zero coefficients, and then

$$f(\theta_f, \varphi_f) \propto \left(w_0^0 Y_{00} + \sqrt{5} w_0^2 Y_{20} \right), \tag{6.6}$$

where the coefficients w_Q^K are classical multipole moments characterizing the differential cross-section of photodissociation and are obtained in the same way as (2.14). The case of weak light action is supposed here, which means that we disregard the above-mentioned alignment of the initial molecular ensemble. If this is not so higher order multipoles may appear with $K > 2$, owing to non-linear light action. Using (D.49) one may turn to the form proposed in [376]:

$$f(\theta_f, \varphi_f) = \frac{\sigma_{ph}}{4\pi} \left[1 + \beta_2 P_2(\cos \theta_f) \right], \tag{6.7}$$

where β_2 is often called the anisotropy parameter.

It is easy to see, with the help of Eqs. (B.1) and (D.49), that in the condition of parallel transition type the angular distribution (6.4) leads to $\beta_2 = 2$, whilst under condition of perpendicular transition type (6.5) it leads to $\beta_2 = -1$ (this corresponds to w_0^2/w_0^0 equal to $2/5$ and $-1/5$ respectively). Note once more that these values take place only in the limiting case of direct prompt dissociation of a diatomic molecule in the axial recoil approximation through one definite electronic state. In practice, photodissociation leads to a more complicated distribution, because photodissociation frequently takes place through two and more states. In the case of more complex molecules the direction of departure is a superposition of axial recoil (along $\mathbf{r}$) and transversal recoil (orthogonal to $\mathbf{r}$), being influenced by the internal motion of the fragments, the relative orientation of the bond and $\mathbf{d}$, and the time delay between the absorption and fragmentation. Note that (6.6) and (6.7) describe only a single photon process when the 'parent' molecule dissociates into only two photofragments. The parameter β_2 reflects the asymmetry of the photofragmentation process and lies in the range $-1 \leq \beta_2 \leq 2$. Measurements of β_2 may yield valuable information on the symmetry type of the dissociative state and the dynamics of the photodissociation, including the time required for photofragmentation. The anisotropy vanishes completely at $\beta_2 = 0$.

The first direct observations of anisotropy in the distribution of photofragments were made in the case of the halogen molecules Br_2 and I_2 [359] by means of the 'marks' left by product atoms on the inner surface of a sphere covered with a sensitive film. Mass spectroscopic registration also made it possible to determine the velocity of photofragments. An alternative method yielding similar information consists of registering the Doppler profile of the fragments. This method was realized considerably later [340] by means of laser induced-fluorescence of H atoms generated in the photolysis of HI molecules.

The second manifestation of anisotropy in photodissociation is the *polarization* of angular momenta $\mathbf{J}_f$ of the photofragments. As can be easily understood, alignment of transition dipole moments $\mathbf{d}$ of a target molecule in the course of adiabatic development of a photodissociation process may lead to a corresponding polarization of angular momenta $\mathbf{J}_f$ in laboratory coordinates due to angular momentum transfer from the light beam to the ensemble of fragments. This means that there exists an angular correlation between the vectors $\mathbf{d}$ and $\mathbf{J}_f$. In cases where the products are in a radiative state, it is most convenient to monitor the $\mathbf{J}_f$ distribution by fluorescence polarization, applying the approach as in Section 2.4 and thus obtaining multipole moments $_b\rho_Q^K$ of the excited state of the fragment. The next step is to produce an interpretation of the fragment multipoles in terms of photofragmentation dynamics; the detailed treatment for arbitrary J_f values based on the angular momentum transfer formulation can be found in [176].

Let us consider some experimental studies. Developing practically the idea proposed as early as 1968 by van Brunt and Zare (see [176, 376] for a review), a number of authors [232, 328, 383] investigated fluorescence polarization on atomic photofragments excited in the process of photodissociation of the simplest diatomic molecules. Observation of linear polarization (of the order of $\mathcal{P} \approx 0.05$) in the fluorescence of Na($^2P_{3/2}$) atoms after the photodissociation of Na$_2$ molecules is reported in [328]. Detailed studies of photodissociation of K$_2$ through the transition $X^1\Sigma_g^+ \to B^1\Pi_u$ have been carried out in Stwalley's laboratory [232]. In this process linear polarization of atomic fluorescence from -0.04 to -0.09 was observed at excitation in the 610–640 nm range. In [383] a quantum mechanical calculation is given for polarizational properties of photofragment fluorescence, as well as experimental results produced for Ca$_2$, confirming the requirements for accounting for coherence between the components $\Lambda = \pm 1$ of the $^1\Pi_u$ state through which the photodissociation

$$Ca_2(X^1\Sigma_g^+) + h\nu \to Ca_2^*(^1\Pi_u) \to Ca(^1S_0) + Ca^*(^1P_1) \qquad (6.8)$$

takes place. This leads to coherent superposition of the $m = \pm 1$ sublevels of the $^1P_1$ state, which, in turn, must cause linear polarization of degree 0.78 in the $^1P_1 - {}^1S_0$ fluorescence. Calculations performed without taking account of coherence yield a considerably lower degree of polarization – only 0.14. The experimental value of 0.64–0.68 obtained in [383] pointed towards coherence effects; the discrepancy from the value 0.78 is largely due to disregard of saturation effects caused by alignment of remaining non-dissociated molecules, as already pointed out previously. A general quantum mechanical treatment of coherence effects manifesting themselves in fluorescence polarization, as well as in the distribution of photofragments formed as a result of simultaneous excitation of different electronic states of different symmetry and correlating with one dissociative limit, are presented in [168]. This paper also predicts oscillations of the degree of polarization of the fragments in dependence on photon energy. The oscillations are determined by the relative amplitudes of photoabsorption and phases of vibrational functions of the continuous spectrum at large distances.

An interesting approach has been attempted in the works of Vasyutinskii and coworkers [252, 254, 380] who showed, for the case of CsI, TlBr and RbI, that conservation of angular momentum in the photodissociation process caused by circularly polarized light produces orientation of the atoms formed, the latter enabling measurement by means of magnetic resonance of the atomic products created. This is so if the lifetime of the molecule in transient state, through which photodissociation proceeds, is much shorter than the coupling time between the electronic momentum of the target molecule and its rotation, and if orientation created in the excited state is transferred to the atoms with almost no loss. As it happens, these conditions are frequently satisfied. This can, in some cases make possible orientation of atoms which cannot be achieved by other means.

A much larger number of papers report on the observation of fluorescence polarization in *molecular photofragments*. The emergence of the correlation $\hat{\mathbf{E}} - \mathbf{d} - \mathbf{J}_f$ offers a lucid explanation here, too. Let the photolysis of a triatomic molecule

$$ABC + h\nu \rightarrow AB(\mathbf{J}_f) + C \tag{6.9}$$

proceed through a bent configuration of the excited state. If the recoil axis (which has a fixed angle with respect to $\mathbf{d}$) is not parallel to the internuclear axis AB, then the separation of fragments causes rotation of the fragment AB around the direction at right angles to the plane containing both axes. This leads to anisotropic distribution of the rotational angular momenta $\mathbf{J}_f$ of the AB molecule in laboratory coordinates (determined by the $\hat{\mathbf{E}}$-vector), which manifests itself in fluorescence polarization of the fragment AB if the latter is produced in the excited electronic state. The explicit expression for the measurable degree of polarization may be found in [274], where the calculations are carried out by the classical approach, neglecting rotational angular momenta of the parent ABC molecule relative to the momenta of the AB fragment. The expression allows us to take account of the finite lifetime of the excited parent molecule ABC*.

The first observation of this kind was that of polarization of radiation of OH at UV photolysis of H_2O, presented in the paper by Chamberlain and Simons [99]. Quantitative testing of the predicted theory was carried out by Hussain, Wiesenfeld and Zare [204], who determined the degree of linear polarization $\mathcal{P} = 0.119$ in the radiation of $HgBr(B^2\Sigma^+ \rightarrow X^2\Sigma^+)$ created by photolysis of $HgBr_2$ by means of 193 nm pulses of an ArF laser. The situation here is of particular interest, since the photolysis of $HgBr_2$ takes place through transition from a linear shape in the ground state into a 'bent' state in the excited one. Accordingly, the transition dipole moment is positioned either in the plane of the molecule or at right angles to it. Thus, photolysis 'selects' different orientations of $HgBr_2$ according to Eq. (2.1). The fact that the measured degree of polarization has been found in good agreement with classical calculation ($\mathcal{P} = 1/7$) was obtained by assuming $\mathbf{d}$ to lie in the plane of the molecule and enabled determination of the C_{2v}-symmetry for the $HgBr_2$ molecule, thereby arriving at the important conclusion that one has here direct photodissociation through the term $^1B_2(^1\Sigma^+)$. Of later works one might mention those on the stereodynamics of photodissociation of ICN and ClCN [180, 190, 217], in which both alignment and orientation of the CN fragment were observed [190] and the effects of electronic and nuclear spin were elucidated. Correlation was found between the vectors $\mathbf{v}_f - \mathbf{J}_f - \mathbf{\Lambda}$, where Λ refers to doublet components, as well as with the electronic spin vector of the CN fragment [217].

If the photodissociation produces ground state photofragments which do not emit light, one may obtain information on the spatial aspects of the dynamics of photodissociation through the polarization of laser-induced fluorescence. As can be easily understood, the polarization characteristics of radiation depend both on the angular parameters or the geometry of the experiment (see Section 2.4) and on the parameters of alignment or orientation in the ground state (see Section 3.2), which reflect the stereodynamics of photolysis. An approach permitting us to separate the desired dynamic reaction parameters from angular (geometric) conditions of experiment is clear from our preceding discussions; see Chapters 3 and 5. Let us imagine that we are eager to determine ground state a multipole moments $_a\rho_0^\kappa/_a\rho_0^0$, which have appeared as a result of some reactive process (weak excitation and the $J \to \infty$ limit is supposed). If we substitute them into (3.19), we may turn to excited state polarization moments $_b\rho_Q^K$ and then to expressions for the intensity and polarization of fluorescence (see (2.24), (2.29), (2.30), (2.32), (2.33), (2.37)); an explicit example for linearly polarized Q-excitation is given by (3.36). Data fitting procedures yield the desired $_a\rho_0^\kappa/_a\rho_0^0$ values immediately, from which one may turn, if desired, to Legendre polynomial coefficients a_K; see (D.50). For arbitrary J values one may use, in a similar way, the formula for fluorescence intensity (5.34) and the system (5.22), (5.23) connecting quantum polarization moments f_0^K of the excited state with ground state polarization moments φ_0^κ (see Section 5.3), with necessary simplifications.

It should be mentioned that a similar task has been evolved by Greene and Zare [176, 177], and by Kummel, Sitz and Zare [249, 250, 251] for two-photon processes. For the latter case the anisotropy of the ground state is characterized by the above authors in terms of real components of the density matrix (see Appendix D). In some recent investigations registration of fluorescence is replaced by registration of charged particles, generated in resonance-enhanced multiphoton ionization (REMPI). Separation of ground state population and alignment parameters in such processes is given in [207].

Information on a considerable number of experimental and theoretical studies concerning photodissociation processes may be also found, e.g., in [66, 67, 199, 262, 351, 403].

2 Photoionization

Photoionization of a molecule may also be considered as belonging to the simplest kind of reactive collisions between a molecule and a photon, leading to the formation of a molecular ion and a photoelectron. As with any other photoprocess, photoionization is of a pronounced anisotropic nature. In analogy with photodissociation we may also have here polarization of molecular angular momentum in the photofragments, namely in the *molecular ion*. We also obtain anisotropic angular distribution of the directions of departure of the products – the departing ions and electrons. Since the phenomena has much in common with the photodissociation considered above [176], we will restrict ourselves to mentioning some concrete examples of investigations.

Let us consider an example of the first case. If the ions are formed in the excited state, the natural characteristic of anisotropy of the process consists of the polarization of their fluorescence. This method was first applied by Poliakoff, Zare, et al. [315] in the reaction

$$N_2(X^1\Sigma_g^+) + h\nu \rightarrow N_2^+(B^2\Sigma_u^+) + e \tag{6.10}$$

with subsequent emission ($\lambda = 391$ nm) in the parallel transition $B^2\Sigma_u^+ \rightarrow X^2\Sigma_g^+$ of the N_2^+ ion. The polarization of the emission is connected with different statistical weights of the degenerate 'parallel' (σ_g) and 'perpendicular' (π_g) photoionization channels. For the parallel channel, both transition dipole moments (of ionization and of emission) lie in the rotation plane of the molecule. Hence, the expected degree of linear polarization, just as in photoexcitation, equals $1/7$ (assuming $J \rightarrow \infty$). For a 'purely' π_g photoionization where P-, R- and Q-transitions are possible, the expected value is $\mathcal{P} = -1/13$. The polarization observed reflects the ratio between the dipole strength of respective ionization channels $r = D_\parallel^2/D_\perp^2$ and is

$$\mathcal{P} = \frac{1-r}{7+13r}. \tag{6.11}$$

The authors of [315] applied linearly polarized synchrotron radiation (45–66 nm) for ionization, which corresponds to photon energy from 18.76 eV (threshold) + 0.7 eV up to ≈ 27 eV. The measured $\mathcal{P}$ values, as dependent on photon energy, changed correspondingly from 0.052 down to approximately half the value, which made it possible to determine the value of r within the range 0.4–0.7. Further improvement of the experiment and refinement of the theoretical description was carried out in [179]. Accounting for the hyperfine and spin–rotational interaction effect made it possible to refine the photoionization channel relation r, which yielded values of 0.2–0.4 for photon energies between threshold and 32 eV.

The use of two-step photoionization to analyze the degree of molecular alignment was suggested by Sinha *et al.* [329]. They applied the method to the frequently studied Na$_2$ molecule, using linearly polarized 488.0 nm (see Table 3.7) and 476.5 nm Ar$^+$-laser excitation. It is interesting that whilst the probability $G(\theta, \varphi)$ of the first step, namely the $X^1\Sigma_u^+ \rightarrow B^1\Pi_u$ photoexcitation, is dependent on θ, φ, as described in Chapter 2, the ionization of Na$_2(B^1\Pi_u)$ to the $^2\Sigma_g^+$ ground state of Na$_2^+$ is found to be independent of the angle between the plane of molecular rotation and the $\hat{E}_i$-vector of ionizing laser beam. This result was confirmed in [329] for both Q- and (P, R)-excitation and was used to analyze the degree of optical alignment in specific v'', J'' states of Na$_2(X^1\Sigma_g^+)$. Besides, the authors demonstrated the preparation of the optically aligned beams of Na$_2^+$ fragments in the ground electronic state (with either $J'' = 43$ or 27) and observed a large anisotropy in the cross-section for collision-induced dissociation of aligned ions Na$_2^+$ at a beam energy near 25 eV [330].

Detection of the angular distribution function $W(\theta, \varphi)$ of the photoionization products is mostly performed with the aid of electrons carrying away the main part of the kinetic energy. The *photoelectron angular distribution* (PAD), similarly to (6.6), is characterized by expansion coefficients over spherical functions Y_{KQ}, or over other quantities proportional to them. In the case of a single-photon process one may also introduce the anisotropy parameter $-1 \leq \beta_2 \leq 2$, similarly to (6.7) [176, 286]. A number of investigations were carried out using a two-photon ionization scheme $(1 + 1$ REMPI). The description of the process for different types of polarization of both exciting and ionizing beams was given by the groups of Berry [186] and McKoy *et al.* [122, 123]. Detailed results have been obtained by Leahy, Reid and Zare on NO molecules [263] in a (1+1) two-photon REMPI process. The advantage in this case consists of the fact that the first act of photoexcitation fixes unequivocally the rovibronic quantum numbers $v' = 1$, $N' = 22$ of the intermediate state NO($A^2\Sigma^+$), from which ionization by the second photon takes place. Selection of individual rotational levels of the intermediate molecular state leads to simplification of the resulting kinetic energy spectrum of the photoelectrons. The energy resolved PADs were registered by the time of flight method (in the absence of an external field), with high energy resolution permitting registration of rotational levels ($N_i = 20 - 24$) of the NO$^+$ ion. The greatest interest is due to the observation of disturbance of cylindrical symmetry in the distribution $W(\theta_f, \varphi_f)$ of electrons in cases where the angle ε between the $\hat{E}$-vectors of both light beams differs from zero. The distribution function $W(\theta_f, \varphi_f)$ shows both θ_f and φ_f dependence, and one cannot describe PAD by the simple expressions (6.6) or (6.7). The non-zero coefficients w_Q^2 and w_Q^4 where $Q = \pm 1, \pm 2$ appear, which is the signature of 'tilted alignment' in PAD. The authors succeeded in producing a full description of the dynamics of the process, including alignment of the angular momenta of the NO^+ ion and three-dimensional distribution of photoelectrons, as well as the signs of relative phase shift of the partial wave functions of the electron.

The angular distribution of photoelectrons is able to supply information on the alignment or orientation of the precursor molecules. Dubs, Dixit and McKoy [123] designed a method of probing the alignment of linear molecules created at light absorption by means of circular dichroism in PAD in the case where photoionization from the intermediate state is caused by the second photon which is circularly polarized. For symmetric top molecules a theory was developed by Leahy, Reid and Zare [264] allowing extraction of the parameters characterizing the orientation of the symmetry axis of the molecule from PAD following two-photon ionization.

Thus, the polarization aspects of such a wide class of photoprocesses, as discussed in the present section, namely photodissociation and photoionization, make it possible to obtain information both on the stereodynamics of the process and on the properties (for instance, symmetry types) of the states through which the transition takes place. It ought to be mentioned that photodissociation can be considered not only as a 'reaction' of a photon with a molecule, but as a 'halfcollision', in which only the second stage of a collision is present, namely the departure of the products without their previous approach. In the following section we will dwell on the polarization of molecules in 'full' collision, both reactive and non-reactive.

6.2 Collisional processes

The material discussed so far was based on the creation of anisotropic distribution of angular momenta under the direct effect of light. At the same time there are other known (and have been for some long time, in some cases) methods of alignment or orientation of the molecular angular momentum. Thus, a different basis for the polarization of the angular momentum consists of *collisions*, both reactive and non-reactive, with other particles in gases or against the surface.

A fundamental property of collisions in the gaseous state consists of the dependence of the probability of an elementary act of collision on the mutual orientation of the angular momenta of the participants. This means, in particular, that in studying the *stereodynamics* of reactions it is necessary to create and register controlled orientations of the angular momenta of the partners and products. On the other hand, it is possible to use collisions as a method for producing such orientation. In order to achieve anisotropy in laboratory coordinates we require a preferred direction in space. This can be obtained by two basic methods: (a) preliminary action upon the reagent by a light beam or by an external field; (b) the use of molecular beams. Far from claiming fullness of argument, we will consider some ideas and examples for realizing the above-mentioned methods in the following sequence: bulk collisions, beam–bulk collisions (or collisions in the formation and propagation of a jet), beam–beam collisions and, finally, beam–surface collisions.

1 Bulk collisions

The simplest and most straightforward idea for producing collisionally polarized molecules in a thermal cell consists of using collisions with the participation of particles which are polarized in the laboratory frame. It seems that the earliest one was the method based on collisions of atoms which have been 'optically pumped' (optically oriented in their ground state) by the Kastler method; see Section 1.1, Fig. 1.1. If the gas constitutes a mixture of a molecular and an atomic component, the conditions being specially created in such a way as to produce such optical orientation of the atoms, we must expect, from considerations of spin conservation in molecular reactions, that polarization of the molecular component must also emerge.

Let us first discuss a system which is traditional for optical pumping in the Kastler sense [106, 224, 226], namely an *optically oriented* alkali atom A (see Fig. 1.1) in a noble gas X buffer surrounding. It is important to take into account the fact that in alkali atoms, owing to hyperfine interaction, nuclear spins are also oriented. However, in a mixture of alkali vapor with a noble gas alkali dimers A_2 which are in the $^1\Sigma_u^+$ electronic ground state are always present. There exist two basic collisional mechanisms which lead to orientation transfer from the optically oriented (spin-polarized) atom A to the dimer A_2: (*a*) creation and destruction of molecules in triple collisions $A + A + X \longleftrightarrow A_2 + X$; (*b*) exchange atom–dimer reaction $A + A_2 \longleftrightarrow A_2 + A$.

The dependence of the dissociation constant k_D and the recombination constants k_R in the reaction (*a*) on the optical polarization of the atoms A was predicted by Bernheim [65] and Kastler [225] and later demonstrated experimentally [7, 363]. The dependence can be understood from the obvious fact that only hydrogen-like atoms with opposite electronic spins may recombine and form a molecule $A_2(X^1\Sigma_g^+)$. Hence we have $k_R = k_R^0(1 - S^2)$, where S is the degree of electron spin polarization of the atoms. A convenient indicator of dimer formation is provided by the kinetics of the laser-induced molecular fluorescence after switching on magnetic resonance which destroys the polarization of the atoms, as performed by Huber and Weber [201] for a $Na - Na_2$ mixture. The series of works [181, 200, 203, 221, 238, 391] was carried out with different vapors developing the method of 'optical pumping of molecules through atom–dimer exchange', connected with detection of the nuclear magnetic resonance (NMR) signal of the dimers, the nuclear orientation of which is created in reactions (*a*) and (*b*), with participation of optically oriented atoms A. A change in orientation of the nuclear spins in the dimers at magnetic resonance leads, in its turn, through decay and exchange reactions which are opposite to (*a*) and (*b*), to a change in transmission of the circularly polarized resonance pumping light; see Fig. 1.1. The complexity of the mechanism was compensated by the simplicity of the experiment in which only resonant changes in pumping light transmission were monitored. The rate constants for reactions (*a*) and (*b*) were determined for Cs_2 and Rb_2 [181, 391], Na_2 [238] and K_2 [221] from the width of the NMR signal. In [200, 203] the values of magnetic moments of alkali dimer nuclei were determined with high precision. This made it possible to determine the difference in magnetic screening $\sigma(A) - \sigma(A_2)$ between atoms and dimers (chemical shift). This, in its turn, permitted estimation of the constant c of the spin–rotational hyperfine interaction H_{IJ}, according to $H_{IJ} = chIJ$, where $c = 3g_I B_e[\sigma(A) - \sigma(A_2)]$, g_I being the nuclear g-factor, and B_e being the rotational constant. We wish to point out that, whilst for $^{23}Na_2$ in [377] much more accurate c values have been obtained by the method of laser–radiofrequency double resonance, the results for the dimers K_2, Rb_2 and Cs_2 from [203] seem to be the only ones available at present, despite large errors.

Serious grounds for the creation of polarized molecules are provided by the results of studies performed by Kartoshkin and Klement'ev [222, 223] on the conservation of spin projection (Wigner's rule) in reactions of spin exchange and chemioionization in collisions between spin polarized metastable helium He or neon Ne atoms and paramagnetic molecules $O_2(^3\Sigma_g^-)$ and $NO(^2\Pi)$. However, the above-mentioned papers traditionally report only on the detection of magnetic resonance signals from atoms, whilst direct experimental data on the polarization of molecular products are lacking. As has already been mentioned in Section 2.5, inelastic non-radiative collision studies with the participation of optically aligned or oriented molecules revealed that the magnetic quantum number M in the laboratory frame is changed very slowly as a result of collisions; see [288] and references therein. It is therefore clear that one may arrive at the polarization of a particular rotation–vibration level which has been populated as a result of rotationally and/or vibrationally inelastic collisions. The same phenomenon is also expected for electronic energy transfer. This can also be considered, in some sense, as a method of obtaining angular momenta polarization for bulk molecules via collisional polarization transfer.

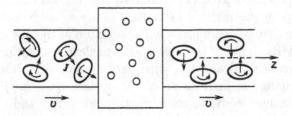

Fig. 6.2.   Alignment of a molecular beam in collisions with a medium of spherically symmetric particles.

2   *Alignment in the beam*

When a molecular beam undergoes scattering on other particles, e.g. on 'spherical' atoms, then alignment of angular momenta may take place in the beam passed 'lower downstream' (Fig. 6.2). Such a collision mechanism of obtaining a 'polarized' beam of diatomic molecules was proposed by Gorter in 1938 [170]. The appearance of alignment in the simplest model is connected with the anisotropy of the intermolecular potential in collisions. For this reason, the scattering cross-section of the 'nonspherical' molecule depends on the orientation of its angular momentum $\mathbf{J}$ with respect to the relative velocity vector $\mathbf{v}_{rel}$ of the colliding partners, assuming maximum value at $\mathbf{J} \parallel \mathbf{v}_{rel}$. The effect was studied in a gas volume in the 1960s and showed a small value, determined by macroscopic transport transfer processes [57, 219]. Since the, subsequently developed, flow of a supersonic jet creates an intensive macroscopic transfer of particles with a fixed axis of transportation, we may expect a considerable amount of angular momenta *alignment* of molecular components; see Fig. 6.2. This was actually realized in [352] in the effusion beam of a Na/Na_2 jet.

To understand the reason for this, one must consider that both processes (*a*) and (*b*) discussed in the preceding subsection proceed simultaneously, making the planes of molecular rotation preferably positioned parallel to the direction of the 'stream', i.e. 'along the flow' of velocity $\mathbf{v}_{rel}$ with respect to the spherically symmetric atoms. Indeed, molecules rotating with $\mathbf{J} \parallel \mathbf{v}_{rel}$ may be considered as 'broadside targets' in collisions with an atom and thus undergo randomizing impact more often than do the molecules with $\mathbf{J} \perp \mathbf{v}_{rel}$. One may therefore say that a *negative alignment* of the ground state angular momenta $\mathbf{J}$ has been created. This alignment is qualitatively similar to the picture in the bottom part of Fig. 3.2(*a*), if the symmetry axis of distribution is formed by the direction of flow $\mathbf{v}_{rel} \parallel \mathbf{z}$; see Fig. 6.2. It follows from Chapter 3 that one of the methods of detecting the effect consists of diminution of the degree of laser-induced fluorescence polarization $\mathcal{P}$ if by $I_\parallel$ in (2.25) we understand the fluorescence which is linearly polarized having the direction of the $\mathbf{E}'$-vector along the beam ($\mathbf{E}' \parallel \mathbf{z}$), and by $I_\perp$ that with $\mathbf{E}' \perp \mathbf{z}$, i.e. at right angles to the beam direction, $\mathbf{v}_{rel} \parallel \mathbf{z}$. The geometry of experiment is that of Fig. 2.5. Here $\mathcal{P}$ depends on the first three even ground state multipole moments $_a\rho_q^\kappa$ of the zero, the second and the fourth rank κ characterizing population and longitudinal alignment ($q = 0$). The direct manifestation of $_a\rho_0^\kappa$ in $\mathcal{P}$ can easily be obtained, as has already been mentioned, by substituting them into (3.19), in order to obtain $_b\rho_0^K$, which leads to $I_\parallel$, $I_\perp$ according to (2.29), (2.30).

Such a method of producing and detecting alignment in a molecular beam was proposed and effected by Sinha, Caldwell and Zare [352] by using $Q \uparrow Q \downarrow$ type fluorescence in the $B^1\Pi_u \longrightarrow X^1\Sigma_g^+$ transition of Na$_2$ excited by an Ar$^+$ laser ($\lambda = 488.0$ nm); see Tab. 3.7. A decrease in $\mathcal{P}$ from 0.48 to 0.44 was observed with increase of pressure p, and one may easily connect these with the degree of alignment $_a\rho_0^2/_a\rho_0^0$ by using (3.36) and neglecting $_a\rho_0^4$. It is important to note that here the laser beam does not *create* any alignment in the ground state, but is only used for its *indication*, i.e., it must act as a probe beam. This enhancement of the alignment with rise in pressure confirms its *collisional* nature. Note that if alignment is caused by non-linear absorption, as considered in Sections 3.2 and 3.3, the opposite is observed, namely an increase in $\mathcal{P}$ with rise in pressure, since the parameter $\chi = \Gamma_p/\gamma_\Sigma$ decreases with pressure owing to an increase in the relaxation rate γ_Σ; see (3.2) and Figs. 3.8 and 3.11.

A direct confirmation of alignment of ground state angular momenta of a Na$_2$ molecular ensemble in the beam also consists of precession of the angular momenta distribution in an external magnetic field $\mathbf{B}$; see Chapter 4. It was observed by the change in fluorescence with magnetic field in the works of Visser *et al.* [384], Treffers and Korving [375], as well as by Pullman and Hershbach on I$_2$ [158, 317, 318]. The magnetic field $\mathbf{B}$, directed at right angles to the molecular beam (Fig. 6.3) acts along a path of length L and creates precession of the aligned molecular ensemble with frequency ω_J, as determined by (4.2). The parameters were chosen such that $\omega_J\tau > 1$, where

$$\omega_J\tau = \frac{g_{J''}\mu_B BL}{v\hbar}. \tag{6.12}$$

Here the time of field action $\tau = L/v$ is determined by the fly-through time of the molecules, the distribution of angular momenta $\mathbf{J}$ being isotropized in a plane at right angles to $\mathbf{B}$. Lower 'downstream' in the beam the relative change in laser-induced fluorescence was registered which took place at switching on of the field $\mathbf{B}$, the $\mathbf{E}$-vector of laser light being directed along the beam direction $\mathbf{v}$. This change may easily be connected

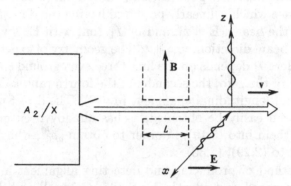

Fig. 6.3. Scheme of registration of alignment in a supersonic jet through precession in a magnetic field **B**, acting along a path length L.

[158, 318, 384] with the anisotropy parameters of the beam molecules
(with polarization moments or magnitudes proportional to them) before
they enter the magnetic field.

The series of works [158, 317, 318] performed recently by the team of
Herschbach (D. Herschbach was the Nobel prize winner in chemistry in
1986 for his investigations of molecular beams) demonstrate promising
prospects in obtaining polarized molecules in a supersonic jet. The authors generalize and normalize to comparable conditions the results of
previous investigations on Na_2 [352, 384], Li_2 [333] and I_2 [334]. One
may draw the conclusion that the proper choice of the similarity factor (product pD of saturated pressure p of the source, and the diameter
D of the nozzle) permits one to obtain a considerably higher degree of
alignment, up to $_a\rho_0^2/_a\rho_0^0 \approx -0.1$. Note that the ratio of polarization
moments $_a\rho_0^2/_a\rho_0^0$ characterizing longitudinal alignment is connected with
the ratio a_2/a_0 between coefficients of Legendre polynomials, presented
in [158, 317, 318] as $a_2/a_0 = 5_a\rho_0^2/_a\rho_0^0$ (see Appendix (D.54)).

Detailed investigations of a I_2/X jet in a wide pressure range and using
the same transitions for registration of fluorescence as given in Fig. 4.13,
enabled [318] two competing mechanisms of alignment formation to be
singled out. The first of them has already been discussed (see Fig. 6.2)
and may be named 'bulk alignment'. Such alignment increases with rise
in initial relative velocity $\mathbf{v}_{rel}$ between the molecule A_2 and the carrier gas
atom X. The relative velocity value, in its turn, increases with increase in
the difference in the masses of A_2 and X. Therefore, the largest *negative
alignment* has been observed in the case of I_2/H_2, and decreases in the
order H_2, D_2, He, Ne, Ar. With further rise in pD, saturation sets in,
followed by a decrease in alignment degree. At first sight this might be

explained by a decrease in v_{rel}, prevailing over a increase in the concentration of the colliding particles. However, the nature of the dependence and, particularly, the change in the sign of alignment, observed at high pD values, necessitated bringing in another mechanism, namely anisotropy in the cooling of rotational motion of the molecules during formation of the jet. As a matter of fact, molecules with $\mathbf{J} \perp \mathbf{v}_{rel}$ cool slightly deeper in the jet than those with $\mathbf{J} \parallel \mathbf{v}_{rel}$, since in this case we have a higher collision probability at a tangent to rotation and thus a reduction of average rotation quantum number value. The rotational temperature of molecules with $\mathbf{J} \perp \mathbf{v}_{rel}$ is correspondingly lower, $T_{r\perp} < T_{r\parallel}$. From the shape of the Boltzmann distribution of rotational population [198, 244] we have the result that at sufficiently low T_r (below 11 K in the case of the I_2/X jet with mean rotational quantum numbers J'' below 15, i.e., on the steeply increasing part of the temperature T_r dependence of the correspondent Boltzmann weight factor P_r) a small difference in $T_{r\perp} < T_{r\parallel}$ leads to a considerable prevalence of the weight factor, $P_{r\perp} < P_{r\parallel}$. This just means that, due to such *anisotropic rotational cooling*, the molecular ensemble as a whole acquires *positive alignment* of $\mathbf{J}$ with a preferred $\mathbf{J}$ direction along the beam.

The concept of the mechanisms of alignment of beam molecules makes it possible to forecast the conditions for attaining the highest alignment effect: more anisotropic molecules, largest difference in their masses with respect to the carrier gas and choice of a suitable nozzle shape which increases in collision frequency. This makes alignment in the jet sufficiently promising for obtaining polarized reagents in stereochemical research. On the other hand, in differently motivated experiments, an understanding of the nature of the effect permits us to exclude, or at least to interpret correctly, its manifestation. Thus, M.P. Chaika [97] succeeded in explaining the way in which such 'selfalignment' (term created by M.P. Chaika) of NO_2 molecules in a jet expansion produced such anomalies in the Hanle signal, which the authors of [392] could not explain.

3 Collisions in crossed beams

A wide variety of works deal with rotational angular momenta polarization of the products in reactive collisions between molecular beams. Usually an atom-transfer reaction of type

$$A + BC \rightarrow AB + C \qquad (6.13)$$

is studied between atom A and a molecule BC (not necessarily diatomic). Four vectors can be specified and measured experimentally, namely the initial and final relative velocities $\mathbf{v}_{rel}$, $\mathbf{v}'_{rel}$, as well as the rotational angular momenta of reactant and product molecules $\mathbf{J}$ and $\mathbf{J}'$. The simplest vector property measured by the molecular beam technique consists of the angular correlation between reagent relative velocity $\mathbf{v}_{rel}$ and molecular product angular momentum $\mathbf{J}'$ (this is the same type of correlation as discussed in the preceding subsection).

The first experimental observations of sufficiently strong polarization (alignment) of rotational angular momenta $\mathbf{J}_{AB}$ of an alkalihalide AB product were achieved by Herschbach and coworkers [197, 283] in exchange reactions (6.13) in which atoms A and B are much heavier than C. In particular, they used A = K or Cs, B = Br or I, C = H or CH_3. The angular momenta $\mathbf{J}'_{AB}$ in this reaction were very large, typically of the order of $100\hbar$. The fundamental cause for so-called 'kinematic polarization' of the products in such a reaction can be understood on the basis of conservation of total angular momentum. In the simplest case

$$\mathbf{l} + \mathbf{J} = \mathbf{l}' + \mathbf{J}', \tag{6.14}$$

where $\mathbf{l}$ and $\mathbf{l}'$ are the orbital angular momenta associated with the relative motion of reactants and products respectively. In the case of a large reaction cross-section $\mathbf{l}$ is large due to large impact parameters and large A+BC reduced mass, whilst $\mathbf{J}_{BC}$ can be considered as relatively small for typical source temperature owing to the small moment of inertia of the BC molecule. As a result, one obtains $l \gg J_{BC}$. The kinematic situation for the products, on the other hand, leads typically to $l' \ll J'_{AB}$, since the AB + C reduced mass is small (for heavy AB and light C) and the AB moment of inertia is large. In such a limit of vanishing exit AB+C interaction (6.14) leads to the simplest condition, $\mathbf{l} \cong \mathbf{J}'$. This means that since the orbital momentum $\mathbf{l}$ due to the reactant approach is perpendicular to the reagent relative velocity $\mathbf{v}_{rel}$ and has azimuthal symmetry around it, the same can be said about product angular momenta $\mathbf{J}'_{AB}$, which must therefore possess negative alignment with respect to $\mathbf{v}_{rel}$. The alignment of $\mathbf{J}'_{AB}$ was determined in [197, 283] by deflection in an inhomogeneous electric field. The anisotropy coefficients a_2 of the Legendre polynomials $P_2(\cos\theta')$, where θ' is the angle between $\mathbf{J}'$ and $\mathbf{v}_{rel}$, characterizing $\mathbf{J}'_{AB}$ alignment were determined as ranging from ≈ -2.1 for the K + HBr case to ≈ -0.95 for Cs + CH_3I, which corresponds, according to (D.54), to ρ_0^2/ρ_0^0 as equal to ≈ -0.43 and -0.19. As can be seen these values mark a stronger alignment than is achieved in jet expansion; see preceding subsection.

The conversion of the orbital angular momentum of the reagents to rotational angular momentum of the products in reaction (6.13) appears not only for heavy atom transfer reactions in heavy + heavy–light A+BC reactive scattering, but is also expected to be the case for other mass combinations and also for lower reactional orbital momentum values. As was shown in [337], the kinematic polarization of the reaction products can be considerable due to the steric requirements of the reaction. From another point of view the polarization, or, more generally, the angular correlation measurements in beam experiments providing single-collision conditions and defining a reference axis (or plane) offer intimate information about the reaction's stereochemistry which is not available from energetic or other scalar properties. The most detailed information is expected when not only the $\mathbf{v}_{rel} - \mathbf{J}'$ correlation, but also three- and four-vector correlations between $\mathbf{v}_{rel}$, $\mathbf{v}'_{rel}$, $\mathbf{J}$ and $\mathbf{J}'$ are registered; the statistical theory for such correlations is developed in [53]

4 Collisions with a surface

Another promising method for producing not only alignment but also orientation in a molecular beam uses spatial effects in *collisions with the surface* of a crystalline solid (see [208, 247, 281, 299, 354, 355] and references therein). These and other spatial effects are being studied intensely in the laboratories of Auerbach, Zare, Stolte and others, beginning in the 1980s. The idea of the method may be visualized as 'the bank shot' of a billiard ball.

One ought to distinguish three basic types of beam–surface impact: *direct non-elastic scattering*, *multiple (indirect) non-elastic scattering*, and *interaction of 'trapping/desorption' type*. They differ in the number of impacts during their residence time on the surface and in the degree of equilibrium attained over rotational degrees of freedom. The third type differs markedly from the Boltzmann one in the first two cases. If we now represent the molecule in the form of a *rigid ellipsoid*, then we may expect, at reflection from a smooth surface, rotational excitation which is largest for angles of incidence and reflection equal to $\theta = 45°$. The direction of the rotational angular momentum $\mathbf{J}$ thus excited must be chiefly orthogonal to the scattering plane (the plane of incidence and reflection).

Thus, considerable alignment of molecular rotation was observed in the scattering of NO molecules from a pure silver crystal Ag(111) [281]. The effect reached maximum value for medium rotational quantum numbers $20.5 < J'' < 35.5$ and dropped to zero for $J'' > 40.5$. The sign of alignment corresponded to the preferred position of $\mathbf{J}$ along the surface, i.e. at right angles to the scattering plane ('cartwheel'-type motion; see Fig. 6.4(a)). Such alignment was observed [354] in the case of N_2/Ag(111) and may be treated as a result of the action of forces normal to the surface. The authors of [354] detected anisotropy of angular momenta of the scattered N_2 molecules by applying $(2 + 2)$ REMPI (two-photon excitation and two-photon resonance-enhanced ionization) with time of flight registration of the N_2^+ ions. They studied changes of the signal with

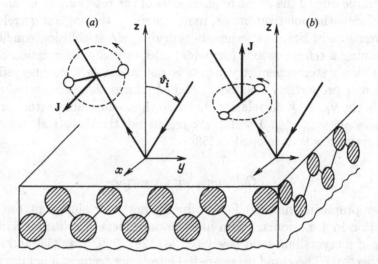

Fig. 6.4.   Alignment of J in interaction between a molecular beam and a surface: (a) 'cartwheel' type with preferred rotation in the scattering plane yz; (b) 'helicopter' type with preferred rotation parallel to the surface plane xy.

change in the polar angle of the $\hat{E}$-vector of the linearly polarized exciting laser beam with respect to the surface normal.

The problem of extracting information on the polarization of the beam molecules under these conditions has been discussed in [207, 354].

However, in conducting this kind of experiment, in the scattering of NO/Pt(111) (applying the method of $(1 + 1)$ REMPI on NO, as mentioned in Section 6.1 [263]), one will also observe, in addition to alignment of 'cartwheel' type at elastic scattering in the desorption process of NO on a Pt crystal, prevalence of alignment of the other sign, i.e. 'helicopter' motion (see Fig. 6.4(b)) with rotation of the molecule parallel to the crystal surface (J positioned orthogonally to the surface $J \parallel z$). This may be explained qualitatively as that during the long-lasting process of trapping/desorption rotation of the molecule is inhibited in the plane orthogonal to the surface, that is, in the yz plane. Simple trajectory calculations [209] of the dynamics of scattering and desorption of NO on Pt confirm this interpretation and demonstrate the possibility of applying angular momentum polarization dependence for studying the dynamics of gas–surface interaction.

In this sense considerable value ought to be attributed to the observation [353, 355] of the emergence of orientation of the angular momentum in the N_2/Ag(111) scattering, described by means of polarization moments of rank $\kappa = 1$ and 3. The orientation was measured along the direction perpendicular to the scattering plane, being dependent on **J** and the scattering angle. This points towards disturbance of cylindrical symmetry which may be interpreted as the role of in-plane forces acting during the collision. It is important that this collision cannot be treated in a model concerning collisions of a rigid rotor with a flat surface, but implies the existence of corrugation of the gas–surface potential and so-called 'surface friction' effects.

For polar polyatomic molecules one has to ascertain the orientation of molecular symmetry axes (x-axis in Fig. 1.4 for a symmetric top molecule). As was established by Novakoski and McClelland [299] it is more likely for a CF_3H molecule to desorb from an Ag(111) surface with the F end of the symmetry axes pointing directly away from the surface, and a degree of orientation of ≈ 0.27 (averaged value of $\langle \cos \theta \rangle$) was observed.

6.3 Orientation in static inhomogeneous fields

So far we have considered various radiational and collisional mechanisms of polarization of molecules. There exist earlier methods applying the action of an external stationary magnetic, later of an external electro-static, field to beam molecules for producing anisotropic distribution of the angular momentum **J** and of the molecular axis.

The first such application refers particularly to the famous *Stern–Gerlach method* from the dawn of quantum mechanics, showing the spatial quantization phenomenon of the angular momentum and confirming the existence of electron spin. The method makes it possible to filter out a beam of oriented atoms or molecules with definite magnetic moment directions and to measure magnetic moment values. In the first experiments the *beam deflection* was used by applying inhomogeneous magnetic fields produced by 'knife'-type magnets (see Fig. 6.5(a)). The deflection appears because of the M-dependent action of a 'magnetic' force along the z-axis, which follows from Eq. (4.7):

$$F_z = -\frac{\partial E_{magn}}{\partial z} = \mu_z \frac{\partial B_z}{\partial z} = -M g_J \frac{\partial B_z}{\partial z}. \tag{6.15}$$

The method was modified in 1933 by Stern and Rabi, together with their coworkers, see [319], which allowed measurement of the nuclear magnetic moments of atoms and molecules. The possibilities and precision of the method were expanded by combining magnetic beam deflection and focussing with magnetic resonance in a homogeneous magnetic field created between the 'deflecting' and 'focussing' magnets, as was first proposed by Rabi in 1938. This *molecular beam magnetic resonance* method (which is often combined with a mass-spectrometer) has been widely used to study different hyperfine interactions in atoms and molecules; see also [294, 374].

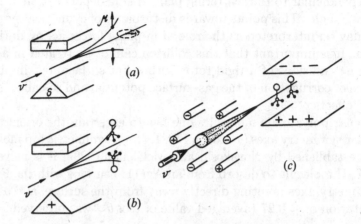

Fig. 6.5. (*a*) Angular momentum orientation via beam deflection in a magnetic field. (*b*) Figure axis orientation for symmetric top molecules via beam deflection in an electric field. (*c*) Figure axis orientation in hexapole focussing field.

From the shape of the resonances the rotational magnetic moment values were determined for the simplest diatomics H_2, D_2, HD and later on for the alkali dimers Li_2, Na_2, Cs_2 (see [87]), and others (see also [155, 294]). The drawbacks of the method were connected with the complexity of the set-up and the lack of selection over vibrational and rotational states.

The application of a *static electric field* is, in principle, also possible for polar molecules due to selection of M projections via the *Stark effect*. An inhomogeneous electric field shown in Fig. 6.5(*b*), which is actually an electric analogue of the Stern–Gerlach field, can also be used to produce deflection of beam molecules [319]. The deflection produced by the *second-order Stark effect* (5.15) was applied, for instance, to observe rotational *alignment* of diatomic molecules, e.g., for alkali halide molecules in such reactions as K + HBr [283]. The application of an electric field, in order to obtain *orientation*, needs the *first-order Stark effect* and is thus a much heavier task for diatomic, linear, or asymmetric top molecules because the rotational motion averages out the electric dipole moment in the first order (there is an exception when, for instance, the degenerate bending vibrational mode is excited, leading to a nearly first-order moment, such as in N_2O and OCS). It is clear that the method based on the first-order Stark effect requires not only polar molecules, but molecules with non-zero angular momentum **K** around the axis of rotation (see Fig. 1.4), leading to a non-zero averaged electric dipole moment $\mathbf{d}_p^K$ along the figure axis. Thus, polar *symmetric top* molecules are most readily oriented. Classically speaking, symmetric top molecules precess in an external electric field rather then 'tumble', and this provides a widely used method of orienting their figure axis $\mathbf{r}_K$ (x-axis in Fig. 1.4). The method was pioneered by the scientific groups of Bernstein [246] and Brooks [85]. A nice pictorial description of the ideas and principles of the method is given in [86], whilst information on the more recent stage can be found in, say, [66, 67, 162, 307].

Let us consider this case in some detail. If collisions are eliminated in a molecular beam, it is possible to orient molecules (their figure axis) by removing the particles possessing unwanted orientation (analogous to the Stern–Gerlach experiment with a magnetic field). Then, classically, the interaction energy with external electric field $\mathcal{E}$ is simply

$$E_{el} = -\mathbf{d}_p\mathcal{E} = -d_p\mathcal{E}\cos\theta, \qquad (6.16)$$

and we have the first-order Stark effect, contrary to the case of rotating diatomics; see (5.15). Fig. 6.5(*b*) shows the idea of deflection of symmetric top molecules with different orientations. Similarly to the case of the Stern–Gerlach method, the field strength (and, hence, the deflecting force) is greatest near the 'knife-edge', and molecules with opposite $\mathbf{d}_p$ are deflected in opposite directions.

For practical use it is appropriate here to use the *electrostatic focussing* technique, Fig. 6.5(*c*), where an inhomogeneous electric field is most often created by six equally spaced rods which are sufficiently long (~ 3 m) and alternately charged. One now obtains a focussing effect for the beam molecules travelling along the symmetry axis where the field is zero, with $\mathbf{d}_p$ directed opposite to the field gradient. In other words, molecules with $\cos\theta < 0$ are focussed towards the array axis, and can be selected, whilst those with $\cos\theta > 0$ are defocussed and those with $\cos\theta = 0$ remain unchanged. Further, the molecules oriented in their local electric field can be oriented in the laboratory frame by applying a homogeneous electric field at the exit of the array; see Fig. 6.5(*c*). The *degree of orientation of figure axis* is often considered as the averaged value $\langle\cos\theta\rangle$. For discrete J we have $\cos\theta = MK/[J(J+1)]$, therefore states with definite $|JKM\rangle$ are selected ('stereoselected' states). Practically, orientation is effective for certain sufficiently low rotational states with, say, $J \leq 3$.

The main goal here is to probe the dependence of chemical reaction probability (reactivity) on the initial orientation of the reagents. Strong beams of molecules in 'stereoselected' states with a sufficiently high orientation degree are most often produced in rotationally cooled symmetric top molecules such as CH_3X, CF_3X, CX_3H, ..., where the X are halide atoms. The main processes investigated were reactive collisions with alkali metal atoms, as well as those with Ar* and Ca*. Direct demonstrations were obtained of the dependence of the reactivity on the 'angle of attack', $\gamma = \widehat{\mathbf{r}_K\mathbf{v}}_{rel}$. The greatest achievement is to measure detailed orientation-dependent reaction cross-sections $f(\theta, E_{tr}, E_{vib}, J, K, M)$ for fixed translational E_{tr} and vibrational E_{vib} energies.

As an example one may again mention a reaction of type (6.13). In particular, the authors of [210] studied the steric effect in the reactive collisions of a beam of excited metastable $Ca^*(^1D_2)$ atoms with oriented symmetric top molecule CH_3F for the reaction:

$$Ca^*(^1D_2) + CH_3F(JKM = 111, 212) \longrightarrow CaF^*(A^2\Pi) + CH_3. \quad (6.17)$$

The orientation degree $\langle \cos\theta \rangle$ was about 0.5 for $|JKM\rangle = |111\rangle$, $T_{rot} = 7$ K. The steric effect manifests itself as the difference $\Delta\sigma_1$ in the chemiluminescence cross-section measured from CaF $(A^2\Pi - X^2\Sigma)$ radiation intensity for the cases where either the fluorine atom or the methyl group of the CH_3F molecule are oriented preferentially along the relative velocity $\mathbf{v}_{rel}$ of the approaching reagents. The ratio $\Delta\sigma_1/\sigma_0$ up to 0.6 was thus obtained, where the cross-section σ_0 refers to random reagent orientation.

6.4   Orientation by homogeneous electric field

Let us now turn to polar diatomic molecules which are the most 'pathological' object for orientation in a static electric field, due to fast rotation of $\mathbf{d}_p \parallel \mathbf{r}$ around the centre of mass (perpendicular to $\mathbf{J}$). Nevertheless, as was already mentioned by Brooks [86], the 'brute electric force' method can be implemented when rotation is slowed, that is when the parameter $\mathbf{d}_p\mathcal{E}/E_{rot}$ is not too small. This can be achieved in practice using deep rotational cooling in a supersonic expansion when rotational temperature T_{rot} can be decreased to 1 K or even less. In this situation, as was first proposed by the research teams of Loesch [273] and Herschbach [158, 159], the favorable ratio $(\mathbf{d}_p\mathcal{E})/[BJ(J+1)]$ can be obtained using strong *homogeneous electric fields*. Then molecules can no longer undergo full rotation (through 360°) but can only *librate* (oscillate about the field direction) over a limited angle, like a pendulum. It can be considered that molecules in the lowest few rotational states are trapped in 'pendular states'. For instance, in a diagnostic experiment [160] which was carried out on $ICl(X^1\Sigma^+)$ molecules seeded in a H_2 jet, the expectation value of $\langle \cos\theta \rangle$ was about 0.3 for $J'' = 1$, $M'' = 1$ at $T_{rot} = 15$ K, $\mathcal{E} = 20$ kV $\cdot$ cm^{-1}. One must take into account, however, the fact that the method cannot be regarded as the orientation of molecular rotation, because the latter is disturbed by the torque of an electric dipole $\mathbf{d}_p \parallel \mathbf{r}$ in a strong electric field. The energy levels correspond to *pendular states* which are the hybrids of field-free rotational states. The registered ro-

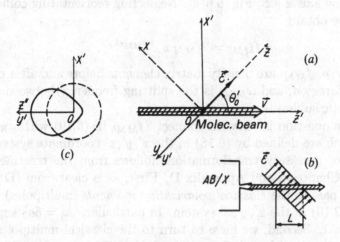

Fig. 6.6. Schematic of realization of alignment–orientation Stark conversion. (a) Choice of coordinate systems. (b) Possible realization scheme for AB molecules seeded in a free jet of X atoms. (c) Symbolic polar plot of **J** distribution (dashed line refers to initial cylindrical symmetry over beam axis z').

tational spectrum then undergoes a dramatic transformation into that reflecting pendular motion [72, 160, 161, 327].

If polar diatomic molecules are previously aligned in a beam (see Section 6.2) there is another possibility, proposed in [43], of producing angular momenta orientation using *alignment-orientation conversion* in a *homogeneous* electric field due to the *second-order Stark effect* (see Section 5.4). We will consider this method in more detail since it is a nice example of how to make use of handling the different approaches presented in Chapter 5 simultaneously.

Let the beam of polar molecules AB with aligned angular momenta **J** and fixed velocity **v** cross an electric field $\mathcal{E}$ region of length L (see Fig. 6.6(a,b)), $\mathcal{E}$ forming an angle $\theta'_0 = 45°$ with respect to **v**. The classical probability density $'\rho(\theta')$ of the spatial **J** distribution (see Fig. 6.6(a)) with respect to the beam direction $z' \parallel$ **v** before entering the field region can be described, using Legendre polynomial expansion coefficients a_K (D.50), as

$$'\rho(\theta') = \frac{n(\theta')}{n_0} = \sum_K a_K P_K(\cos\theta') \approx 1 + a_2 P_2(\cos\theta') \qquad (6.18)$$

if we neglect higher order alignment coefficients with $K > 2$.

Because of the quantum origin of the Stark effect (see Section 5.2), the evolution of the density matrix elements $f_{MM'}$ must be considered for the ground state molecules. In the field region we must choose the quantization axis $\mathbf{z} \parallel \boldsymbol{\mathcal{E}}$, Fig. 6.6($a$). Neglecting reorientating collisions in a beam, we obtain

$$f_{MM'} = {}^0f_{MM'}e^{-i\omega_{MM'}L/v}, \tag{6.19}$$

where ${}^0f_{MM'}$, $f_{MM'}$ are density matrix elements before and after crossing the field $\boldsymbol{\mathcal{E}}$ region, and $\omega_{MM'}$ is the splitting frequency between the M, M' magnetic sublevels.

Now the question is: how to connect ${}^0f_{MM'}$ in (6.19) with a_K coefficients which are defined by (6.18) in the x', y', z' coordinate system. The sequence of necessary transformations follows from the treatment presented in Chapter 5 and Appendix D. First, as is clear from (D.54), we can easily pass to the *classical polarization moments* (multipoles) $'\rho_0^K$ defined by (2.16) in the x', y', z' system. In particular, $'\rho_0^2 = 5a_2$ supposing $'\rho_0^0 = a_0 = 1$. Second, we have to turn to the classical multipoles ρ_Q^K in the x, y, z frame with $\mathbf{z} \parallel \boldsymbol{\mathcal{E}}$; see Fig. 6.6($a$). This can be achieved by means of Wigner D-matrices, and according to (A.12), we obtain

$$\rho_2^Q = {}'\rho_0^K D_{0Q}^{(2)}(0, \theta_0', 0), \tag{6.20}$$

where the $D_{0Q}^{(2)}$ values can be found in Table A.2 from Appendix A. This transformation leads to the appearance of non-zero *transverse alignment* values ρ_Q^2 with $Q = \pm 1, \pm 2$, thus including tilted alignment components $\rho_{\pm 1}^2$. Further, we use the fact that the classical multipoles ρ_Q^K are approaching their quantum analogs ${}^0f_Q^K$ at $J \gg 1$, as follows from the analysis in Sections 5.6 and 5.7. Hence, we obtain ${}^0f_Q^K \approx \rho_Q^K$, and it is now possible to use (5.17), (5.18) and (C.6), in order to pass from ${}^0f_Q^K$ to ${}^0f_{MM'}$:

$$^0f_{MM'} = \sum_{KQ} \frac{2K+1}{2J+1}(-1)^Q {}^0f_Q^K C_{JMKQ}^{JM'}. \tag{6.21}$$

Consequently, as a result of transformation $a_2 \rightarrow {}'\rho_0^2 \rightarrow \rho_Q^2 \rightarrow {}^0f_Q^K \rightarrow {}^0f_{MM'}$, we obtain ${}^0f_{MM'}$ and substitute them into (6.19), thus obtaining the desired density matrix elements $f_{MM'}$.

In order to analyze the electric field induced level splitting effect we will again expand $f_{MM'}$ over *quantum polarization moments* f_Q^K. As may be seen from (5.40), which is the reverse of (6.21), we obtain in the x, y, z frame

$$f_Q^K = (-1)^Q \sum_M C_{JMKQ}^{JM+Q0} f_{MM+Q} e^{-i\omega_{MM+Q}}, \tag{6.22}$$

and if M sublevels are split non-equidistantly in field $\boldsymbol{\mathcal{E}}$, ${}^0f_{MM}$, ${}^0f_{MM+1}$ and ${}^0f_{M-1M+1}$ produce not only transverse alignment f_Q^2, but also *trans-*

verse orientation $f^1_{\pm 1}$ components, in agreement with (5.41). This orientation appears in the direction perpendicular to the $\mathbf{v}$, $\mathcal{E}$ plane, or along the $\mathbf{y}$ axis (see Fig. 6.6(a)), thus breaking the axial symmetry of $\mathbf{J}$ distribution over the beam axis $\mathbf{z'} \parallel \mathbf{v}$, as is schematically shown in Fig. 6.6(c). Note then, because the field $\mathcal{E}$ is not acting on $^0f_{MM}$, owing to $\omega_{MM} = 0$ in (6.19), f^K_0 remains unchanged, and the appearance of longitudinal orientation f^1_0 along the beam axis is excluded, as has already been mentioned in Section 5.4 in connection with (5.40) and (5.41).

In order to calculate what fraction of alignment a_2 is transformed into transverse orientation, we must use a coordinate system with the $\mathbf{z''}$ axis parallel to the direction of orientation created, namely $\mathbf{z''} \parallel \mathbf{y}$; see Fig. 6.6($a$). In such a system the transverse orientation components $\rho^1_{\pm 1}$ (which are, according to Section 5.6, the classical analogue of $f^1_{\pm 1}$) are transformed, according to (A.12), into one 'longitudinal' component $^t\rho^1_0$. This allows one to use (D.54) and to return to a corresponding Legendre polynomial coefficient: $^ta_1 = 3^t\rho^1_0 = 3\sqrt{2}\,\mathrm{Im}\rho^1_1$. In order to obtain ρ^1_1 the form of the Stark effect must be specified. We will use the Stark energy expression in the form (5.15) for a diatomic (or linear) molecule, neglecting hyperfine interaction. This leads to an expression for $^ta_1$ of the form

$$^ta_1 = A \sum_{l=0}^{J-1} (2l+1)(J+l+1)(J-l)\sin\left[(2l+1)\beta\right], \qquad (6.23)$$

where A is a normalizing factor, and

$$\beta = \frac{3\pi}{J(J+1)(2J-1)(2J+3)} \frac{d_p^2 \mathcal{E}^2}{h^2 B} \frac{L}{v}. \qquad (6.24)$$

Calculation allows us to determine the *efficiency factor* for alignment–orientation conversion: $\kappa = {}^ta_1/a_2$; see Figs. 6.7, 6.8 and 6.10. As may be seen from Fig. 6.7, in $\beta \sim \mathcal{E}^2$ coordinates orientation appears in the form of some regular structure. Such a 'Stark orientation grille' (SOG) exhibits equidistantly spaced alternating dispersion type 'principal' orientation signals centered at positions $\beta = n\pi, n = 1, 2, \ldots$, corresponding to field strength $\mathcal{E}^2_n$ values

$$\mathcal{E}^2_n = \frac{nh^2 B v J(J+1)(2J-1)(2J+3)}{3L d_p^2}. \qquad (6.25)$$

Between the centers of adjacent principal signals secondary maxima of similar form can be distinguished. As J increases from $J = 10$ to 100, Fig. 6.8, principal $\kappa(\beta)$ signals become relatively sharper (see Fig. 6.9), and secondary peaks, whilst increasing in number, become negligible in their effect. Such behavior of the orientation reminds one of a diffraction-grating-signal characteristic pattern. In this case the $\kappa(\beta)$ dependence,

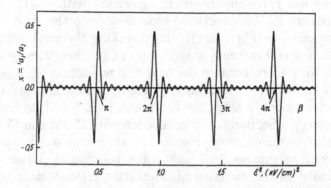

Fig. 6.7. SOG structure for conversion efficiency ${}^t a_1/a_2 = \kappa(\beta)$. $\mathcal{E}^2$ values correspond to an example of NaK molecule with $J = 10$.

however, is linear with respect to $\sin[(2l + 1)\beta]$ (see Eq. (6.23)) (not squared as in the case of a diffraction grating), and this leads to dispersion-type signals. The amplitude values of $|\kappa(\varphi)|$ are large, being ≈ 0.468 for $J \geq 5$.

Let us examine as a concrete example a polar NaK molecule for which noticeable alignment in a supersonic beam can be expected, similar to the case of Na$_2$/Na [352, 384]. A mixed halogen diatomic, such as IBr, could also be considered as a convenient object [158]. Dependences $\kappa(\mathcal{E}^2)$ and $\kappa(\mathcal{E})$ for NaK in its electronic ground $X^1\Sigma^+$ state obtained from Eqs. (6.23), (6.24) and (6.25), assuming [202] that $d_p = 2.667\,\text{D}$, $B = 0.0905\,\text{cm}^{-1}$, $L = 5\,\text{cm}$ and $v = 10^5\,\text{cm}\cdot\text{s}^{-1}$, are given in Figs. 6.7 and 6.8 for $J = 10, 11, 100$. Values of field strength $\mathcal{E}_1$ for $n = 1$ correspond-ing to the location of the zero point for the first principal signal (taking place for $\beta = \pi$) are given in Fig. 6.9. The signal width $\Delta\mathcal{E}_1$, defined as the distance between positions of maximum and minimum orientation values for the first principal signal, is also shown in Fig. 6.9. As may be seen, the relative signal width $\Delta\mathcal{E}_1/\mathcal{E}_1$ diminishes with J growth. It is interesting that the $\mathcal{E}_1$ difference for different J allows, in principle, the production of orientation only for a selected J level; see Figs. 6.8(a, b) for $J = 10$ and 11. Moreover, selectivity of the SOG structure may allow one to resolve (orient separately) different isotope molecules. Figure 6.10 gives an example allowing one to compare orientation signals for ${}^{23}\text{Na}^{39}\text{K}$ and ${}^{23}\text{Na}^{41}\text{K}$. The values of $B = 0.0886\,\text{cm}^{-1}$ and $v = 9.8425 \cdot 10^4\text{cm}\cdot\text{s}^{-1}$ were taken for a ${}^{23}\text{Na}^{41}\text{K}$ molecule, assuming that the changes in numer-ical value of B are caused by changes in the moments of inertia of the molecule [156] and that changes in v are caused by changes in molecu-lar mass. The isotopic shift is in principle resolvable, even for $J = 10$

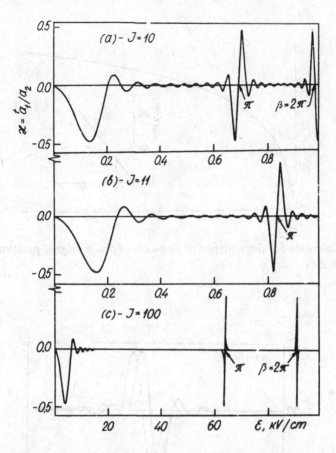

Fig. 6.8. Alignment–orientation conversion parameter κ dependence on electric field strength $\mathcal{E}$ as calculated for a $^{23}\mathrm{Na}^{39}\mathrm{K}$ $(X^1\Sigma)$ molecule: (*a*) $J = 10$; (*b*) $J = 11$; (*c*) $J = 100$.

(see Fig. 6.10(*a*)), growing dramatically for $J = 100$; see Fig. 6.10(*b*). Of course, the expected resolution takes place for a monoenergetic beam of molecules with fixed v value. From another point of view, however, for a given J value beam molecules with definite v can be oriented selectively. For a beam with a given, say, Maxwellian v distribution it may turn out to be more convenient to use the zero-order ($\beta = 0$) SOG signal, namely, the first minimum in Fig. 6.8, despite the loss in J selectivity. In fact, this requires considerably smaller electric field values of only about 135 V/cm for $J = 10$; even a field strength of 4 kV/cm needed for $J = 100$ does not seem too excessive. It must be emphasized here that we need much smaller electric field values than in the 'brute force' method described above, and the higher J values can be oriented. In contrast to the brute

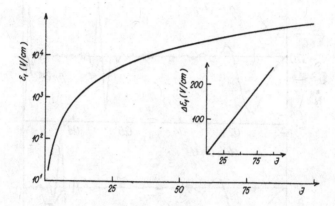

Fig. 6.9. Calculated J dependence of first-order ($\beta = \pi$) signal position $\mathcal{E}_1$ and signal width $\Delta\mathcal{E}_1$.

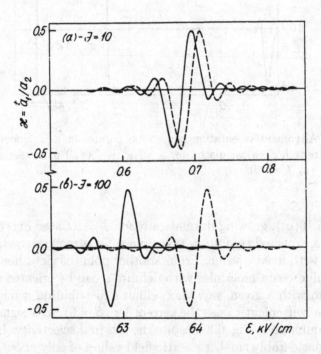

Fig. 6.10. First-order ($\beta = \pi$) orientation maxima calculations for $^{23}\text{Na}^{41}\text{K}$ isotopic molecule: (a) for $J = 10$; (b) for $J = 100$. Dashes reproduce signals for $^{23}\text{Na}^{39}\text{K}$.

force method, the rotational movement remains undisturbed here because of the small value of $d_p \mathcal{E}_1 / B$, being about 0.34 for $J = 10$ and 31.7 for $J = 100$, which leads to $d_p \mathcal{E}_1 / [BJ(J+1)] < 4 \cdot 10^{-3}$.

In this book we have hopefully demonstrated that the theory of angular momentum is a remarkable tool for analyzing light–molecule interaction revealing a bundle of pictorial and valuable features of ordered angular momentum ensembles in the polarization of emission, collisional processes, external field influence and the creation and destruction of coherence. It is to wonder how much 'geometric' factors can tell about the structure of an isolated molecule and about the collisional processes including chemical reactions which are of significant interest to the chemical physicists. We hope that, although the research area is developing quickly, the strong emphasis on theory given in the book will prevent the material from dating too rapidly.

Appendix A
Vector calculus in cyclic coordinates

In physical calculations containing angular momenta, cyclic coordinates are frequently employed [379]. *Cyclic unit vectors* are introduced according to the following rules:

$$
\left.
\begin{aligned}
\mathbf{e}_{+1} &= -(1/\sqrt{2})(\mathbf{e}_x + i\mathbf{e}_y), \\
\mathbf{e}_0 &= \mathbf{e}_z, \\
\mathbf{e}_{-1} &= (1/\sqrt{2})(\mathbf{e}_x - i\mathbf{e}_y),
\end{aligned}
\right\}
\tag{A.1}
$$

where $\mathbf{e}_x$, $\mathbf{e}_y$, $\mathbf{e}_z$ are the base unit vectors in cartesian coordinates. Cyclic unit vectors $\mathbf{e}_\mu$, $\mu = 0, \pm 1$ satisfy the following symmetry properties:

$$
\mathbf{e}_\mu = (-1)^\mu \mathbf{e}^*_{-\mu}.
\tag{A.2}
$$

Any vector $\mathbf{A}$ may be resolved over cyclic unit vectors $\mathbf{e}_\mu$:

$$
\mathbf{A} = \sum_\mu A^\mu \mathbf{e}_\mu,
\tag{A.3}
$$

where the components A^μ of the vector $\mathbf{A}$ are connected to the cartesian components of this vector as

$$
\left.
\begin{aligned}
A^{+1} &= -(1/\sqrt{2})(A_x - iA_y), \\
A^0 &= A_z, \\
A^{-1} &= (1/\sqrt{2})(A_x + iA_y).
\end{aligned}
\right\}
\tag{A.4}
$$

The quantities A^μ thus defined are called *contragredient* components of the vector $\mathbf{A}$.

In addition to Eq. (A.3) one may also apply *cogredient* expansion of $\mathbf{A}$. Then

$$
\mathbf{A} = \sum_\mu (-1)^\mu A_\mu \mathbf{e}_{-\mu}.
\tag{A.5}
$$

In this case the connections between the cogredient A_μ, contragredient

A^μ and the cartesian components of the vector are of the form:

$$\left.\begin{aligned}
A_{+1} = -A^{-1} &= -(1/\sqrt{2})(A_x + iA_y), \\
A_0 = A^0 &= A_z, \\
A_{-1} = -A^{+1} &= (1/\sqrt{2})(A_x - iA_y).
\end{aligned}\right\} \qquad (A.6)$$

Using cogredient vector components, it is necessary to keep in mind the fact that the component A_μ characterizes the contribution to the vector **A** of unit vector $\mathbf{e}_{-\mu}$ with the sign $(-1)^\mu$, and not of unit vector $\mathbf{e}_\mu$, which may occasionally lead to misunderstanding.

Considering the *orthogonality properties* of the cyclic unit vectors:

$$(\mathbf{e}_\mu \mathbf{e}_\nu^*) = \delta_{\mu\nu}, \qquad (A.7)$$

one may obtain the components of vector **A** as

$$\left.\begin{aligned}
A^\mu &= (\mathbf{A}\mathbf{e}_\mu^*) = (-1)^\mu (\mathbf{A}\mathbf{e}_{-\mu}), \\
A_\mu &= (\mathbf{A}\mathbf{e}_\mu).
\end{aligned}\right\} \qquad (A.8)$$

The *scalar product* of vectors **A** and **B** may be expressed in cyclic coordinates as

$$(\mathbf{AB}) = \sum_\mu A^\mu B_\mu = \sum_\mu A_\mu B^\mu = \sum_\mu (-1)^\mu A_\mu B_{-\mu} = \sum_\mu (-1)^\mu A^\mu B^{-\mu}. \qquad (A.9)$$

If one performs a turn of the coordinate system $x'y'z'$ by Euler angles α, β, γ, thus transferring it to the position xyz, then the components A_μ and A^μ of the vector **A** in the new coordinate system may be expressed through the components $A_{\mu'}$, $A^{\mu'}$ in the initial coordinate system by means of *Wigner D-functions*

$$\left.\begin{aligned}
A^\mu &= \sum_{\mu'} A^{\mu'} D_{\mu'\mu}^{(1)\,*}(\alpha, \beta, \gamma), \\
A_\mu &= \sum_{\mu'} A_{\mu'} D_{\mu'\mu}^{(1)}(\alpha, \beta, \gamma).
\end{aligned}\right\} \qquad (A.10)$$

The turn of coordinates ought to be performed in the following sequence: first around the z'-axis by the angle $\alpha(0 \le \alpha < 2\pi)$, then around the new y'-axis by the angle $\beta(0 \le \beta < \pi)$ and, last, around the new z'-axis by the angle $\gamma(0 \le \gamma < 2\pi)$. The positive turning direction is the counterclockwise one, if viewed from the side of the positive direction of the coordinate axis. The explicit form of the matrix $D_{\mu'\mu}^{(1)}(\alpha, \beta, \gamma)$ is presented in the Table A.1 and that of matrix $D_{\mu'\mu}^{(2)}(\alpha, \beta, \gamma)$ in Table A.2.

The D-matrix possesses a whole range of symmetry properties, of which we only present a few here:

$$\left.\begin{aligned}
D_{\mu\mu'}^{(1)}(\alpha, \beta, \gamma) &= D_{\mu\mu'}^{(1)\,*}(-\alpha, \beta, -\gamma), \\
D_{\mu\mu'}^{(1)}(\alpha, \beta, \gamma \pm n\pi) &= (-i)^{\pm 2n\mu'} D_{\mu\mu'}^{(1)}(\alpha, \beta, \gamma).
\end{aligned}\right\} \qquad (A.11)$$

Table A.1.   Wigner D-functions $D^{(1)}_{\mu'\mu}(\alpha, \beta, \gamma)$

$\mu'\backslash\mu$	1	0	-1
1	$e^{-i\alpha}\frac{1+\cos\beta}{2}e^{-i\gamma}$	$-e^{-i\alpha}\frac{\sin\beta}{\sqrt{2}}$	$e^{-i\alpha}\frac{1-\cos\beta}{2}e^{i\gamma}$
0	$\frac{\sin\beta}{\sqrt{2}}e^{-i\gamma}$	$\cos\beta$	$-\frac{\sin\beta}{\sqrt{2}}e^{i\gamma}$
-1	$e^{i\alpha}\frac{1-\cos\beta}{2}e^{-i\gamma}$	$e^{i\alpha}\frac{\sin\beta}{\sqrt{2}}$	$e^{i\alpha}\frac{1+\cos\beta}{2}e^{i\gamma}$

The D-matrices make it possible to write down the expressions for the transformation at turning the coordinate system not only for the vector, but also for the *tensor* T^K_Q of arbitrary rank K:

$$T^K_Q = \sum_{Q'} T^K_{Q'} D^{(K)}_{Q'Q}(\alpha, \beta, \gamma). \tag{A.12}$$

It is frequently useful to use an expression which permits us to find the integral of the product of three D-functions:

$$\int_0^{2\pi} \int_0^{\pi} \int_0^{2\pi} D^{(K_3)}_{Q_3 Q'_3}{}^{*}(\alpha, \beta, \gamma) D^{(K_2)}_{Q_2 Q'_2}(\alpha, \beta, \gamma) D^{(K_1)}_{Q_1 Q'_1}(\alpha, \beta, \gamma) \sin\beta \, d\alpha \, d\beta \, d\gamma$$

$$= \frac{8\pi^2}{2K_3 + 1} C^{K_3 Q_3}_{K_1 Q_1 K_2 Q_2} C^{K_3 Q'_3}_{K_1 Q'_1 K_2 Q'_2}, \tag{A.13}$$

where $K_1 + K_2 + K_3$ must be an integer and $C^{c\gamma}_{a\alpha b\beta}$ is the Clebsch–Gordan coefficient (see Appendix C).

Table A.2. Wigner D-functions $D^{(2)}_{\mu'\mu}(\alpha,\beta,\gamma)$

$\mu'\backslash\mu$	2	1	0	-1	-2
2	$e^{-2i\alpha}\dfrac{(1+\cos\beta)^2}{4}e^{-2i\gamma}$	$-e^{-2i\alpha}\sin\beta\dfrac{(1+\cos\beta)}{2}e^{-i\gamma}$	$e^{-2i\alpha}\dfrac12\sqrt{\dfrac32}\sin^2\beta$	$-e^{-2i\alpha}\sin\beta\dfrac{(1-\cos\beta)}{2}e^{i\gamma}$	$e^{-2i\alpha}\dfrac{(1-\cos\beta)^2}{4}e^{2i\gamma}$
1	$e^{-i\alpha}\sin\beta\dfrac{(1+\cos\beta)}{2}e^{-2i\gamma}$	$e^{-i\alpha}\dfrac{2\cos^2\beta+\cos\beta-1}{2}e^{-i\gamma}$	$-e^{-i\alpha}\sqrt{\dfrac32}\sin\beta\cos\beta$	$e^{-i\alpha}\dfrac{2\cos^2\beta-\cos\beta-1}{2}e^{i\gamma}$	$-e^{-i\alpha}\sin\beta\dfrac{(1-\cos\beta)}{2}e^{2i\gamma}$
0	$\dfrac12\sqrt{\dfrac32}\sin^2\beta\;e^{-2i\gamma}$	$\sqrt{\dfrac32}\sin\beta\cos\beta\;e^{-i\gamma}$	$\dfrac{3\cos^2\beta-1}{2}$	$-\sqrt{\dfrac32}\sin\beta\cos\beta\;e^{i\gamma}$	$\dfrac12\sqrt{\dfrac32}\sin^2\beta\;e^{2i\gamma}$
-1	$e^{i\alpha}\sin\beta\dfrac{(1-\cos\beta)}{2}e^{-2i\gamma}$	$e^{i\alpha}\dfrac{2\cos^2\beta-\cos\beta-1}{2}e^{-i\gamma}$	$e^{i\alpha}\sqrt{\dfrac32}\sin\beta\cos\beta$	$e^{i\alpha}\dfrac{2\cos^2\beta+\cos\beta-1}{2}e^{i\gamma}$	$e^{i\alpha}\sin\beta\dfrac{(1+\cos\beta)}{2}e^{2i\gamma}$
-2	$e^{2i\alpha}\dfrac{(1-\cos\beta)^2}{4}e^{-2i\gamma}$	$e^{2i\alpha}\sin\beta\dfrac{(1-\cos\beta)}{2}e^{-i\gamma}$	$e^{2i\alpha}\dfrac12\sqrt{\dfrac32}\sin^2\beta$	$e^{2i\alpha}\sin\beta\dfrac{(1+\cos\beta)}{2}e^{i\gamma}$	$e^{2i\alpha}\dfrac{(1+\cos\beta)^2}{4}e^{2i\gamma}$

Appendix B
Spherical functions $Y_{KQ}(\theta, \varphi)$ and some of theirproperties

The *spherical functions* $Y_{KQ}(\theta, \varphi)$ with $K \leq 4, -K \leq Q \leq K$ may be expressed by means of the following analytical functions:

$$
\begin{aligned}
Y_{00} &= \sqrt{\tfrac{1}{4\pi}}; \\
Y_{11} &= -\tfrac{1}{2}\sqrt{\tfrac{3}{2\pi}} \sin\theta e^{i\varphi}, \\
Y_{10} &= \tfrac{1}{2}\sqrt{\tfrac{3}{\pi}} \cos\theta, \\
Y_{1-1} &= \tfrac{1}{2}\sqrt{\tfrac{3}{2\pi}} \sin\theta e^{-i\varphi}; \\
Y_{22} &= \tfrac{1}{4}\sqrt{\tfrac{15}{2\pi}} \sin^2\theta e^{i2\varphi}, \\
Y_{21} &= -\tfrac{1}{2}\sqrt{\tfrac{15}{2\pi}} \cos\theta \sin\theta e^{i\varphi}, \\
Y_{20} &= \tfrac{1}{4}\sqrt{\tfrac{5}{\pi}} \left(3\cos^2\theta - 1\right), \\
Y_{2-1} &= \tfrac{1}{2}\sqrt{\tfrac{15}{2\pi}} \cos\theta \sin\theta e^{-i\varphi}, \\
Y_{2-2} &= \tfrac{1}{4}\sqrt{\tfrac{15}{2\pi}} \sin^2\theta e^{-i2\varphi}; \\
Y_{33} &= -\tfrac{1}{8}\sqrt{\tfrac{35}{\pi}} \sin^3\theta e^{i3\varphi}, \\
Y_{32} &= \tfrac{1}{4}\sqrt{\tfrac{105}{2\pi}} \cos\theta \sin^2\theta e^{i2\varphi}, \\
Y_{31} &= -\tfrac{1}{8}\sqrt{\tfrac{21}{\pi}} \left(5\cos^2\theta - 1\right) \sin\theta e^{i\varphi}, \\
Y_{30} &= \tfrac{1}{4}\sqrt{\tfrac{7}{\pi}} \left(5\cos^2\theta - 3\right) \cos\theta, \\
Y_{3-1} &= \tfrac{1}{8}\sqrt{\tfrac{21}{\pi}} \left(5\cos^2\theta - 1\right) \sin\theta e^{-i\varphi}, \\
Y_{3-2} &= \tfrac{1}{4}\sqrt{\tfrac{105}{2\pi}} \cos\theta \sin^2\theta e^{-i2\varphi}, \\
Y_{3-3} &= \tfrac{1}{8}\sqrt{\tfrac{35}{\pi}} \sin^3\theta e^{-i3\varphi}; \\
Y_{44} &= \tfrac{3}{16}\sqrt{\tfrac{35}{2\pi}} \sin^4\theta e^{i4\varphi}, \\
Y_{43} &= -\tfrac{3}{8}\sqrt{\tfrac{35}{\pi}} \sin^3\theta \cos\theta e^{i3\varphi}, \\
Y_{42} &= \tfrac{3}{8}\sqrt{\tfrac{5}{2\pi}} \sin^2\theta \left(7\cos^2\theta - 1\right) e^{i2\varphi}, \\
Y_{41} &= -\tfrac{3}{8}\sqrt{\tfrac{5}{\pi}} \sin\theta \left(7\cos^3\theta - 3\cos\theta\right) e^{i\varphi}, \\
Y_{40} &= \tfrac{3}{16}\sqrt{\tfrac{1}{\pi}} \left(35\cos^4\theta - 30\cos^2\theta + 3\right), \\
Y_{4-1} &= \tfrac{3}{8}\sqrt{\tfrac{5}{\pi}} \sin\theta \left(7\cos^3\theta - 3\cos\theta\right) e^{-i\varphi}, \\
Y_{4-2} &= \tfrac{3}{8}\sqrt{\tfrac{5}{2\pi}} \sin^2\theta \left(7\cos^2\theta - 1\right) e^{-i2\varphi}, \\
Y_{4-3} &= \tfrac{3}{8}\sqrt{\tfrac{35}{\pi}} \sin^3\theta \cos\theta e^{-i3\varphi}, \\
Y_{4-4} &= \tfrac{3}{16}\sqrt{\tfrac{35}{2\pi}} \sin^4\theta e^{-i4\varphi}.
\end{aligned}
\tag{B.1}
$$

The spherical functions satisfy the following property of complex conjugation:

$$Y_{KQ}(\theta, \varphi) = (-1)^Q Y_{K-Q}^*(\theta, \varphi).$$
(B.2)

The *orthogonality* and *normalization* conditions of spherical functions may be represented as

$$\int_0^\pi \int_0^{2\pi} Y_{KQ}^*(\theta, \varphi) Y_{K'Q'}(\theta, \varphi) \sin\theta \, d\theta \, d\varphi = \delta_{KK'}\delta_{QQ'}.$$
(B.3)

The spherical functions may be expressed through the Wigner D-function:

$$Y_{KQ}(\theta, \varphi) = \sqrt{\frac{2K+1}{4\pi}} D_{0-Q}^{(K)}(\chi, \theta, \varphi) = (-1)^Q \sqrt{\frac{2K+1}{4\pi}} D_{0Q}^{(K)*}(\chi, \theta, \varphi).$$
(B.4)

Appendix C
The Clebsch–Gordan coefficients. $6j$- and $9j$-symbols

The *Clebsch–Gordan* coefficients $C^{jm}_{j_1 m_1 j_2 m_2}$ denote the quantum mechanical probability amplitude that the angular momenta $\mathbf{j}_1$ with projection m_1, and $\mathbf{j}_2$ with projection m_2 will add to form a total momentum $\mathbf{j}$ with projection m. Hence, only those coefficients differ from zero for which we have validity of the *triangle rule*, namely

$$|j_1 - j_2| \leq j \leq j_1 + j_2, \tag{C.1}$$

and of the *projection condition*

$$m_1 + m_2 = m, \tag{C.2}$$

that is, the magnetic quantum numbers m_1 and m_2 add algebraically while the angular momenta $\mathbf{j}_1$ and $\mathbf{j}_2$ add vectorially. The following *symmetry properties* hold for the Clebsch–Gordan coefficients:

$$
\begin{aligned}
C^{c\gamma}_{a\alpha b\beta} &= (-1)^{a+b-c} C^{c\gamma}_{b\beta a\alpha} = (-1)^{a-\alpha} \sqrt{\frac{2c+1}{2b+1}} C^{b-\beta}_{a\alpha c-\gamma} \\
&= (-1)^{a-\alpha} \sqrt{\frac{2c+1}{2b+1}} C^{b\beta}_{c\gamma a-\alpha} = (-1)^{b+\beta} \sqrt{\frac{2c+1}{2a+1}} C^{a-\alpha}_{c-\gamma b\beta} \\
&= (-1)^{b+\beta} \sqrt{\frac{2c+1}{2a+1}} C^{a\alpha}_{b-\beta c\gamma}, \tag{C.3} \\
&= (-1)^{a+b-c} C^{c-\gamma}_{a-\alpha b-\beta}. \tag{C.4}
\end{aligned}
$$

The following relations are useful for calculations:

$$C^{c\gamma}_{a\alpha 00} = \delta_{ac}\delta_{\alpha\gamma}, \quad C^{00}_{a\alpha b\beta} = (-1)^{a-\alpha} \frac{\delta_{ab}\delta_{\alpha-\beta}}{\sqrt{2a+1}}. \tag{C.5}$$

Sometimes, instead of Clebsch–Gordan coefficients, one uses the *Wigner 3j-symbols* which possess simpler symmetry properties. These are connected with the Clebsch–Gordan coefficients in the following manner:

$$\begin{pmatrix} a & b & c \\ \alpha & \beta & \gamma \end{pmatrix} = (-1)^{c+2a+\gamma} \frac{C^{c\gamma}_{a-\alpha b-\beta}}{\sqrt{2c+1}}. \tag{C.6}$$

It is frequently useful to employ formulae for sums of $3j$-symbols, connecting them with $6j$-symbols (given in curly brackets, $\{ \begin{smallmatrix} \cdot & \cdot & \cdot \\ \cdot & \cdot & \cdot \end{smallmatrix} \}$):

$$\sum_{\kappa} (-1)^{q-\kappa} \begin{pmatrix} a & b & q \\ \alpha & \beta & -\kappa \end{pmatrix} \begin{pmatrix} q & d & c \\ \kappa & \delta & \gamma \end{pmatrix}$$

$$= (-1)^{2a} \sum_{x\xi} (-1)^{x-\xi}(2x+1) \begin{pmatrix} a & c & x \\ \alpha & \gamma & -\xi \end{pmatrix} \begin{pmatrix} x & d & b \\ \xi & \delta & \beta \end{pmatrix} \begin{Bmatrix} b & d & x \\ c & a & q \end{Bmatrix}. \tag{C.7}$$

$$\sum_{\kappa\psi\rho} (-1)^{p-\psi+q-\kappa+r-\rho} \begin{pmatrix} p & a & q \\ \psi & \alpha & -\kappa \end{pmatrix} \begin{pmatrix} q & b & r \\ \kappa & \beta & -\rho \end{pmatrix} \begin{pmatrix} r & c & p \\ \rho & \gamma & -\psi \end{pmatrix}$$

$$= \begin{pmatrix} a & b & c \\ -\alpha & -\beta & -\gamma \end{pmatrix} \begin{Bmatrix} a & b & c \\ r & p & q \end{Bmatrix}, \tag{C.8}$$

$$\sum_{\kappa\psi\rho\sigma\tau} (-1)^{p-\psi+q-\kappa+r-\rho+s-\sigma+t-\tau} \begin{pmatrix} p & a & q \\ \psi & -\alpha & \kappa \end{pmatrix} \begin{pmatrix} q & r & t \\ -\kappa & \rho & \tau \end{pmatrix}$$

$$\times \begin{pmatrix} r & a' & s \\ -\rho & \alpha' & \sigma \end{pmatrix} \begin{pmatrix} s & p & t \\ -\sigma & -\psi & -\tau \end{pmatrix} = \frac{(-1)^{a-\alpha}}{2a+1} \begin{Bmatrix} q & p & a \\ s & r & t \end{Bmatrix} \delta_{aa'} \delta_{\alpha\alpha'}, \tag{C.9}$$

and also with $9j$-symbols $\{ \begin{smallmatrix} \cdot & \cdot & \cdot \\ \cdot & \cdot & \cdot \end{smallmatrix} \}$:

$$\sum_{\psi\kappa} \begin{pmatrix} a & b & p \\ \alpha & \beta & \psi \end{pmatrix} \begin{pmatrix} p & c & q \\ -\psi & -\gamma & -\kappa \end{pmatrix} \begin{pmatrix} q & d & e \\ \kappa & \delta & \varepsilon \end{pmatrix}$$

$$= \sum_{x\xi y\eta} (2x+1)(2y+1) \begin{pmatrix} a & d & x \\ \alpha & \delta & \xi \end{pmatrix} \begin{pmatrix} x & c & y \\ -\xi & -\gamma & -\eta \end{pmatrix}$$

$$\times \begin{pmatrix} y & b & e \\ \eta & \beta & \varepsilon \end{pmatrix} \begin{Bmatrix} a & b & p \\ d & e & q \\ x & y & c \end{Bmatrix}. \tag{C.10}$$

Table C.1 presents the numerical values of some Clebsch–Gordan coefficients. Table C.2 contains formulae for the, often necessary, Clebsch–

Table C.1.   Numerical values of some Clebsch–Gordan coeffcients

a	α	b	β	$C^{0\alpha+\beta}_{a\alpha b\beta}$	a	α	b	β	$C^{0\alpha+\beta}_{a\alpha b\beta}$
0	0	0	0	1	2	2	2	-2	$\frac{1}{\sqrt{5}}$
1	1	1	-1	$\frac{1}{\sqrt{3}}$	2	1	2	-1	$-\frac{1}{\sqrt{5}}$
1	0	1	0	$-\frac{1}{\sqrt{3}}$	2	0	2	0	$\frac{1}{\sqrt{5}}$

a	α	b	β	$C^{1\alpha+\beta}_{a\alpha b\beta}$	a	α	b	β	$C^{1\alpha+\beta}_{a\alpha b\beta}$
1	1	0	0	1	2	0	1	1	$\frac{1}{\sqrt{2\cdot5}}$
1	0	0	0	1	2	0	1	0	$-\sqrt{\frac{2}{5}}$
1	1	1	0	$\frac{1}{\sqrt{2}}$	2	2	2	-1	$\frac{1}{\sqrt{5}}$
1	1	1	-1	$\frac{1}{\sqrt{2}}$	2	2	2	-2	$\sqrt{\frac{2}{5}}$
1	0	1	0	0	2	1	2	0	$-\sqrt{\frac{3}{2\cdot5}}$
2	2	1	-1	$\sqrt{\frac{3}{5}}$	2	1	2	-1	$-\frac{1}{\sqrt{2\cdot5}}$
2	1	1	0	$-\sqrt{\frac{3}{2\cdot5}}$	2	0	2	0	0
2	1	1	-1	$\sqrt{\frac{3}{2\cdot5}}$					

a	α	b	β	$C^{2\alpha+\beta}_{a\alpha b\beta}$	a	α	b	β	$C^{2\alpha+\beta}_{a\alpha b\beta}$
1	1	1	1	1	2	1	1	-1	$\frac{1}{\sqrt{2}}$
1	1	1	0	$\frac{1}{\sqrt{2}}$	2	0	1	1	$-\frac{1}{\sqrt{2}}$
1	1	1	-1	$\frac{1}{\sqrt{2\cdot3}}$	2	0	1	0	0
1	0	1	0	$\sqrt{\frac{2}{3}}$	2	2	2	0	$\sqrt{\frac{2}{7}}$
2	2	0	0	1	2	2	2	-1	$\sqrt{\frac{3}{7}}$
2	1	0	0	1	2	2	2	-2	$\sqrt{\frac{2}{7}}$
2	0	0	0	1	2	1	2	1	$-\sqrt{\frac{3}{7}}$
2	2	1	0	$\sqrt{\frac{2}{3}}$	2	1	2	0	$-\frac{1}{\sqrt{2\cdot7}}$
2	2	1	-1	$\frac{1}{\sqrt{3}}$	2	1	2	-1	$\frac{1}{\sqrt{2\cdot7}}$
2	1	1	1	$-\frac{1}{\sqrt{3}}$	2	0	2	0	$-\sqrt{\frac{2}{7}}$
2	1	1	0	$\frac{1}{\sqrt{2\cdot3}}$					

a	α	b	β	$C^{4\alpha+\beta}_{a\alpha b\beta}$	a	α	b	β	$C^{4\alpha+\beta}_{a\alpha b\beta}$
2	2	2	-2	$\frac{1}{\sqrt{2\cdot5\cdot7}}$	4	2	2	0	$-\frac{4}{\sqrt{5\cdot7\cdot11}}$
2	0	2	0	$3\sqrt{\frac{2}{5\cdot7}}$	4	4	2	-2	$\sqrt{\frac{6}{5\cdot11}}$
2	2	2	0	$\sqrt{\frac{3}{2\cdot7}}$	4	2	2	2	$\sqrt{\frac{6}{5\cdot11}}$
4	0	2	0	$-2\sqrt{\frac{5}{7\cdot11}}$	4	4	2	0	$2\sqrt{\frac{7}{5\cdot11}}$
4	2	2	-2	$3\sqrt{\frac{3}{5\cdot11}}$					

Table C.2. The explicit form of the $C^{c\gamma}_{a\alpha 1\beta}$

c	$\beta = 1$	$\beta = 0$	$\beta = -1$
$a+1$	$\sqrt{\dfrac{(c+\gamma-1)(c+\gamma)}{(2c-1)2c}}$	$\sqrt{\dfrac{(c+\gamma)(c-\gamma)}{(2c-1)c}}$	$\sqrt{\dfrac{(c-\gamma-1)(c-\gamma)}{(2c-1)2c}}$
a	$-\sqrt{\dfrac{(c+\gamma)(c-\gamma+1)}{2c(c+1)}}$	$\dfrac{\gamma}{\sqrt{c(c+1)}}$	$\sqrt{\dfrac{(c-\gamma)(c+\gamma+1)}{2c(c+1)}}$
$a-1$	$\sqrt{\dfrac{(c-\gamma+1)(c-\gamma+2)}{(2c+2)(2c+3)}}$	$-\sqrt{\dfrac{(c+\gamma+1)(c-\gamma+1)}{(c+1)(2c+3)}}$	$\sqrt{\dfrac{(c+\gamma+1)(c+\gamma+2)}{(2c+2)(2c+3)}}$

Table C.3. Numerical values of $6j$-symbols for integral indices not exceeding 2

a	b	c	d	e	f	$\left\{\begin{matrix} a & b & c \\ d & e & f \end{matrix}\right\}$
1	1	1	1	1	1	$\frac{1}{6}$
2	1	1	1	1	1	$\frac{1}{6}$
2	1	1	2	1	1	$\frac{1}{30}$
2	2	1	1	1	1	$-\frac{1}{2\sqrt{5}}$
2	2	1	1	1	2	$\frac{1}{6\sqrt{5}}$
2	2	1	2	1	1	$-\frac{1}{10}$
2	2	1	2	2	1	$\frac{1}{6}$
2	2	2	1	1	1	$\frac{\sqrt{7}}{10\sqrt{3}}$
2	2	2	2	1	1	$\frac{\sqrt{7}}{10\sqrt{3}}$
2	2	2	2	2	1	$-\frac{1}{10}$
2	2	2	2	2	2	$-\frac{3}{70}$

Gordon coefficients $C^{c\gamma}_{a\alpha 1\beta}$ with $b = 1$. Table C.3 contains the numerical values of $6j$-symbols for integral indices ≤ 2. More complete tables may be found in [379].

The magnitude of the $6j$-symbols does not change at permutation of columns, as well as at permutation of any two elements of the top line with two elements of a lower line positioned under the former.

The following relation is useful for practical calculations:

$$\left\{\begin{matrix} a & b & 0 \\ d & e & f \end{matrix}\right\} = (-1)^{a+e+f}\,\frac{\delta_{ab}\delta_{de}}{\sqrt{(2a+1)(2d+1)}}. \qquad (C.11)$$

In cases where some of the angular momenta entering into the $6j$-symbols become large, one may apply *asymptotic formulae* for the $6j$-symbols.

In particular:

$$\left\{ \begin{array}{ccc} a & b & c \\ R+d & R+e & R+f \end{array} \right\} \approx \frac{(-1)^{a+b+d+e}}{\sqrt{2R(2c+1)}} C^{c\gamma}_{a\alpha b\beta},\qquad (C.12)$$

where $\alpha = f - e$, $\beta = d - f$, $\gamma = d - e$, $R \gg 1$, and

$$\left\{ \begin{array}{ccc} a & R+b & R+c \\ d & R+e & R+f \end{array} \right\}$$

$$\approx (-1)^{\varphi} \left(\frac{(a-b+c)!(a-e+f)!(c+d-e)!(-b+d+f)!}{(a+b-c)!(a+e-f)!(-c+d+e)!(b+d-f)!} \right)^{(1/2)\mathrm{sgn}(c+f-b-e)}$$

$$\times \frac{(2R)^{-1-|b+e-c-f|}}{|b+e-c-f|!} \left[1 + O\left(R^{-2}\right) \right],\qquad (C.13)$$

where $\varphi = a + d + \min(b+e, c+f)$, $\mathrm{sgn}\,x = \left\{ \begin{array}{cc} 1 & \text{at}\quad x \geq 0 \\ -1 & \text{at}\quad x < 0 \end{array} \right.$. The calculation of the $9j$-symbols may be reduced to the calculation of a number of $6j$-symbols:

$$\left\{ \begin{array}{ccc} a & b & c \\ d & e & f \\ g & h & i \end{array} \right\}$$

$$= \sum_y (-1)^{2y}(2y+1) \left\{ \begin{array}{ccc} a & b & c \\ f & i & y \end{array} \right\} \left\{ \begin{array}{ccc} d & e & f \\ b & y & h \end{array} \right\} \left\{ \begin{array}{ccc} g & h & i \\ y & a & d \end{array} \right\} .$$

$$(C.14)$$

Appendix D
Various methods of introducing polarization moments

In cases where we consider various processes with the participation of atoms and molecules and when the state of the particle ensemble is characterized with the aid of the *quantum density operator* $\hat{\rho}$, it is frequently convenient to perform the expansion of this operator over *irreducible tensor operators* $\hat{T}_Q^K$. Further, in the process of forming the theory, one may use the coefficients of this expansion f_Q^K, i.e. the *polarization moments*. This, in fact, is what we have attempted to demonstrate in the present book. The basic advantage of such an approach lies in the extreme simplicity of transforming the tensor operators $\hat{T}_Q^K$ and, correspondingly, also the polarization moments f_Q^K in the process of turning the coordinate system. This makes it possible to gain maximum advantage in simplifying the equation of motion for f_Q^K by making use of the symmetry properties of the processes under treatment.

All the aforesaid also applies to the *classical limit* of the quantum density operator $\hat{\rho}$, namely to the *probability density* $\rho(\theta, \varphi)$ if one performs its expansion over the classical analogue of tensor operators to which the *spherical functions* $Y_{KQ}(\theta, \varphi)$ belong. Unfortunately, different ways of expansion are used by different authors, both in the quantum approach and in the classical limit. This complicates considerably comparison of the results obtained by these authors, including experimental data.

In this appendix we aim briefly to compare and analyze a number of frequently used approaches to the problem of introducing polarization moments into the description of atomic and molecular angular momenta polarization.

D.1   Polarization moments in quantum physics

For two levels J' and J'' which are linked to each other by optical transition the density operator $\hat{\rho}$ may be represented by the matrix

$$\langle J_1 M_1 | \hat{\rho} | J_2 M_2 \rangle = \rho_{J_1 M_1, J_2 M_2}, \qquad (D.1)$$

which is called the *density matrix*. Here J_1 and J_2 have the values J' or J''. This matrix contains *four submatrices* and may be written as follows [110]:

$$\rho = \begin{pmatrix} {}_{J''J''}\rho & {}_{J''J'}\rho \\ {}_{J'J''}\rho & {}_{J'J'}\rho \end{pmatrix}. \qquad (D.2)$$

The *non-diagonal submatrices* ${}_{J''J'}\rho$ and ${}_{J'J''}\rho$ describe the *optical coherences* between the magnetic sublevels of the states J'' and J'. The submatrices ${}_{J''J''}\rho$ and ${}_{J'J'}\rho$ describe the particles on levels J'' and J' respectively. Their *diagonal elements* characterize the *populations* of the respective sublevels M'' and M', whilst the *non-diagonal elements* describe the *Zeeman coherences*.

The expansion of $\hat{\rho}$ may be performed in two ways, using as bases either the operators ${}_{J_1 J_2}\hat{T}_Q^K$, or their conjugates $({}_{J_1 J_2}\hat{T}_Q^K)^\dagger$:

$$_{J_1 J_2}\hat{\rho} = \sum_{KQ} {}_{J_1 J_2}^{c} f_Q^K \, {}_{J_1 J_2}\hat{T}_Q^K, \qquad (D.3)$$

$$_{J_1 J_2}\hat{\rho} = \sum_{KQ} {}_{J_1 J_2}^{g} f_Q^K \, ({}_{J_1 J_2}\hat{T}_Q^K)^\dagger. \qquad (D.4)$$

The difference between these expansions consists of the following. In the first case the operators ${}_{J_1 J_2}\hat{T}_Q^K$ and the moments are *contragredient* quantities. This means that a turn of the coordinate system ${}_{J_1 J_2}\hat{T}_Q^K$ causes transformation according to (A.12), i.e.

$$_{J_1 J_2}\hat{T}_Q^K = \sum_{Q'} {}_{J_1 J_2}\hat{T}_{Q'}^K D_{Q'Q}^{(K)}(\alpha, \beta, \gamma), \qquad (D.5)$$

whilst the polarization moments ${}_{J_1 J_2}^{c} f_Q^K$ transform slightly differently:

$$_{J_1 J_2}^{c} f_Q^K = \sum_{Q'} D_{QQ'}^{(K)}(\alpha, \beta, \gamma) \, {}_{J_1 J_2}^{c} f_{Q'}^K = \sum_{Q'} {}_{J_1 J_2}^{c} f_{Q'}^K D_{Q'Q}^{(K)^*}(\alpha, \beta, \gamma). \quad (D.6)$$

In the second case the expansion is *cogredient*, and both ${}_{J_1 J_2}\hat{T}_Q^K$ and polarization moments ${}_{J_1 J_2}^{g} f_Q^K$ transform, at the turn of coordinates, in a similar way to the form of (D.5). The form of expansion (D.3) is applied, e.g., in [46, 73, 95, 110, 187, 304, 309, 402]), and also in other papers by these authors, whilst form (D.4) was first introduced by Dyakonov in

[133] and used by the authors of such papers as [39, 96, 242, 303], and Chapter 2 of the present book.

To determine the explicit form of the matrix elements of the tensor operators we employ the Wigner–Eckart theorem [136, 379, 402]

$$\langle J_1 M_1 |_{J_1 J_2} \hat{T}_Q^K | J_2 M_2 \rangle = (-1)^{J_1 - M_1} \begin{pmatrix} J_1 & K & J_2 \\ -M_1 & Q & M_2 \end{pmatrix} \langle J_1 \|_{J_1 J_2} \hat{T}_Q^K \| J_2 \rangle.$$
(D.7)

Different authors introduce the above matrix element $\langle J_1 \|_{J_1 J_2} \hat{T}_Q^K \| J_2 \rangle$ in different ways. Thus the authors of [46, 73, 95, 110, 304] define it as

$$\langle J_1 \|_{J_1 J_2}^{1} \hat{T}_Q^K \| J_2 \rangle = \sqrt{2K + 1},$$
(D.8)

which provides for orthonormality of the operators:

$$\mathrm{tr} \left[_{J_1 J_2}^{1} \hat{T}_Q^K \left(_{J_1 J_2}^{1} \hat{T}_{Q'}^{K'} \right)^\dagger \right]$$

$$= \sum_{M_1 M_2} \langle J_1 M_1 |_{J_1 J_2}^{1} \hat{T}_Q^K | J_2 M_2 \rangle \langle J_2 M_2 |_{J_1 J_2}^{1} \hat{T}_{Q'}^{K'}{}^\dagger | J_1 M_1 \rangle = \delta_{KK'} \delta_{QQ'},$$
(D.9)

where [73, 304]

$$\langle J_1 M_1 |_{J_1 J_2}^{1} \hat{T}_{Q'}^{K'}{}^\dagger | J_2 M_2 \rangle = \langle J_2 M_2 |_{J_1 J_2}^{1} \hat{T}_{Q'}^{K'} | J_1 M_1 \rangle^*.$$
(D.10)

In [39, 133, 242, 303] the above matrix element is assumed equal to

$$\langle J_1 \|_{J_1 J_1}^{2} \hat{T}_Q^K \| J_1 \rangle = (-1)^Q \frac{2K + 1}{\sqrt{2J_1 + 1}},$$
(D.11)

which destroys normalization of the operators. In this case the Kronecker symbols in the righthand part of Eq. (D.9) must in addition be multiplied by $(2K + 1)/(2J_1 + 1)$, and the formula must be regarded as that for $J_1 = J_2$. Such normalization, as will be shown subsequently, yields certain advantages in interpreting the polarization moments. At the same time the additional factor $(-1)^Q$ in (D.11) requires modification of the transformation formula (D.5) by an additional factor $(-1)^{Q+Q'}$ (under the summation symbol). The orthogonality property of the tensor operators makes it possible to find inverse relations to (D.3) and (D.4). For normalization, as in papers [46, 73, 95, 110, 304], we obtain, from (D.3) and (D.9),

$$_{J_1 J_2}^{c} f_Q^K = \sum_{M_1 M_2} (-1)^{J_1 - M_1} \sqrt{2K + 1} \begin{pmatrix} J_1 & K & J_2 \\ -M_1 & Q & M_2 \end{pmatrix} \rho_{J_1 M_1, J_2 M_2}.$$
(D.12)

Applying the normalization of the tensor operator used in [39, 133, 242,

303], as well as the expansion (D.4), we obtain

$$_{J_1 J_1}^{g} f_Q^K = \sum_{M_1 M_2} (-1)^{J_1 - M_1} \sqrt{2J_1 + 1} \begin{pmatrix} J_1 & K & J_1 \\ -M_1 & -Q & M_2 \end{pmatrix} \rho_{J_1 M_1, J_1 M_2}.$$

(D.13)

We see that the two most widely used methods of polarization moment definition, namely (D.3) and (D.4), differ both in their transformation properties at turn of the coordinate system (which is not always fully realized), and in normalization. In addition, the definition of the matrix element of the operator $_{J_1 J_1} \hat{T}_Q^K$, as assumed in [133], does not permit us to use (D.13) directly for the description of optical coherences.

Let us now dwell in greater detail on the *physical meaning* of polarization moments. It is simplest to interpret them for states with a definite angular momentum, where we have $J_1 = J_2 = J$. We wish to stress that in our subsequent discussion we are going to consider only states with a definite angular momentum. From the expansion formulae (D.3) and (D.4), using the accepted normalization, we obtain:

$$^c f_Q^K = \text{tr}\left(\rho^1 \hat{T}_Q^{K\dagger}\right) = \langle {}^1 \hat{T}_Q^{K\dagger} \rangle,$$

(D.14)

$$^g f_Q^K = \frac{2J+1}{2K+1} \text{tr}\left(\rho^2 \hat{T}_Q^K\right) = \frac{2J+1}{2K+1} \langle {}^2 \hat{T}_Q^K \rangle.$$

(D.15)

It is important here to note that in Eq. (D.14) it is assumed that the given matrix element is employed in the form of (D.8), and in Eq. (D.15) it is employed in the form of (D.11). Then, applying the explicit form for the 3j-symbols, we obtain directly for zero rank elements

$$^c f_0^0 = \frac{\text{tr}\rho}{\sqrt{2J+1}}, \qquad ^g f_0^0 = \text{tr}\rho,$$

(D.16)

where $\text{tr}\rho = n_J$ is the full *population* of level J. The *zero rank* moments are, in essence, *normalizing factors* which also set the scale of measurements for all the other moments. Sometimes, for the sake of convenience in calculations [303], the population of level J is normalized to unity, $n_J = 1$, which is particularly convenient using the cogredient form of polarization moments – $^g f_Q^K$.

In order to determine the physical meaning of moments of higher rank, one may keep in mind the fact that the tensor operators $\hat{T}_Q^K$ are proportional to the angular momenta operator $\hat{J}^{(K)}$ [73]. Thus, for instance, the three components of the operator $\hat{J}^{(1)}$ are defined as

$$J_{\pm 1}^{(1)} = \mp \frac{1}{\sqrt{2}}(J_x \pm iJ_y), \qquad J_0^{(1)} = J_z,$$

(D.17)

and the corresponding matrix elements are defined as

$$\langle JM_1|J_0^{(1)}|JM_2\rangle = M_2\delta_{M_1M_2},$$
$$\langle JM_1|J_{\pm1}^{(1)}|JM_2\rangle = \mp\frac{1}{\sqrt{2}}\sqrt{(J\mp M_2)(J\pm M_2+1)}\delta_{M_1M_2\pm1}.$$

$$(D.18)$$

On the other hand, we have:

$$\langle JM_1|^1\hat{T}_0^1|JM_2\rangle = (-1)^{J-M_1}\sqrt{3}\begin{pmatrix} J & 1 & J \\ -M_1 & 0 & M_2 \end{pmatrix}$$

$$= \sqrt{\frac{3}{(2J+1)(J+1)J}}M_2\delta_{M_1M_2},$$

$$\langle JM_1|^2\hat{T}_0^1|JM_2\rangle = (-1)^{J-M_1}\frac{3}{\sqrt{2J+1}}\begin{pmatrix} J & 1 & J \\ -M_1 & 0 & M_2 \end{pmatrix}$$

$$= \frac{3}{(2J+1)\sqrt{(J+1)J}}M_2\delta_{M_1M_2}, \qquad (D.19)$$

$$\langle JM_1|^1\hat{T}_{\pm1}^1|JM_2\rangle = (-1)^{J-M_1}\sqrt{3}\begin{pmatrix} J & 1 & J \\ -M_1 & \pm1 & M_2 \end{pmatrix}$$

$$= \mp\sqrt{\frac{3}{2}}\sqrt{\frac{(J\mp M_2)(J\pm M_2+1)}{(2J+1)(J+1)J}}\delta_{M_1M_2\pm1},$$

$$\langle JM_1|^2\hat{T}_{\pm1}^1|JM_2\rangle = (-1)^{J-M_1}\frac{3}{\sqrt{2J+1}}\begin{pmatrix} J & 1 & J \\ -M_1 & \pm1 & M_2 \end{pmatrix}$$

$$= \mp\frac{3}{2J+1}\sqrt{\frac{(J\mp M_2)(J\pm M_2+1)}{2(J+1)J}}\delta_{M_1M_2\pm1},$$

$$(D.20)$$

Comparing (D.18) and (D.20), we obtain

$$^1\hat{T}_Q^1 = \sqrt{\frac{3}{(2J+1)(J+1)J}}J_Q^{(1)} = {}^1N_1 J_Q^{(1)}, \qquad (D.21)$$

$$^2\hat{T}_Q^1 = (-1)^Q\frac{3}{(2J+1)\sqrt{(J+1)J}}J_Q^{(1)} = {}^2N_1 J_Q^{(1)}. \qquad (D.22)$$

Similarly, we have for second rank angular momentum operators [95]

$$\left.\begin{array}{l} J_0^{(2)} = \frac{1}{\sqrt{6}}(3J_z^2 - \mathbf{J}^2), \\ J_{\pm1}^{(2)} = \mp\frac{1}{2}[(J_xJ_z + J_zJ_x)\pm i(J_yJ_z + J_zJ_y)], \\ J_{\pm2}^{(2)} = \frac{1}{2}[J_x^2 - J_y^2 \pm i(J_xJ_y + J_yJ_x)], \end{array}\right\} \qquad (D.23)$$

and obtain

$$
{}^1\hat{T}_Q^2 = \sqrt{\frac{30}{(2J+3)(2J+1)J(2J-1)(J+1)}}J_Q^{(2)} = {}^1N_2 J_Q^{(2)}, \quad \text{(D.24)}
$$

$$
{}^2\hat{T}_Q^2 = (-1)^Q \frac{5}{2J+1}\sqrt{\frac{6}{(2J+3)J(J+1)(2J-1)}}J_Q^{(2)} = {}^2N_2 J_Q^{(2)}.
$$
$$
\text{(D.25)}
$$

Hence, one may interpret polarization moments as

$$
{}^c f_Q^K = {}^1N_K \langle J_Q^{(K)\dagger} \rangle = {}^1N_K \langle J_Q^{(K)} \rangle^*, \quad \text{(D.26)}
$$

$$
{}^g f_Q^K = \frac{2J+1}{2K+1} {}^2N_K \langle J_Q^{(K)} \rangle. \qquad \text{(D.27)}
$$

Such an interpretation is convenient for a number of reasons. Thus, for instance, it is well known that the magnetic moment, averaged over a particle ensemble, may be defined as

$$
\langle \boldsymbol{\mu} \rangle = g\mu_B \langle \mathbf{J} \rangle, \quad \text{(D.28)}
$$

and its cyclic components, accordingly, are

$$
\langle \mu_Q \rangle = \frac{2K+1}{(2J+1)\,({}^2N_1)} g\mu_B {}^g f_Q^1. \quad \text{(D.29)}
$$

It is also possible to show that

$$
\frac{{}^g f_0^1}{{}^g f_0^0} = \frac{2J+1}{\sqrt{3}n_J} {}^2N_1 \langle J_0^{(1)} \rangle = \langle J_z \rangle \frac{1}{n_J \sqrt{(J+1)J}} = \langle \cos\theta \rangle, \quad \text{(D.30)}
$$

$$
\frac{{}^c f_0^1}{{}^c f_0^0} = \frac{\sqrt{2J+1}}{n_J} {}^1N_1 \langle J_0^{(1)} \rangle = \frac{\sqrt{3}\langle J_z \rangle}{n_J \sqrt{(J+1)J}} = \sqrt{3}\langle \cos\theta \rangle, \quad \text{(D.31)}
$$

where $\cos\theta = M_J/\sqrt{(J+1)J}$. In general, one may define the normalizing factors in Eqs. (D.26) and (D.27) as the ratio between reduced matrix elements

$$
\langle {}^i N_K \rangle = \frac{\langle J \| {}^i\hat{T}_Q^K \| J \rangle}{\langle J \| \hat{J}_Q^{(K)} \| J \rangle}, \quad \text{(D.32)}
$$

where the quantities $\langle J \| {}^i\hat{T}_Q^K \| J \rangle$ are given by Eqs. (D.8) and (D.11), whilst the quantities $\langle J \| {}^i\hat{J}_Q^K \| J \rangle$ up to range $K = 4$ are presented in Table D.1. Use of (D.26), together with (D.32), (D.8) and (D.11) makes it possible to obtain a general formula connecting both methods for introducing polarization moments:

$$
\left({}^c f_Q^K \right)^* = (-1)^Q \sqrt{\frac{2K+1}{2J+1}} {}^g f_Q^K. \quad \text{(D.33)}
$$

Table D.1. Reduced matrix elements of the angular momentum operator

$$
\begin{aligned}
\langle J\|J_Q^{(1)}\|J\rangle &= \sqrt{J(J+1)(2J+1)} \\
\langle J\|J_Q^{(2)}\|J\rangle &= \sqrt{\tfrac{1}{6}J(J+1)(2J+1)(2J+3)(2J-1)} \\
\langle J\|J_Q^{(3)}\|J\rangle &= \sqrt{\tfrac{1}{10}(J-1)J(J+1)(J+2)(2J+1)} \\
&\quad \times\sqrt{(2J+3)(2J-1)} \\
\langle J\|J_Q^{(4)}\|J\rangle &= \sqrt{\tfrac{1}{70}(J-1)J(J+1)(J+2)(2J+1)} \\
&\quad \times\sqrt{(2J+3)(2J+5)(2J-1)(2J-3)}
\end{aligned}
$$

Table D.2. Angular momentum operators $\hat{J}_Q^{(K)}$

$$
\begin{aligned}
J_0^{(3)} &= \tfrac{1}{\sqrt{10}}\left(2J_z^3 - J_{x^2z} - J_{y^2z}\right) \\
J_{\pm1}^{(3)} &= \mp\tfrac{1}{\sqrt{30}}\left[\left(2J_{xz^2} - \tfrac{3}{2}J_x^3 - \tfrac{1}{2}J_{xy^2}\right) \pm i\left(2J_{yz^2} - \tfrac{3}{2}J_y^3 - \tfrac{1}{2}J_{x^2y}\right)\right] \\
J_{\pm2}^{(3)} &= \tfrac{1}{2\sqrt{3}}\left(J_{x^2z} - J_{y^2z} \pm iJ_{xyz}\right) \\
J_{\pm3}^{(3)} &= \mp\tfrac{1}{2\sqrt{2}}\left[\left(J_x^3 - J_{xy^2}\right) \pm i\left(J_{x^2y} - J_y^3\right)\right] \\
J_0^{(4)} &= \tfrac{2}{\sqrt{70}}\left[\left(2J_z^4 - J_{x^2z^2} - J_{y^2z^2}\right) + \tfrac{1}{4}\left(3J_x^4 + 3J_y^4 + J_{x^2y^2}\right)\right] \\
J_{\pm1}^{(4)} &= \mp\tfrac{1}{4\sqrt{14}}\left[\left(4J_{xz^3} - 3J_{x^3z} - J_{xy^2z}\right) \pm i\left(4J_{yz^3} - 3J_{y^3z} - J_{x^2yz}\right)\right] \\
J_{\pm2}^{(4)} &= \tfrac{1}{2\sqrt{7}}\left[\left(J_{x^2z^2} - J_{y^2z^2} + J_y^4 - J_x^4\right) \pm \tfrac{i}{2}\left(2J_{xyz^2} - J_{x^3y} - J_{xy^3}\right)\right] \\
J_{\pm3}^{(4)} &= \mp\tfrac{1}{4\sqrt{2}}\left[\left(J_{x^3z} - J_{xy^2z}\right) \pm i\left(J_{x^2yz} - J_{y^3z}\right)\right] \\
J_{\pm4}^{(4)} &= \tfrac{1}{4}\left[\left(J_x^4 + J_y^4 - J_{x^2y^2}\right) \pm i\left(J_{x^3y} - J_{xy^3}\right)\right]
\end{aligned}
$$

Operators with several indices are symmetric sums of all permutations,
e.g. $J_{xy^2} = J_x J_y^2 + J_y^2 J_x + J_y J_x J_y$.

Definitions of operators for angular momenta of range higher than two are presented in Table D.2. A similar table given in [95] contains misprints which are corrected in [250]. Various papers use various notations for tensor operators and polarization moments. A summary of the various ways of denoting tensor operators may be found in [73, 136, 304], and of denoting polarization moments may be found in [304].

Both methods used in practice for introducing polarization moments $^c f_Q^K$ and $^g f_Q^K$ possess certain disadvantages. The disadvantage of using the contragredient form of definition is that the polarization moments are proportional not to the mean value of the corresponding operator of angular momentum, but to the complex conjugate magnitude. This property is a direct consequence of the contragrediency at turning the coordinate system.

The disadvantages of the cogredient form of definition are connected with the presence of the factor $(-1)^Q$ in the normalizing coefficient ^2N_K (D.32), which is a result of defining the reduced matrix element of the tensor operator (D.11) into which this factor is introduced, in our opinion, without particular necessity.

However, both forms have gained wide application in practice, and it is, probably, hardly any use to introduce yet a 'third' way of defining polarization moments, which would be free of the above disadvantages. As it is, having at one's disposal two forms is already a 'luxury' that makes comparison of experimental results difficult.

D.2   Real components of the density matrix

The components of polarization moments, as defined by Eqs. (D.14) and (D.15), are, in the general case, complex quantities. This results in a certain inconvenience in their application, since the equations of motion are also complex.

In the paper by Fano [141] real irreducible tensor operators were first obtained:

$$\hat{T}_{Q+}^{\{K\}} = (-1)^Q\sqrt{2 - \delta_{Q0}}\operatorname{Re}\hat{T}_Q^K, \quad Q \geq 0, \tag{D.34}$$

$$\hat{T}_{Q-}^{\{K\}} = (-1)^Q(1 - \delta_{Q0})\sqrt{2}\operatorname{Im}\hat{T}_Q^K, \quad Q \geq 0, \tag{D.35}$$

which allow one to perform the expansion of the density matrix by the introduction of *real polarization moments* $f_{Q\pm}^{\{K\}}$. Although the problem of such an expansion has been discussed repeatedly [58, 59, 141, 304], it nevertheless did not gain widespread application until recently. This is chiefly due to the fact that the theory of real irreducible operators $\hat{T}_{Q\pm}^{\{K\}}$ has not been elaborated. In particular, freely accessible matrix tables (similar to the Wigner D-matrices) have not been produced for them or for their conjugate values at turn of the coordinate system.

Nevertheless, in a number of papers [177, 208, 249, 250, 251, 354] the authors have been developing the theory and applying 'real polarization moments' in practice. In the first two papers [177, 249] so-called *distribution moments* in the ground state are introduced as

$$A_Q^K = c(K)\frac{\operatorname{Re}\langle(JM|\hat{J}_Q^{(K)}|JM')\rangle}{[(JM|\mathbf{J}^2|JM)]^{K/2}}, \tag{D.36}$$

where the normalizing factor $c(K)$ is given in the form of a table which we present here (Table D.3). The zero rank moment is normalized to unity, $A_0^{(0)} = \operatorname{tr}\rho = 1$. Regrettably, the general rule for defining $c(K)$ is not given in [177, 249] (the table gives $c(K)$, $K \leq 4$). Eq. (D.36) corresponds

Table D.3.   Normalizing factors $c(K)$ and $v(K)$

K	0	1	2	3	4
$c(K)$	1	1	$\sqrt{6}$	$\frac{\sqrt{10}}{2}$	$\sqrt{\frac{35}{8}}$
$v(K)$	$\sqrt{4\pi}$	$\sqrt{\frac{4\pi}{3}}$	$\sqrt{\frac{8\pi}{15}}$	$\sqrt{\frac{8\pi}{35}}$	$4\sqrt{\frac{2\pi}{315}}$

to an expansion of the density matrix over $\hat{J}_Q^{(K)\dagger}$. Accordingly, Eq. (D.36) may be rewritten in the form

$$A_Q^{(K)} = c(K)\frac{\mathrm{Re}\left[\mathrm{tr}\left(\rho\hat{J}_Q^{(K)}\right)\right]}{[(J+1)J]^{K/2}n_J}, \tag{D.37}$$

which permits comparison between the distribution moments $A_Q^{(K)}$ and the polarization moments previously considered. Thus, from (D.27), (D.32), (D.11) and (D.37) we have:

$$\frac{\mathrm{Re}\left(^g f_Q^K\right)}{A_Q^{(K)}} = (-1)^Q\frac{\sqrt{2J+1}[(J+1)J]^{K/2}}{c(K)\langle J\|\hat{J}_Q^{(K)}\|J\rangle}n_J. \tag{D.38}$$

Unfortunately, the authors of [177, 249] do not consider the question of how $A_Q^{(K)}$ transforms on turning the coordinate system. Attempts to obtain matrices which effect such a transformation from D-matrices meet with considerable difficulty. In addition, a number of relations for $A_Q^{(K)}$, for instance Eq. (17) in [249], are valid only for $Q = 0$, and not in the general case. Despite the fact that distribution moments are successfully used for the interpretation of experimental results [208, 209], their theory requires further refinement.

In [249, 250, 251] the magnitudes $A_{Q\pm}^{\{K\}}$ are defined. They have been named *real tensor moments* of ground state distribution:

$$A_{Q\pm}^{\{K\}} = c(K)\frac{\langle\langle JM|\hat{J}_{Q\pm}^{\{K\}}|JM'\rangle\rangle}{(JM|\mathbf{J}^2|JM)^{K/2}}, \tag{D.39}$$

where the real angular momentum operators $\hat{J}_Q^{\{K\}}$ are defined by the general rules (D.34) and (D.35), taking $\hat{J}_Q^{\{K\}}$ as a base. The general normalization of $A_{Q\pm}^{\{K\}}$ is based on the agreement that $A_0^{\{0\}} = 1$. On turn of the coordinate system these magnitudes transform with the aid of matrices which are tabulated in [141] for rank $K \leq 3$. Eq. (D.39) may be rewritten as

$$A_{Q\pm}^{\{K\}} = c(K)\frac{\text{tr}\left(\rho \hat{J}_{Q\pm}^{\{K\}}\right)}{[(J+1)J]^{K/2}n_J}, \tag{D.40}$$

which makes it possible to compare them with the polarization moments. Simultaneous application of Eqs. (D.27), (D.32), (D.11) and (D.40), (D.34), (D.35) yields

$$\frac{\text{Re}\left({}^g f_Q^K\right)}{A_{Q+}^{\{K\}}} = \frac{\sqrt{2J+1}[(J+1)J]^{K/2}}{c(K)(2-\delta_{Q0})^{1/2}\langle J\|\hat{J}_Q^{(K)}\|J\rangle}n_J, \quad Q \geq 0, \tag{D.41}$$

$$\frac{\text{Im}\left({}^g f_Q^K\right)}{A_{Q-}^{\{K\}}} = \frac{\sqrt{2J+1}[(J+1)J]^{K/2}}{c(K)\sqrt{2}\langle J\|\hat{J}_Q^{(K)}\|J\rangle}n_J, \quad Q > 0. \tag{D.42}$$

Real tensor moments are successfully applied to the interpretation of experimental results [354].

D.3 Expansion of probability density

In the expansion of the *classical probability density* $\rho(\theta, \varphi)$ over spherical functions $Y_{KQ}(\theta, \varphi)$, one may, just as in the quantum mechanical approach, use both *cogredient* and *contragredient* forms of expansion, similarly to Eqs. (D.3) and (D.4), i.e. in the form

$$\rho(\theta, \varphi) = \sum_{KQ} {}^c\rho_Q^K Y_{KQ}(\theta, \varphi), \tag{D.43}$$

$$\rho(\theta, \varphi) = \sum_{KQ} {}^g\rho_Q^K Y_{KQ}^*(\theta, \varphi). \tag{D.44}$$

In chemical physics research, both forms have been successfully employed. Thus, for instance, the authors of [128, 129, 304] have applied the contragredient form of $\rho(\theta, \varphi)$ expansion, whilst those of [19, 95] have applied the cogredient form. Both methods lead to corresponding ways of polarization moment transformation at turn of the coordinate system.

In the standard expansion form, i.e. without applying additional normalizing factors in (D.3) and (D.4), this leads to the following simple interpretation of the physical meaning of the introduced moments:

$$^c\rho_Q^K = \langle Y_{KQ}^* \rangle, \tag{D.45}$$

$$^g\rho_Q^K = \langle Y_{KQ} \rangle. \tag{D.46}$$

In particular, using the explicit form of spherical functions (see Appendix B) we may write:

$$^c\rho_0^0 = {}^g\rho_0^0 = \frac{W}{\sqrt{4\pi}},\tag{D.47}$$

$$^c\rho_0^1 = {}^g\rho_0^1 = \sqrt{\frac{3}{4\pi}}W\langle\cos\theta\rangle.\tag{D.48}$$

Here W is the probability of finding the molecule on the level under examination, but $\langle\cos\theta\rangle$ is the mean value of the cosine defining angular momentum orientation for a separate particle on this level.

The above interpretation is, in essence, the result of the limiting case of Eqs. (D.15) and (D.14), as may be confirmed by comparison between (D.47), (D.48) and (D.31). It will emerge that in the classical limit the *total population* n_J will be replaced by the *total probability* W, and the normalizing factor $(2J + 1)^{-1/2}$ will be replaced by $(4\pi)^{-1/2}$. The latter replacement is due to the fact that, in passing from the quantum to the classical approach, the number of possible spatial orientations of the angular momentum, equalling $2J + 1$, must be replaced by the full solid angle 4π.

Using the explicit form of spherical functions (Appendix B), it is also possible to obtain an interpretation, analogous to (D.47) and (D.48) for *classical polarization moments* $^c\rho_Q^K$, $^g\rho_Q^K$ with other values of K, Q. It may be pointed out that, sometimes, for instance in [19, 95], normalization $W = 1$ is used.

In the present book we have used the cogredient expansion form (2.14), where, as distinct from the standard form, an additional normalizing factor has been introduced, namely $(-1)^Q\sqrt{(2K + 1)/4\pi}$. Our expansion of the classical probability density $\rho(\theta, \varphi)$ thus differs from the standard one in exactly the same way as the expansion of the quantum mechanical density matrix $\hat{\rho}$ over $^2T_Q^K$ differs from the expansion over $^1T_Q^K$. In Section 5.3 we present a comparison between the physical meaning of the classical polarization moments ρ_Q^K, as used in the present book, and the quantum mechanical polarization moments f_Q^K, as determined by the cogredient method using normalization (D.11).

It was shown in Section 2.3 that for the description of axially symmetric angular momenta distribution the use of only the polarization moments ρ_Q^K where $Q = 0$ are sufficient. In this case $\rho(\theta, \varphi)$ is frequently [9, 111, 191, 289] expanded over simpler functions connected with $Y_{K0}(\theta, \varphi)$, namely *Legendre polynomials* [183] $P_K(\cos\theta)$ which may be easily obtained in explicit form using the formula

$$P_K(\cos\theta) = \sqrt{\frac{4\pi}{2K + 1}}Y_{K0}(\theta, \varphi)\tag{D.49}$$

and the spherical functions (B.1) from Appendix B. The expansion of the form

$$n(\theta) = n_J \sum_{K=0}^{\infty} a_K P_K(\cos\theta) \qquad (D.50)$$

is most frequently used with normalization

$$\frac{1}{4\pi} \int_0^\pi \frac{n(\theta)}{n_J} \sin\theta \, d\theta \, d\varphi = 1, \qquad (D.51)$$

where $n(\theta)$ is the molecular angular momenta distribution function over the angle θ, whilst n_J is the total number of molecules on the level under investigation. The Legendre polynomials are orthogonal functions, although not normalized to unity:

$$\int_0^\pi P_K(\cos\theta) P_L(\cos\theta) \sin\theta \, d\theta = \frac{2}{2K+1} \delta_{KL}, \qquad (D.52)$$

which leads to the following form of expressions for a_K:

$$a_K = \frac{2K+1}{2n_J} \int_0^\pi n(\theta) P_K(\cos\theta) \sin\theta \, d\theta. \qquad (D.53)$$

From (D.51) we obtain the normalization $a_0 = 1$. Comparing Eq. (D.53) with the definition of ρ_Q^K (2.16), we may obtain, taking account of the normalization for $n(\theta)$ and $\rho(\theta, \varphi)$:

$$a_K = \frac{2K+1}{W} \rho_0^K = (2K+1) \frac{\rho_0^K}{\rho_0^0}. \qquad (D.54)$$

In this particular case of axially symmetric distribution both coefficients a_K and ρ_0^K are real magnitudes.

For some other particular cases of $\rho(\theta, \varphi)$ distribution symmetry the methods of obtaining real expansion coefficients and their symmetry properties are analyzed in [289]. In the general case of arbitrary $\rho(\theta, \varphi)$ symmetry it is also possible to introduce real expansion coefficients over real functions $Y_{Q\pm}^{\{K\}}(\theta, \varphi)$ which are defined in analogy with the operators $T_{Q\pm}^{\{K\}}$:

$$Y_{Q+}^{\{K\}} = (-1)^Q \sqrt{2 - \delta_{Q0}} \operatorname{Re} Y_Q^K, \qquad Q \geq 0, \qquad (D.55)$$

$$Y_{Q-}^{\{K\}} = (-1)^Q (1 - \delta_{Q0}) \sqrt{2} \operatorname{Im} Y_Q^K, \qquad Q \geq 0. \qquad (D.56)$$

The real functions $Y_{Q\pm}^{\{K\}}$, introduced in this way, are proportional to the zonal, sectorial and tesserial harmonics sometimes used [379]. The most detailed treatment of the expansion of $\rho(\theta, \varphi)$ over $Y_{Q\pm}^{\{K\}}$ may be found in

[249, 251], in which an expansion is used with the following normalization:

$$A_{Q\pm}^{\{K\}} = c(K)v(K)\langle Y_{Q\pm}^{\{K\}}(\theta, \varphi)\rangle, \tag{D.57}$$

$$v(K) = \frac{\mathrm{Re}\left(J_Q^K\right)}{(\mathbf{J}^2)^{K/2}\mathrm{Re}(Y_{KQ})}. \tag{D.58}$$

Numerical values of $c(K)$, $v(K)$, for $K \leq 4$ are given in Table D.3. For expansion coefficients we use, following [249, 251], the same notation $A_{Q\pm}^{\{K\}}$ as for the expansion coefficients of the quantum mechanical density matrix over real $T_{Q\pm}^{\{K\}}$. Such a choice of notation may be justified by the fact that classical $A_{Q\pm}^{\{K\}}$ constitute an asymptotic limit of quantum mechanical $A_{Q\pm}^{\{K\}}$ in their transition to the limit $J \to \infty$.

Comparing Eqs. (2.16), (D.45), (D.46) and (D.57), one may see that

$$\frac{A_{Q+}^{\{K\}}}{\mathrm{Re}^g\rho_Q^K} = \frac{A_{Q+}^{\{K\}}}{\mathrm{Re}^c\rho_Q^K} = \frac{A_{Q+}^{\{K\}}}{(-1)^Q\sqrt{(2K+1)/4\pi}\mathrm{Re}\,\rho_Q^K}$$

$$= c(K)v(K)(-1)^Q\sqrt{2 - \delta_{Q0}}, \quad Q \geq 0, \tag{D.59}$$

$$\frac{A_{Q-}^{\{K\}}}{\mathrm{Im}^g\rho_Q^K} = -\frac{A_{Q-}^{\{K\}}}{\mathrm{Im}^c\rho_Q^K} = \frac{A_{Q-}^{\{K\}}}{(-1)^Q\sqrt{(2K+1)/4\pi}\mathrm{Im}\,\rho_Q^K}$$

$$= c(K)v(K)(-1)^Q(1 - \delta_{Q0})\sqrt{2}, \quad Q \geq 0. \tag{D.60}$$

Thus, we have attempted to give, in the present appendix, an idea of the various methods of determining classical and quantum mechanical polarization moments and some related coefficients. We have considered only those methods which are most frequently used in atomic, molecular and chemical physics. An analysis of a great variety of different approaches creates the impression that sometimes the authors of one or other investigation find it easier to introduce new definitions of 'their own' multipole moments, rather than find a way in the rather muddled system of previously used ones. This situation complicates comparison between the results obtained by various authors considerably. We hope that the material contained in the present appendix might, to some extent, simplify such a comparison.

References

[1] Alber, G. and Zoller, P. (1991). Laser excitation of electronic wave packets in Rydberg atoms, *Physics Reports*, **199**, 231–280.

[2] Aleksandrov, E.B. (1963). Quantum beats of luminescence under modulated light excitation, *Optika i Spektroskopiya*, **14**, 436–438. [*Opt. Spectrosc. (USSR)*, **14**, 233–234].

[3] Aleksandrov, E.B. (1964). Luminescence beats induced by pulsed excitation of coherent states, *Optika i Spektroskopiya*, **17**, 957–960. [*Opt. Spectrosc. (USSR)*, **17**, 522–523].

[4] Aleksandrov, E.B. (1972). Optical manifestation of interference of non-degenerated atomic states, *Uspekhi Fizicheskikh Nauk*, **107**, 595–622. [*Sov. Phys.—Usp.*, **15**, 436–451].

[5] Aleksandrov, E.B. (1979). Quantum beats, *Proc. VI Intern. Conf. Atomic Physics, Zinatne, Plenum Press, Riga, New York, London*, pp. 521–534.

[6] Aleksandrov, E.B., Chaika, M.P. and Khvostenko, G.I. (1993). *Interference of Atomic States* (Springer–Verlag, New York).

[7] Alzetta, G., Gozzini, A. and Moi, L. (1972). Effet de l'orientation atomique par pompage optique sur la formation des molecules K_2, *C. R. Acad. Sci., Paris. Ser. B*, **274**, 39–42.

[8] Aryutyunan, V.M., Adonts, G.G. and Kanatsyan, E.G. (1982). Determination of atomic-level polarization decay times by a pulse delay technique, *Optika i Spektroskopiya*, **53**, 792–795. [*Opt. Spectrosc. (USSR)*, **53**, 472–474].

[9] Atkorn, R. and Zare, R.N. (1984). Effects of saturation on laser-induced fluorescence measurements of population and polarization, *Ann. Rev. Phys. Chem.*, **35**, 265–289.

[10] Atutov, S.N., Jermolayev, I.M. and Shalagin, A.M. (1986). Light-induced current in sodium vapor, *Zhurnal Eksperimental'noi i*

Teoreticheskoi Fiziki, **90**, 1963–1971. [*Sov. Phys.—JETP* **63**, 1149–1154].

[11] Auzinsh, M.P., Pirags, I.Ya., Ferber, R.S. and Shmit, O.A. (1980). Direct measurement of thermalization rate of the ground state of K_2 molecules, *JETP Lett.*, **31**, 554–557.

[12] Auzinsh, M.P. and Ferber, R.S. (1983). Manifestation of the sixth-order polarization moment in the Hanle-effect signal of the electronic ground state of dimers, *Opt. Spectrosc. (USSR)*, **55**, 674–675.

[13] Auzinsh, M.P., Ferber, R.S. and Pirags, I.Ya. (1983). K_2 ground-state relaxation studies from transient process kinetics, *J. Phys. B: At. Mol. Phys.*, **16**, 2759–2771.

[14] Auzinsh, M.P. and Ferber, R.S. (1984). Observation of a quantum-beat resonance between magnetic sublevels with $\Delta M = 4$, *JETP Lett.*, **39**, 452–455.

[15] Auzinsh, M.P. (1984). On the solution of rate equations for polarization momenta in case of large angular moments, *Izv. Akad. Nauk Latv. SSR Ser. Fiz. Tekh. Nauk*, **1**, 9–15.

[16] Auzinsh, M.P. and Ferber, R.S. (1984). On the manifestation of ground state polarization moments in transient process kinetics of diatomic molecules, *Izv. Akad. Nauk Latv. SSR Ser. Fiz. Tekh. Nauk*, **1**, 16–20.

[17] Auzinsh, M.P. and Ferber, R.S. (1985). Exceeding the classical limit of the degree of polarization under the nonlinear Hanle effect of diatomic molecules, *Opt. Spectrosc. (USSR)*, **59**, 6–9.

[18] Auzinsh, M.P., Tamanis, M.Ya. and Ferber, R.S. (1985). Simultaneous determination of the Landé g factor and relaxation rate of the ground state for diatomic molecules using the Hanle effect, *Opt. Spectrosc. (USSR)*, **59**, 828–829.

[19] Auzinsh, M.P. and Ferber, R.S. (1985). Classical treatment of optical pumping by expansion to multipoles, *Izv. Akad. Nauk Latv. SSR Ser. Fiz. Tekh. Nauk*, **3**, 3–9.

[20] Auzinsh, M.P., Tamanis, M.Ya. and Ferber, R.S. (1985). Observation of quantum beats in the kinetics of the thermalization of diatomic molecules in the electronic ground state, *JETP Lett.*, **42**, 160–163.

[21] Auzinsh, M.P., Suvorov, A.E. and Ferber, R.S. (1985). Description of nonlinear resonance beats of diatomic molecules in the model of dipole, *Izv. Akad. Nauk Latv. SSR Ser. Fiz. Tekh. Nauk*, **6**, 49–52.

[22] Auzinsh, M.P. and Ferber, R.S. (1985). Oriented gas of diatomic molecules in a magnetic field, *Sov. Phys. Tech. Phys.*, **30**, 923–927.

[23] Auzinsh, M.P., Tamanis, M.Ya. and Ferber, R.S. (1986). Zeeman quantum beats during the transient process after optical depopulation of the ground electronic state of diatomic molecules, *Sov. Phys.—JETP*, **63**, 688–693.

[24] Auzinsh, M.P., Ferber, R.S., Harya, Ya.A. and Pirags, I.Ya. (1986). The effect of collisions on the intensity and polarization of laser-induced $D^1\Pi \to X^1\Sigma^+$ fluorescence from NaK, *Chem. Phys. Lett.*, **124**, 116–120.

[25] Auzinsh, M.P. (1986). Polarization moments in a state with high angular momentum, *Opt. Spectrosc. (USSR)*, **60**, 248–250.

[26] Auzinsh, M.P., Tamanis, M.Ya. and Ferber, R.S. (1987). Determination of the sign of the Landé factor of diatomic molecules in the ground and excited states by the Hanle effect, *Opt. Spectrosc. (USSR)*, **63**, 582–588.

[27] Auzinsh, M.P. (1987). Polarization of laser-excited fluorescence of diatomic molecules and the magnetic-field effect, *Opt. Spectrosc. (USSR)*, **63**, 721–725.

[28] Auzinsh, M.P., Nasyrov, K.A., Tamanis, M.Ya., Ferber, R.S. and Shalagin, A.M. (1987). Resonance of quantum beats in a system of magnetic sublevels of the ground electronic state of molecules, *Sov. Phys.—JETP*, **65**, 891–897.

[29] Auzinsh, M.P. (1988). Hanle effect in the ground electronic state of dimers with allowance for the finite value of angular momentum, *Opt. Spectrosc. (USSR)*, **65**, 153–156.

[30] Auzinsh, M.P. and Ferber, R.S. (1989). Optical orientation and alignment of high-lying vibrational–rotational levels of diatomic molecules under their fluorescence population, *Opt. Spectrosc. (USSR)*, **66**, 158–163.

[31] Auzinsh, M.P. (1989). Study of the electronic ground state of molecules by polarization-spectroscopy methods, *Opt. Spectrosc. (USSR)*, **67**, 616–619.

[32] Auzinsh, M.P., Tamanis, M.Ya., Ferber, R.S. and Kharya, Ya.A. (1989). Inclusion of thermal motion in laser fluorescence lifetime measurement using ^{80}Se$_2$ as an example, *Opt. Spectrosc. (USSR)*, **67**, 750–753.

[33] Auzinsh, M.P. (1989). Nonlinear parametric and relaxation resonance of quantum beats, *Izv. Akad. Nauk Latv. SSR Ser. Fiz. Tekh. Nauk*, **6**, 3–7.

[34] Auzinsh, M.P., Nasyrov, K.A., Tamanis, M.Ya., Ferber, R.S. and Shalagin, A.M. (1990). Determination of the ground-state Landé factor for diatomic molecules by a beat-resonance method, *Chem. Phys. Lett.*, **167**, 129–136.

[35] Auzinsh, M.P. and Ferber, R.S. (1990). Orientation of atoms and molecules under excitation by elliptically polarized light, *Opt. Spectrosc. (USSR)*, **68**, 149–152.

[36] Auzinsh, M.P. (1990). Orientation of molecules in the ground state by linearly polarized light with a wide spectral profile, *Opt. Spectrosc. (USSR)*, **68**, 695–697.

[37] Auzinsh, M.P. (1990). Nonlinear phase resonance of quantum beats in the dimer ground state, *Opt. Spectrosc. (USSR)*, **68**, 750–752.

[38] Auzinsh, M.P. (1990). On the possibility of experimental observation of the dynamic Stark effect influence on the laser induced fluorescence of dimers, *Optika i Spektroskopiya*, **69**, 302–306. [*Opt. Spectrosc. (USSR)*, **69**, 182–185].

[39] Auzinsh, M.P. and Ferber, R.S. (1991). Optical pumping of diatomic molecules in the electronic ground state: Classical and quantum approaches, *Phys. Rev. A*, **43**, 2374–2386.

[40] Auzinsh, M.P. (1991). The influence of optical pumping of molecules on the polarization of laser induced fluorescence, *Latvian Journal of Physics and Technical Sciences*, **1**, 3–10.

[41] Auzinsh, M.P. (1992). Dynamic Stark effect action on optical pumping of atoms in an external magnetic field, *Phys. Lett. A*, **169**, 463–468.

[42] Auzinsh, M.P. (1992). General restrictions for the relaxation constants of the polarization moments of the density matrix, *Chem. Phys. Lett.*, **198**, 305–310.

[43] Auzinsh, M.P. and Ferber, R.S. (1992). *J*-Selective Stark orientation of molecular rotation in a beam, *Phys. Rev. Lett.*, **69**, 3463–3466.

[44] Auzinsh, M.P. and Ferber, R.S. (1993). Emergence of circularity at linear polarized excitation of molecules, *J. Chem. Phys.*, **99**, 5742–5747.

[45] Bagaeva, I.N. and Chaika, M.P. (1989). Lifetimes of highly excited states of Argon, *Optika i Spektroskopiya*, **66**, 984–989. [*Opt. Spectrosc. (USSR)*, **66**, 575–578].

[46] Bain, A.J. and McCaffery, A.J. (1984). Complete determination of the state multipoles of rotationally resolved polarized fluorescence using a single experimental geometry, *J. Chem. Phys.*, **80**, 5883–5892.

[47] Bain, A.J. and McCaffery, A.J. (1984). Multipolar evolution in two- and *N*-photon excitation, *Chem. Phys. Lett.*, **108**, 275–282.

[48] Bain, A.J., McCaffery, A.J., Proctor, M.J. and Whitaker, B.J. (1984). Laser-interrogated dichroic spectroscopy: A sensitive probe of molecular anisotropies, *Chem. Phys. Lett.*, **110**, 663–665.

[49] Bain, A.J. and McCaffery, A.J. (1985). On the measurement of molecular anisotropies using laser techniques. I. Polarized laser fluorescence, *J. Chem. Phys.*, **83**, 2627–2631.

[50] Bain, A.J. and McCaffery, A.J. (1985). On the measurements of molecular anisotropies using laser techniques. II. Differential absorption of circularly and linearly polarized light, *J. Chem. Phys.*, **83**, 2632–2640.

[51] Bain, A.J. and McCaffery, A.J. (1985). On the measurements of molecular anisotropies using laser techniques. III. Detection of the higher multipoles, *J. Chem. Phys.*, **83**, 2641–2645.

[52] Band, Y.B. and Yulienne, P.S. (1992). Complete alignment and orientation of atoms and molecules by stimulated Raman scattering with temporally shifted lasers, *J. Chem. Phys.*, **96**, 3339–3341.

[53] Barnwell, J.D., Loeser, J.G. and Herschbach, D.R. (1983). Angular
      correlations in chemical reactions. Statistical theory for four-vector
      correlations, *J. Phys. Chem.*, **87**, 2781–2786.

[54] Baskin, J.S., Felker, P.M. and Zewail, A.H. (1987). Purely rotational
      coherence effect and time-resolved sub-Doppler spectroscopy of large
      molecules. II. Experimental, *J. Chem. Phys.*, **86**, 2483–2499.

[55] Becker, M., Gaubatz, U., Bergmann, K. and Jones, P.L. (1987). Efficient
      and selective population of high vibrational levels by stimulated near
      resonance Raman scattering, *J. Chem. Phys.*, **87**, 5064–5076.

[56] Bell, W.E. and Bloom, A.L. (1961). Optically driven spin precession,
      Phys. Rev. Lett., **6**, 280–281.

[57] Bennaker, J.J.M., Scoles, G., Knaap, H.F.P. and Jonkman, R.M. (1962).
      The influence of a magnetic field on the transport properties of diatomic
      molecules in the gaseous state, *Phys. Lett.*, **2**, 5–6.

[58] Benreuven, A. (1966). Symmetry considerations in pressure broadening
      theory, *Phys. Rev.*, **141**, 34–40.

[59] Benreuven, A. (1966). Impact broadening of microwave spectra, *Phys.
      Rev.*, **145**, 7–22.

[60] Bergmann, K. and Demtröder, W. (1972). Inelastic cross section of
      excited molecules. III. Absolute cross section for rotational and
      vibrational transitions in the $Na_2(B^1\Pi_u)$ state, *J. Phys. B: At. Mol.
      Phys.*, **5**, 2098–2106.

[61] Bergmann, K., Engelhardt, R., Hefter, U. and Witt, J. (1978).
      State-resolved differential cross sections for rotational transitions in
      $Na_2 + Ne(He)$ collisions, *Phys. Rev. Lett.*, **40**, 1446–1450.

[62] Bergmann, K., Engelhardt, R., Hefter, U. and Witt, J. (1979).
      State-to-state differential cross sections for rotational transitions in
      $Na_2 + Ne$ collisions, *J. Chem. Phys.*, **71**, 2726–2739.

[63] Bergmann, K., Hefter, U. and Witt, J. (1980). State-to-state differential
      cross sections for rotationally inelastic scattering of Na_2 by He, *J.
      Chem. Phys.*, **72**, 4777–4790.

[64] Berman, P.R. (1985). Current research in optical pumping, *Ann. Phys.
      (Paris)*, **10**, 985–993.

[65] Bernheim, R.A. (1965). *Optical Pumping* (Benjamin, New York).

[66] Bernstein, R.B., Herschbach, D.R. and Levine, R.D. (1987). Dynamic
      aspects of stereochemistry, *J. Phys. Chem.*, **91**, 5365–5377.

[67] (1991). Bernstein memorial issue on molecular dynamics, *J. Phys.
      Chem.*, **95**, 7961–8421.

[68] Bersohn, R. and Lin, S.H. (1969). Orientation of targets by beam
      excitation, *Adv. Chem. Phys.*, **16**, 67–100.

[69] Bersohn, R. (1980). Molecular photodissociation by an ultraviolet photon, *IEEE J. Quantum Electron.*, **QE-16**, 1208–1218.

[70] Beyer, H.J. and Kleinpoppen, H. (1979). Applications of anticrossing spectroscopy, *Proc. VI Intern. Conf. Atomic Phys.*, *Zinatne, Plenum Press, Riga, New York, London*, pp. 435–461.

[71] Bitto, H. and Huber, J.R. (1992). Molecular quantum beats. High-resolution spectroscopy in time domain, *Acc. Chem. Res.*, **25**, 65–71.

[72] Block, P.A., Bohac, E.J. and Miller, R.E. (1992). Spectroscopy of pendular states: The use of molecular complexes in achieving orientation, *Phys. Rev. Lett.*, **68**, 1303–1306.

[73] Blum, K. (1981). *Density Matrix. Theory and Applications* (Plenum Press, New York, London).

[74] Bonch-Bruevich, A.M. and Khodovoi, V.A. (1967). Present-day methods of Stark effect investigation in atoms, *Uspekhi Fizicheskikh Nauk*, **93**, 71–110. [*Sov. Phys.—Usp.* **10**, 637—657].

[75] Born, M. and Wolf, E. (1968). *Principles of Optics* (Pergamon Press, Oxford, London, Edinburgh, New York, Paris, Frankfurt).

[76] Bouchiat, M.A. and Grossetete, F. (1966). Methodé expérimentales d'etude de la relaxation d'une vapeur alcaline orienteé optiquement, *J. Phys. (Paris)*, **27**, 353–366.

[77] Brandt, S. (1970). *Statistical and Computational Methods in Data Analysis* (North Holland Publishing Company, Amsterdam, London, American Elsevier Publishing Company, Inc., New York).

[78] Bras, N., Butaux, J., Jeannet, J.C. and Perrin, D. (1985). Diffusion effects on relaxation measurements. Example: vibrational relaxation of $NO(v = 5)$ molecules studied by laser-induced fluorescence, *J. Phys. B: At. Mol. Phys.*, **18**, 3901–3908.

[79] Braun, P.A. and Rebane, T.K. (1982). Semiclassical theory of molecular hydromagnetism, *Khimicheskaya Fizika*, **1**, 447–456.

[80] Braun, P.A., Volodicheva, M.I. and Rebane, T.K. (1988). Effect of nuclear core vibrations on the magnitude and sign of the rotational g factor of a molecule with closed electron shell, *Optika i Spektroskopiya*, **65**, 306–310. [*Opt. Spectrosc. (USSR)*, **65**, 183–186].

[81] Brebrick, R.F. (1968). Tellurium vapor pressure and optical density at $350° - 615°$, *J. Phys. Chem.*, **72**, 1032–1036.

[82] Brechignac, Ph., Picard-Bersellini, A., Charneau, R. and Launay, J.M. (1980). Rotational relaxation of CO by collisions with H_2 molecules: a comparison between theory and experiment, *Chem. Phys.*, **53**, 165–183.

[83] Breit, G. (1933). Quantum theory of dispersion, *Rev. Mod. Phys.*, **5**, 91–141.

[84] Brieger, M., Hese, A., Renn, A. and Sodeik, A. (1980). The dipole moment of ^{7}LiH in the electronically excited $A^1\Sigma^+$ state, *Chem. Phys. Lett.*, **76**, 465–468.

[85] Brooks, P.R. and Jones, E.M. (1966). Reactive scattering of K atoms from oriented CH_3I molecules, *J. Chem. Phys.*, **45**, 3449–3450.

[86] Brooks, P.R. (1976). Reactions of oriented molecules, *Science*, **193**, 11–16.

[87] Brooks, R.A., Anderson, C.H. and Ramsey, N.F. (1964). Rotational magnetic moments of the alkali molecules, *Phys. Rev. A*, **136**, 62–68.

[88] Broyer, M., Vigué, J. and Lehmann, J.-C. (1972). Rotational Landé factors in the $B^3\Pi_{0_u^+}$ state of iodine, *Phys. Lett. A*, **40A**, 43–49.

[89] Broyer, M., Lehmann, J.-C. and Vigué, J. (1975). g-factors and lifetimes in the B state of molecular iodine, *J. Phys. (Paris)*, **36**, 235–241.

[90] Brucat, P.J. and Zare, R.N. (1983). $NO_2\tilde{A}^2B_2$ state properties from Zeeman quantum beats, *J. Chem. Phys.*, **73**, 100–111.

[91] Brunner, T.A., Smith, N., Karp, A.W. and Pritchard, D.E. (1981). Rotational energy transfer in $Na_2^*(A^1\Sigma)$ colliding with Xe, Kr, Ar, Ne, He, H_2, CH_4 and N_2: Experiment and fitting laws, *J. Chem. Phys.*, **74**, 3324–3341.

[92] Büchler, A. and Meschi, D.J. (1975). The magnetic moment of Se_2, *J. Chem. Phys.*, **63**, 3586–3590.

[93] Cagnac, B. (1961). Orientation nucléaire par pompage optique des isotopes impairs du mercure, *Ann. Phys. (Paris)*, **6**, 467–526.

[94] Carver, T.R. (1963). Optical pumping, *Science*, **141**, 599–608.

[95] Case, D.A., McClelland, G.M. and Herschbach, D.R. (1978). Angular momentum polarization in molecular collisions: Classical and quantum theory for measurements using resonance fluorescence. *Mol. Phys.*, **35**, 541–573.

[96] Chaika, M.P. (1975). *Interference of degenerated atomic states (in Russian)* (Nauka, Leningrad).

[97] Chaika, M.P. (1986). Self aligment of molecular beam, *Optika i Spektroskopiya*, **60**, 1103–1106. [*Opt. Spectrosc. (USSR)*, **60**, 681–683].

[98] Chaiken, J., Benson, T., Gurnick, M. and McDonald, J.D. (1979). Quantum beats in single rovibronic state fluorescence of biacetyl, *Chem. Phys. Lett.*, **61**, 195–198.

[99] Chamberlain, G.A. and Simons, J.P. (1975). Polarized photofluorescence excitation spectroscopy: a new technique for the study of molecular photodissociation. Photolysis of H_2O in the vacuum ultraviolet, *Chem. Phys. Lett.*, **32**, 355–358.

[100] Chapman, G.D. (1967). Magnetic resonance in mercury vapour induced by frequency-modulated light, *Proc. Phys. Soc.*, **92**, 1070–1073.

[101] Chu, S. and Wieman, C. (1989). Laser cooling and trapping of atoms, *J. Opt. Soc. Am. B*, **6**, 2020.

[102] Clark, R. and McCaffery, A.P. (1978). Laser fluorescence studies of molecular iodine. II. Relaxation of oriented ground and excited state molecules, *Mol. Phys.*, **35**, 617–637.

[103] Cohen, E.R. and Taylor, B.N. (1986). The 1986 Adjustment of the Fundamental Physical Constants, report of the CODATA Task Group on Fundamental Constants, *CODATA Bulletin* **63**, Pergamon, Elmsford, New York; (1987). The 1986 adjustment of the fundamental constants, *Rev. Mod. Phys.*, **59**, 1121–1148; (1993). The fundamental physical constants, *Phys. Today* August, Part 2, 9–13.

[104] Cohen-Tannoudji, C. (1962). Théorie quantique du cycle de pompage optique. Verification experimentale des nouveaux effets prevus (1-re partie), *Ann. de Phys. (Paris)*, **7**, 423–461.

[105] Cohen-Tannoudji, C. (1962). Théorie quantique du cycle de pompage optique. Verification experimentale des nouveaux effets prevus (2-e partie), *Ann. de Phys. (Paris)*, **7**, 469–504.

[106] Cohen-Tannoudji, C. and Kastler, A. (1966) *Progress in Optics* (North-Holland, Amsterdam).

[107] Colegrove, F.D., Franken, P.A., Lewis, P.R. and Sands, R.H. (1959). Novel method of spectroscopy with application to precision fine structure measurements, *Phys. Rev. Lett.*, **3**, 420–422.

[108] Corney, A. and Series, G.W. (1964). Theory of resonance fluorescence excited by modulated or pulsed light, *Proc. Phys. Soc.*, **83**, 207–216.

[109] Dalby, F.W., Vigué, J. and Lehmann, J.C. (1975). On Hanle effect in the band system of the Se_2 molecule, *Can. J. Phys.*, **53**, 140–144.

[110] Decomps, B., Dumont, M. and Ducloy, M. (1976). Linear and nonlinear phenomena in laser optical pumping. In *Laser Spectroscopy of Atoms and Molecules. Topics in Applied Physics*, vol. 2, ed. H. Walther, pp. 283–347 (Springer-Verlag, Berlin, Heidelberg, New York).

[111] Deech, J.S. and Baylis, W.E. (1971). Double exponential decay of imprisoned resonance radiation, *Can. J. Phys.*, **49**, 90–101.

[112] Dehmelt, H.G. and Jefferts, K.B. (1962). Alignment of the H_2^+ molecular ion by selective photodissociation. I, *Phys. Rev.*, **125**, 1318–1322.

[113] Demkov, Yu.N. and Rebane, T.K. (1991). Atomic and molecular magnetic moments related to the anomalous electron magnetic moments, *Optika i Spektroskopiya*, **71**, 714–716. [*Opt. Spectrosc. (USSR)*, **71**, 414–415].

[114] Demtröder, W., McClintock, M. and Zare, R.N. (1969). Spectroscopy of Na_2 using laser-induced fluorescence, *J. Chem. Phys.*, **51**, 5495–5508.

[115] Demtröder, W., Stetzenbach, W. and Stock, W. (1976). Lifetimes and Franck–Condon factors for the $B^1\Pi_u$–$X^1\Sigma_g^+$ system of Na_2, *J. Mol. Spectrosc.*, **61**, 382–394.

[116] Demtröder, W. (1982). *Laser Spectroscopy. Basic Concepts and Instrumentation* (Springer-Verlag, Berlin, Heidelberg, New York).

[117] De Vries, M.S., Srdanov, V.I., Hanrahan, C.P. and Martin, R.M. (1982). Observation of steric effects in the reaction of Xe* with photodissociation polarized IBr, *J. Chem. Phys.*, **77**, 2688–2689.
De Vries, M.S., Srdanov, V.I., Hanrahan, C.P. and Martin, R.M. (1983). Orientation dependence in the reaction of Xe* with photodissociation polarized IBr, *J. Chem. Phys.*, **78**, 5582–5589.

[118] Dodd, J.N. and Series, G.W. (1961). Theory of modulation of light in a double resonance experiment, *Proc. R. Soc. London, Ser. A*, **263**, 353–370.

[119] Dodd, J.N., Kaul, R.D. and Warington, D.M. (1964). The modulation of resonance fluorescence excited by pulsed light, *Proc. Phys. Soc. London, Sect. A*, **84**, 176–178.

[120] Dodd, J.N. and Series, G.W. (1978). Time-resolved fluorescence spectroscopy, In: *Progress in Atomic Spectroscopy, PT A, Plenum Publ. Corp.*, pp. 639–677.

[121] Dohnt, G., Hese, A., Renn, A. and Schweda, H.S. (1979). Molecular electric field level crossing spectroscopy: The dipole moment in the $A^1\Sigma^+$ state of BaO, *Chem. Phys.*, **42**, 183–190.

[122] Dubs, R.L., McKoy, V. and Dixit, S.N. (1988). Atomic and molecular alignment from photoelectric angular distribution in $(n+1)$ resonantly enhanced multiphoton ionization, *J. Chem. Phys.*, **88**, 968–974.

[123] Dubs, R.L., Dixit, S.N. and McKoy, V. (1986). Circular dichroism in photoelectron angular distribution as a probe of atomic and molecular alignment, *J. Chem. Phys.*, **85**, 656–663.
Dubs, R.L., Dixit, S.N. and McKoy, V. (1986). Extraction of alignment parameters from circular dichroic photoelectron angular distribution (CDAD) measurements, *J. Chem. Phys.*, **85**, 6267–6269.
Dubs, R.L. and McKoy, V. (1989). Molecular alignment from circular dichroic photoelectrons angular distribution in $(n+1)$ resonance enhanced multiphoton ionization, *J. Chem. Phys.*, **91**, 5208–5211.

[124] Drullinger, R.E. and Zare, R.N. (1969). Optical pumping of molecules, *J. Chem. Phys.*, **51**, 5532–5542.

[125] Drullinger, R.E. and Zare, R.N. (1973). Optical pumping of molecules. II. Relaxation studies, *J. Chem. Phys.*, **59**, 4225–4234.

[126] Ducas, T.W., Littman, M.G. and Zimmerman, M.L. (1975). Observation of oscillations in resonance absorption from a coherent superposition of atomic states, *Phys. Rev. Lett.*, **35**, 1752–1754.

[127] Ducloy, M. (1973). Nonlinear effects in optical pumping of atoms by a high intensity multimode gas laser. General theory, *Phys. Rev. A*, **8**, 1844–1859.

[128] Ducloy, M. (1975). Application du formalisme des états cohérents de moment angulaire a quelques problémes de physique atomique, *J. Phys. (Paris)*, **36**, 927–941.

[129] Ducloy, M. (1976). Non-linear effects in optical pumping with lasers. I. General theory of the classical limit for levels of large angular momenta, *J. Phys. B: At. Mol. Phys.*, **9**, 357–381.

[130] Dufayard, J. and Nedelec, O. (1982). Collision transfers between CdH $A^2\Pi\, v' = 0$ rotational states induced by He or Ar, *Chem. Phys.*, **71**, 279–288.

[131] Dumont, A.M. (1968). Influence de l'agitation thermique sur la mesure d'une duree de vie d'atomes excités, *J. Quant. Spectrosc. and Radiat. Transfer*, **8**, 1551–1554.

[132] Dyakonov, M.I. and Perel, V.I. (1964). Coherence relaxation during difusion of resonance radiation, *Zhurnal Eksperimental'noi i Teoreticheskoi Fiziki*, **47**,1483–1495. [*Sov. Phys.—JETP* **20**, 997–1004].

[133] Dyakonov, M.I. (1964). Theory of resonance scattering of light by a gas in the presence of a magnetic field, *Zhurnal Eksperimental'noi i Teoreticheskoi Fiziki*, **47**, 2213–2221. [(1965). *Sov. Phys.—JETP*, **20**, 1484–1489].

[134] Dyakonov, M.I. and Perel, V.I. (1965). Coherence relaxation of excited atoms in collisions, *Zhurnal Eksperimental'noi i Teoreticheskoi Fiziki*, **48**, 345–352. [*Sov. Phys.—JETP*, **21**, 227–231].

[135] Dyakonov, M.I. and Perel, V.I. (1972). General inequalities for the relaxation constants of a spin density matrix, *Phys. Lett.*, **41 A**, 451–452.

[136] Edmonds, A.R. (1974). *Angular Momentum in Quantum Mechanics*, (Princeton University Press, Princeton, New Jersey).

[137] Elbel, M., Simon, M. and Strauss, Th. (1990). Transformation of an atomic alignment into an orientation in a discharge, *Ann. Phys. (Leipzig)*, **47**, 467–474.

[138] Ennen, G. and Ottinger, Ch. (1974). Rotation–vibration–translation energy transfer in laser excited $Li_2(B^1\Pi_u)$, *Chem. Phys.*, **3**, 404–412.

[139] Fano, U. (1957). Description of states in quantum mechanics by density matrix operator techniques, *Rev. Mod. Phys.*, **29**, 74–93.

[140] Fano, U. and Racah, G. (1959). *Irreducible Tensorial Sets* (Academic Press, New York).

[141] Fano, U. (1960). Real representations of coordinate rotations, *J. Math. Phys.*, **1**, 417–423.

[142] Fano, U. (1964). Precession equation of a spinning particle in nonuniform fields, *Phys. Rev.*, **133**, B828–B830.

[143] Feinberg, R., Teets, R.E., Rubbmark, J. and Schawlow, A.L. (1977). Ground state relaxation measurements by laser-induced depopulation, *J. Chem. Phys.*, **66**, 4330–4333.

[144] Felker, P.M. and Zewail, A.H. (1985). Dynamics of intramolecular vibrational–energy redistribution (IVR). I. Coherence effects, *J. Chem. Phys.*, **82**, 2961–2974.
Dynamics of intramolecular vibrational–energy redistribution (IVR). II. Excess energy dependence, *ibid*, 2975–2993.
Dynamics of intramolecular vibrational-energy redistribution (IVR). III. Role of molecular rotations, *ibid*, 2994–3010.

[145] Felker, P.M. and Zewail, A.H. (1987). Purely rotational coherence effect and time-resolved sub-Doppler spectroscopy of large molecules. I. Theoretical, *J. Chem. Phys.*, **86**, 2460–2482.

[146] Felker, P.M. and Zewail A.H. (1988). Picosecond time-resolved dynamics of vibrational-energy redistribution and coherence in beam-isolated molecules, *Adv. Chem. Phys.*, **70**, 265–363.

[147] Fell, C.P., McCaffery, A.J. and Ticktin, A. (1989). Measurements of collisional broadening and shift parameters in $X^1\Sigma_g^+ \to A^1\Sigma_u^+$ Li_2 using Doppler scanning techniques, *J. Chem. Phys.*, **90**, 852–861.

[148] Feofilov, P.P. (1961) *The Physical Basis of Polarized Emission* (Consultants Bureau, New York).

[149] Ferber, R.S., Shmit, O.A. and Tamanis, M.Ya. (1978). Ground state Hanle effect in optically aligned diatomic molecules, *Abstr. VI Intern. Conf. Atomic Physics, Zinatne, Plenum Press, Riga, New York, London*, pp. 345–346.

[150] Ferber, R.S. (1978). Optical alignment and orientation of diatomic molecules in the ground electronic state, *Izvestiya Akademii Nauk Latviyskoi SSR*, **8**, 85–99.

[151] Ferber, R.S. (1979). Determination of rate constants and cross sections in ground state Na_2 and K_2 by the method of laser optical pumping, *Izvestiya Akademii Nauk SSSR, Seriya Fizicheskaya* **43**, 419–423.

[152] Ferber, R.S., Shmit, O.A. and Tamanis, M.Ya. (1979). Ground state Hanle effect in optically aligned diatomic molecules Na_2 and K_2, *Chem. Phys. Lett.*, **61**, 441–444.

[153] Ferber, R.S., Okunevich, A.I., Shmit, O.A. and Tamanis, M.Ya. (1982). Landé factor measurements for the $^{130}Te_2$ electronic ground state, *Chem. Phys. Lett.*, **90**, 476–480.

[154] Ferber, R.S., Shmit, O.A. and Tamanis, M.Ya. (1982). Lifetimes and Landé factors in the $A0_u^+$ and $B0_u^+$ states of $^{130}Te_2$, *Chem. Phys. Lett.*, **92**, 393–397.

[155] Flygare, W.H. and Benson, R.C. (1971). The molecular Zeeman effect in diamagnetic molecules and the determination of molecular magnetic moments (g values), magnetic susceptibilities, and molecular quadrupole moments, *Mol. Phys.*, **20**, 225–250.

[156] Flygare, W.H. (1978). *Molecular Structure and Dynamics* (Prentice-Hall, Inc., New Jersey).

[157] Fonda, L., Mankoč-Borštnik, N. and Rosina, M. (1988). Coherent rotational states, *Phys. Rep.*, **158**, 161–204.

[158] Friedrich, B., Pullman, D.P. and Herschbach, D.R. (1991). Alignment and orientation of rotationally cool molecules, *J. Phys. Chem.*, **95**, 8118–8129.

[159] Friedrich, B. and Herschbach, D.R. (1991). On the possibility of orienting rotationally cooled polar molecules in an electric field, *Z. Phys. D—Atoms, Molecules and Clusters*, **18**, 153–161.

[160] Friedrich, B. and Herschbach, D.R. (1991). Spatial orientation of molecules in strong electric fields and evidence for pendular states, *Nature (London)*, **353**, 412–414.

[161] Friedrich, B., Herschbach, D.R., Rost, J.–M., Rubahn, H.–G., Reger, M. and Verbeek, M. (1993). Optical spectra of spatially oriented molecules: ICl in a strong electric field, *J. Chem. Soc. Faraday Trans.*, **89**, 1539–1549.

[162] Gandhi, S.R., Xu, Q.X., Curtis, T.J. and Bernstein, R.B. (1987). Oriented molecular beams: focused beam of rotationally cold polar polyatomic molecules, *J. Chem. Phys.*, **91**, 5437–5441.

[163] Gaubatz, U., Rudecki, P., Becker, M., Schiemann, S., Külz, M. and Bergmann, K. (1988). Population switching between vibrational levels in molecular beams, *Chem. Phys. Lett.*, **149**, 463–468.

[164] Gericke, K.–H., Gläser, H.G., Maul, C. and Comes, F.J. (1990). Joint product state distribution of coincidently generated photofragment pairs, *J. Chem. Phys.*, **92**, 411–419.

[165] German, K.R. and Zare, R.N. (1969). Measurement of the Hanle effect for the OH radical, *Phys. Rev.*, **186**, 9–13.

[166] German, K.R., Bergman, T.H., Weinstock, E.M. and Zare, R.N. (1973). Zero-field level crossing and optical radio-frequency double resonance studies of the $A^2\Sigma^+$ states of OH and OD, *J. Chem. Phys.*, **58**, 4304–4318.

[167] Gerrard, A. and Burch, J.M. (1975). *Introduction to Matrix Methods in Optics* (A Wiley-Interscience Publication, London, New York, Sydney, Toronto).

[168] Glass-Maujean, M. and Beswick, J.A. (1989). Coherence effects in the polarization of photofragments, *J. Chem. Soc., Faraday Trans. 2*, **85**, 983–1002.

[169] Gornik, W., Kaiser, D., Lange, W., Luther, J. and Schulz, H.H. (1972). Quantum beats under pulsed dye laser excitation, *Opt. Commun.*, **6**, 327–329.

[170] Gorter, C.J. (1938). Zur Interpretierung des Sentleben-Effektes, *Naturwissenschaften*, **26**, 140.

[171] Gouedard, G. and Lehmann, J.-C. (1970). Resonance optique et effect Hanle sur la molécule, *C. R. Acad. Sci. Ser. B*, **270**, 1664–1678.

[172] Gouedard, G. and Lehmann, J.C. (1975). Effet Hanle et résonance en lumiére modulée sur le niveau $B1_u, v' = 0, J' = 105$ de la molécule ($^{80}Se_2$) exitée par la raie 4727 Å, *C. R. Acad. Sci. Ser. B*, **280**, 471–474.

[173] Gouedard, G. and Lehmann, J.-C. (1977). Landé factors measurements in the $^3\Sigma_u^+$ state of $^{80}Se_2$, *J. Phys. (Paris)*, *Lettres*, **38**, L85–L86.

[174] Gouedard, G. and Lehmann, J.C. (1981). Pulsed dye-laser studies in the $B^3\Sigma_u^-$ state of Se_2 Landé factors and perturbations, *Faraday Disc. Chem. Soc.*, **71**, 143–149.

[175] Gray, J.A., Lin, M., and Field, R.W. (1990). Zeeman spectroscopy and deperturbation of the low-lying states of NiH, *J. Chem. Phys.*, **92**, 4651–4659.

[176] Greene, C.H. and Zare, R.N. (1982). Photofragment alignment and orientation, *Ann. Rev. Phys. Chem.*, **33**, 119–150.

[177] Greene, C.H. and Zare, R.N. (1983). Determination of product population and alignment using laser-induced fluorescence, *J. Chem. Phys.*, **78**, 6741–6753.

[178] Grushevskii, V.B., Tamanis, M.Ya., Ferber, R.S. and Shmit, O.A. (1977). Relaxation of K_2 molecules, optically pumped by the radiation from a He–Ne laser in collisions with atoms, *Opt. Spectrosc. (USSR)*, **42**, 572–573.

[179] Guest, J.A., Jackson, K.H. and Zare, R.N. (1983). Determination of rotational alignment of $N_2 B^2\Sigma_u^+$ and $CO^+ B^2\Sigma^+$ following photoionization of N_2 and CO, *Phys. Rev. A*, **28**, 2217–2228.

[180] Guest, J.A., O'Halloran, M.A. and Zare, R.N. (1984). Influence of electron and nuclear spin on photofragment: Application to ClCN dissociation at 157.6 nm, *Chem. Phys. Lett.*, **103**, 261–265.

[181] Gupta, R., Happer, W., Moe, G. and Park, W. (1974). Nuclear magnetic resonance of diatomic molecules in optically pumped alkali vapors, *Phys. Rev. Lett.*, **32**, 574–577.

[182] Hack, E. and Huber, J.R. (1991). Quantum beat spectroscopy of molecules, *International Reviews in Physical Chemistry*, **10**, 287–317.

[183] (1964). *Handbook of Mathematical Functions*, edited by Abramowitz, M. and Stegun, I.A. (National Bureau of Standards, Washington).

[184] Hanle, W. (1924). Über magnetische Beeinflussung der Polarization der Resonanzfluoreszenz, *Z. Phys.*, **30**, 93–105.

[185] Hanle, W. (1989). *Memorien* (Justus-Liebig-Universität, Giessen).

[186] Hansen, J.C. and Berry, R.S. (1984). Angular distribution of electrons from resonant two-photon ionization of molecules, *J. Chem. Phys.*, **80**, 4078–4096.

[187] Happer, W. (1972). Optical pumping, *Rev. Mod. Phys.*, **44**, 169–249.

[188] Haroche, S. (1976). Quantum beats and time resolved fluorescence spectroscopy, *In: High-resolved Laser Spectroscopy*, Ed. K. Shimoda, *Springer, Berlin*, pp. 253–311.

[189] Harya, Ya.A., Ferber, R.S., Kuz'menko, N.E., Shmit, O.A. and Stolyarov, A.V. (1987). Intensities of the laser-induced fluorescence of $^{130}Te_2$ and electronic transition strengths for the $A0_u^+ - X0_g^+$ and $B0_u^+ - X0_g^+$ systems, *J. Mol. Spectrosc.*, **125**, 1-13.

[190] Hasselbrink, E., Waldeck, J.R. and Zare, R.N. (1988). Orientation of the CN $X^2\Sigma^+$ fragment following photolysis of ICN by circularly polarized light, *Chem. Phys.*, **126**, 191–200.

[191] Hefter, U., Ziegler, G., Mattheus, A., Fischer, A. and Bergmann, K. (1986). Preparation and detection of alignment with high $|m|$ selectivity by saturated laser optical pumping in molecular beams, *J. Chem. Phys.*, **85**, 286–302.

[192] Heitler, W. (1954). *The Quantum Theory of Radiation* (The Clarendon Press, Oxford).

[193] Henke, H., Selzle, H.L., Hays, T.R., Lin, S.H. and Schlag, E.W. (1981). Effect of collisions and magnetic field on quantum beats in biacetyl, *Chem. Phys. Lett.*, **77**, 448–452.

[194] Herzberg, G. (1957). *Molecular Spectra and Molecular Structure. I. Spectra of Diatomic Molecules* (D. Van Nostrand Company, Princeton, New Jersey, Toronto, London, New York)

[195] Hinchen, J.J. and Hobbs, R.H. (1976). Rotation relaxation studies of HF using double resonance, *J. Chem. Phys.*, **65**, 2732–2739.

[196] Hinchen, J.J. and Hobbs, R.H. (1979). Rotational population transfer in HF, *J. Appl. Phys.*, **50**, 628–636.

[197] Hsu, D.S.Y., Weinstein, N.D. and Herschbach, D.R. (1975). Rotational polarization of reaction products. Analysis of electric deflection profiles, *Mol. Phys.*, **29**, 257–278.

[198] Hougen, J.T. (1970). *The Calculation of Rotational Energy Levels and Rotational Line Intensities in Diatomic Molecules* (National Bureau of Standards (U. S.), Monograph 115, Washington D.C.).

[199] Houston, P.L. (1987). Vector correlations in photodissociation dynamics, *J. Phys. Chem.*, **91**, 5388–5397.

[200] Huber, R., König, F. and Weber, H.G. (1977). Magnetic shielding studies on the alkali molecules Na_2 and Cs_2 by NMR, *Z. Physik A*, **281**, 25–33.

[201] Huber, R. and Weber, H.G. (1978). Collision induced dissociation of Na_2: Dependence of the dissociation rate on the vibrational distribution, *Chem. Phys.*, **35**, 461–467.

[202] Huber, K.P. and Herzberg, G. (1979). *Molecular Spectra and Molecular Structure, IV. Constants of Diatomic Molecules* (Van Nostrand Reinhold Company, New York, Cincinati, Atlanta, Dallas, San Francisco, London, Toronto, Melbourne).

[203] Huber, R., Knapp, M., König, F., Reinhard, H. and Weber, H.G. (1980). Magnetic shielding and spin-rotation interaction in ground state alkali molecules, *Z. Physik A–Atoms and Nuclei*, **296**, 95–99.

[204] Hussain, J., Wiesenfeld, J.R. and Zare, R.N. (1980). Photofragment fluorescence polarization following photolysis of $HgBr_2$ at 193 nm, *J. Chem. Phys.*, **72**, 2479–2483.

[205] Ibbs, K.G. and McCaffery, A.J. (1981). Polarised laser fluorescence of Se_2, *J. Chem. Soc., Faraday Trans. 2*, **77**, 631–635.

[206] Ivanco, M., Hager, J., Sharfin, W. and Wallace, S.C. (1983). Quantum interference phenomena in the radiative decay of the $\tilde{C}(^1B_2)$ state of SO_2, *J. Chem. Phys.*, **78**, 6531–6540.

[207] Jacobs, D.C. and Zare, R.N. (1986). Reduction of $1 + 1$ resonance enhanced MPI spectra to populations and alignment factors, *J. Chem. Phys.*, **85**, 5457–5468.

[208] Jacobs, D.C., Kolasinski, K.W., Shane, S.F. and Zare, R.N. (1989). Rotational population and alignment distribution for inelastic scattering and trapping/desorption of NO on Pt(111), *J. Chem. Phys.*, **91**, 3182–3195.

[209] Jacobs, D.C. and Zare, R.N. (1989). Simplified trajectory method for modeling gas-surface scattering: The NO/Pt(111) system, *J. Chem. Phys.*, **91**, 3196–3207.

[210] Jannsen, M.H.M., Parker, D.H. and Stolte, S. (1991). Steric properties of the reactive system Ca $(^1D_2) + CH_3F(JKM) \rightarrow CaF(A) + CH_3$, *J. Phys. Chem.*, **95**, 8142–8153.

[211] Jefferts, K.B. (1968). Rotational HFS spectra of H_2^+ molecular ions, *Phys. Rev. Lett.*, **20**, 39–41.

[212] Jefferts, K.B. (1969). Hyperfine structure in the molecular ion H_2^+, *Phys. Rev. Lett.*, **23**, 1476–1478.

[213] Jeyes, S.R., McCaffery, A.J., Rowe, M.D. and Kato, H. (1977). Selection rules for collisional energy transfer in homonuclear diatomics. Rotationally inelastic collisions, *Chem. Phys. Lett.*, **48**, 91–94.

[214] Jeyes, S.R., McCaffery, A.J. and Rowe, M.D. (1978). Energy and angular momentum transfer in homonuclear diatomic molecules from polarized laser fluorescence, *Mol. Phys.*, **36**, 845–867.

[215] Jeyes, S.R., McCaffery, A.J. and Rowe, M.D. (1978). Energy and angular moment transfer in atom–diatom collisions from polarized laser fluorescence, *Mol. Phys.*, **36**, 1865–1891.

[216] Jones, P.L., Hefter, U., Mattheus, A., Witt, J., Bergmann, K., Müller, W., Meyer, W and Schinke, R. (1982). Angular resolved rotationally inelastic scattering of Na_2–Ne: Comparision between experiment and theory, *Phys. Rev. A*, **26**, 1283–1301.

[217] Joswig, H., O'Halloran, M.A. and Zare, R.N. (1986). Photodissociation Dynamics of ICN, *Faraday Discuss. Chem. Soc.*, **82**, Paper 3, 1–10.

[218] Judd, B. (1975). *Angular Momentum Theory for Diatomic Molecules* (Academic Press, New York).

[219] Kagan, Yu.M. and Afanasiev, A.M. (1961). Contribution to the kinetic theory of gases with rotational degrees of freedom, *Zhurnal Eksperimental'noi i Teoreticheskoi Fiziki*, **41**, 1536–1545. [*Sov. Phys.—JETP*, **14**, 1096–1101].

[220] Kais, S. and Levine, R.D. (1987). Directed states of molecules, *J. Phys. Chem.*, **91**, 5462–5465.

[221] Kamke, W. (1975). Nuclear magnetic resonance of K_2 in optically pumped potassium vapor, *Phys. Lett. A*, **55**, 15–16.

[222] Kartoshkin, V.A. and Klement'ev, G.V. (1987). Spin exchange in the He (2^3S_1)–$O_2(^3\Sigma_g^-)$ system, *Optika i Spektroskopiya* **63**, 465–469. [*Opt. Spectrosc. (USSR)*, **63**, 271–272].

[223] Kartoshkin, V.A. and Klement'ev, G.V. (1988). Conservation of total spin in a collision of an atom with molecule in the Π-state, *Optika i Spektroskopiya* **64**, 1198–1200. [*Opt. Spectrosc. (USSR)* **64**, 715–716].

[224] Kastler, A. (1950). Quelques suggestion concentrant la production optique et la détection optique d'une inégalité de population des niveaux, *J. Phys. Radium*, **11**, 225–265.

[225] Kastler, A. (1968). Optical pumping and molecule formation, *Acta Phys. Pol.*, **34**, 693–694.

[226] Kastler, A. (1972). *New Directions in Atomic Physics* (Yale Univ. Press, New Haven, Conn.).

[227] Kastler, A. (1978). *Coherence in Spectroscopy and Modern Physics* (Plenum Press, New York).

[228] Kazantsev, S.A., Polinovskaya, N.Ya., P'atnickii, L.I. and Edel'man, S.A. (1988). Polarization of atomic ensembles in ionized gases, *Uspekhi Fizicheskikh Nauk*, **156**, 3–46. [*Sov. Phys.—Usp.*, **31**, 785–809].

[229] Khare, V., Kouri, D.J. and Hoffman, D.K. (1981). On j_z-preserving properties in molecular collisions. I. Quantal coupled states and classical impulsive approximation, *J. Chem. Phys.*, **74**, 2275–2286.

[230] Khare, V., Kouri, D.J. and Hoffman, D.K. (1982). On j_z-preserving properties in molecular collisions. II. Close-coupling study of state-to-state differential cross sections, *J. Chem. Phys.*, **76**, 4493–4501.

[231] Killinger, D.K., Wang, C.C. and Hanabusa, M. (1976). Intensity and pressure dependence of resonance fluorescence of OH induced by a tunable uv laser, *Phys. Rev. A*, **13**, 2145–2152.

[232] Kleiber, P.D., Wang, J.-X., Sando, K.M., Zaviropulos, V. and Stwalley, W.C. (1991). Photodissociation of $K_2(X^1\Sigma_g^+ - B^1\Pi_u)$, *J. Chem. Phys.* **95**, 4168–4176.

[233] Klintsare, I.P., Tamanis, M.Ya. and Ferber, R.S. (1989). Quantum beats in the fluorescence kinetics of $Te_2(A0_u^+)$ molecules, *Opt. Spektrosc. (USSR)*, **66**, 484–485.

[234] Klintsare, I.P., Stolyarov, A.V., Tamanis, M.Ya., Ferber, R.S. and Kharya, Ya.A. (1989). Anomalous behavior of Landé factors of the $Te_2(B0_u^+)$ molecule and of the intensities of the $B0_u^+-X1_g^+$ transition, *Opt. Spectrosc. (USSR)*, **66**, 595–597.

[235] Klintsare, I.P., Tamanis, M.Ya, Stolyarov, A.V., Auzinsh, M.P. and Ferber, R.S. (1993). Alignment–orientation conversion by quadratic Zeeman effect: Analysis and observation for Te_2, *J. Chem. Phys.*, **99**, 5748–5753.

[236] Klyucharev, A.N. and Janson, M.L. (1988). *Elementary Processes in Alkali Metal Plasma (in Russian)* (Energoizdat, Moscow).

[237] Kolwas, M. and Szonert, J. (1986). Reorientation of Na_2 by He under multiple-collision conditions, *Chem. Phys. Lett.*, **130**, 498–503.

[238] Kompitsas, M. and Weber, H.G. (1975). A study of the nuclear spin polarization of Na_2 by optical pumping, *Chem. Phys. Lett.*, **35**, 277–279.

[239] König, F. and Weber, H.G. (1980). Relaxation studies of ground state Na_2 by optical pumping transient, *Chem. Phys.*, **45**, 91–100.

[240] Kono, H., Fujimura, Y. and Lin, S.H. (1981). Theory of quantum beats: Master equation approach, *J. Chem. Phys.*, **75**, 2569–2576.

[241] Konstantinov, O.V. and Perel, V.I. (1964). Coherence of states at scattering of modulated light, *Zhurnal Eksperimental'noi i Teoreticheskoi Fiziki*, **45**, 279–284. [*Sov. Phys.—JETP*, **18**, 195–198].

[242] Kotlikov, E.N. and Kondratjeva, V.A. (1980). Effect of a strong electromagnetic field on the shape of the crossing signals in zero magnetic field, *Opt. Spectrosc. (USSR)*, **48**, 367–371.

[243] Kotlikov, E.N. and Chaika, M.P. (1983). Origin of orientation irradiation by linearly polarized nonresonant light. I. Description in the limits of classical-oscillator model, *Optika i Spektroskopiya*, **55**, 242–245. [*Opt. Spectrosc. (USSR)*, **55**, 142–144].

[244] Kovács, I. (1967). *Rotational Structure in the Spectra of Diatomic Molecules* (American Elsevier, New York)

[245] Kowalski, F.W., Hale, P.D. and Shattil, S.J. (1988). Broadband continuous-wave laser, *Opt. Lett.*, **13**, 622–624.

[246] Kramer, K.H. and Bernstein, R.B. (1965). Focusing and orientation of symmetric-top molecules with the electric six-pole field, *J. Chem. Phys.*, **42**, 767–770.

[247] Kuipers, E.W., Tenner, M.G., Kleyn, A.W. and Stolte, S. (1989). Dependence of the NO/Ag(111) trapping probability on molecular orientation, *Chem. Phys.*, **138**, 451–460.

[248] Kuklinski, J.R., Gaubatz, U., Hioe, F.T. and Bergmann, K. (1989). Adiabatic population transfer in a three-level system driven by delayed laser pulses, *Phys. Rev. A*, **40**, 6741–6744.

[249] Kummel, A.C., Sitz, G.O. and Zare, R.N. (1986). Determination of population and alignment of the ground state using two-photon nonresonance excitation, *J. Chem. Phys.*, **85**, 6874–6897.

[250] Kummel, A.C., Sitz, G.O. and Zare, R.N. (1988). Determination of orientation of the ground state using two-photon nonresonant excitation, *J. Chem. Phys.*, **88**, 6707–6732.

[251] Kummel, A.C., Sitz, G.O. and Zare, R.N. (1988). Determination of population, alignment, and orientation using laser induced fluorescence with unresolved emission, *J. Chem. Phys.*, **88**, 7357–7368.

[252] Kupriyanov, D.V., Sevastianov, B.N. and Vasyutinskii O.S. (1990). Polarization of thallium atoms produced in molecular photodissociation: experiment and theory, *Z. Phys. D–Atoms, Molecules and Clusters*, **15**, 105–115.

[253] Kupriyanov, D.V. and Sokolov, I.M. (1991). Generation of squeezed electromagnetic field states on interaction between radiation and optically oriented atoms, *Zhurnal Eksperimental'noi i Teoreticheskoi Fiziki*, **99**, 93–106. [*Sov. Phys.—JETP*, **72**, 50–57].

[254] Kupriyanov, D.V., Picheyev, B.V. and Vasyutinskii, O.S. (1993). Photodissociation of RbI at 266 nm: spin orientation of ground state Rb atoms, *J. Phys. B*, **26**, L803–L810.

[255] Kurzel, R.B., Steinfeld, J.I., Hatzenbuhler, D.A. and Leroi, G.E. (1971). Energy transfer processes in monochromatically excited iodine molecules, *J. Chem. Phys.*, **55**, 4822–4831.

[256] Kuz'menko, N.E., Kuznetsova, L.A. and Kuz'akov, Yu.Ya. (1984). *Franck–Condon Factors in Diatomic Molecules (in Russian)* (Moscow University, Moscow).

[257] Kuz'menko, N.E., Pirags, I.Ya., Pritkov, S.E., Stolyarov, A.V. and Ferber, R.S. (1987). Determination of strength of electronic transition in $B^1\Pi_u$–$X^1\Sigma_g^+$ system of potassium dimers $^{39}K_2$ from intensities of laser-induced fluorescence, *Izv. Akad. Nauk Latv. SSR, Ser. Fiz. Tekh. Nauk*, **4**, 3–10.

[258] Kuznetsova, L.A., Kuz'menko, N.E. and Kuz'akov, Yu.Ya. (1980). *Probabilities of Optical Transitions in Diatomic Molecules (in Russian)* (Nauka, Moscow).

[259] Landau, L.D. and Lifshitz, E.M. (1988). *Theoretical Physics. V. 2. Field Theory (in Russian)* (Nauka, Moscow).

[260] Lange, W. and Mlynek, J. (1978). Quantum beats in transmission by time-resolved polarization spectroscopy, *Phys. Rev. Lett.*, **40**, 1373–1375.

[261] *Laser and Coherence Spectroscopy* (1978). Edited by Steinfeld, J.I. (Plenum Press, New York, London).

[262] Lavi, R., Schwartz-Lavi, D., Bar, I. and Rosenwaks, S. (1987). Directional properties in photodissociation: A probe for the symmetry and geometry of excited states of dimethylnitrosamine and *tert*-butyl nitrite, *J. Phys. Chem.*, **91**, 5398–5402.

[263] Leahy, D.J., Reid, K.L. and Zare, R.N. (1991). Effect of breaking cylindrical symmetry on photoelectron angular distribution resulting from resonance-enhanced two-photon ionization, *J. Chem. Phys.*, **95**, 1746–1756., Complete description of two-photon (1 + 1) ionization of NO deduced from rotationally resolved photoelectron angular distribution, *ibid*, 1757–1767.

[264] Leahy, D.J., Reid, K.L. and Zare, R.N. (1991). Determination of molecular symmetry axis ($\hat{z}$) orientation via photoelectron angular distribution measurements, *J. Phys. Chem.*, **95**, 8154–8158.

[265] Lebed', I.V., Nikitin, E.E. and Umansky, S.Ya. (1977). Collision-induced transitions between the rotational levels of diatomic molecules in Π state, *Optika i Spektroskopiya*, **43**, 636–644. [*Opt. Spectrosc. (USSR)*, **43**, 378–383].

[266] Lebow, P., Raab, F. and Metcalf, H. (1979). Measurements of *g*-factor by quantum beats in the OH free radicals, *Phys. Rev. Lett.*, **42**, 85–88.

[267] Lefebvre-Brion, H. and Field, R.W. (1986). *Perturbations in the Spectra of Diatomic Molecules* (Academic Press, New York, London).

[268] Letokhov, V.S. and Chebotayev, V.P. (1977). *Nonlinear Laser Spectroscopy* (Springer-Verlag, Heidelberg).

[269] Levenson, M.D. (1982). *Introduction to Nonlinear Laser Spectroscopy* (Academic Press, New York).

[270] Levi, D.H. (1972). Molecular level anticrossing in the CN radicals, *J. Chem. Phys.*, **56**, 5493–5499.

[271] Levine, R.D. (1990). The chemical shape of molecules: An introduction to dynamical stereochemistry, *J. Phys. Chem.*, **94**, 8872–8880.

[272] Ling, J.H. and Wilson, K.R. (1976). Molecular alignment and photofragment spectroscopy, *J. Chem. Phys.*, **65**, 881–893.

[273] Loesch, H.J. and Remscheid, A. (1990). Brute force in molecular reaction dynamics: A novel technique for measuring steric effects, *J. Chem. Phys.*, **93**, 4779–4790.

[274] Loge, G.W. and Zare, R.N. (1981). Dependence of diatomic photofragment fluorescence polarization on triatomic predissociation lifetime, *Mol. Phys.*, **43**, 1419–1428.

[275] Lombardi, M. (1967). Note sur la possibilité d'orient un atome par superposition de deux interactions séparément non orientantes en particulier par alignement électronique et relaxation anisotrope, *C. R. Acad. Sci., Ser. B*, **265B**, 191–194.

[276] Lombardi, M. (1969). Création d'orientation par combinaison de deux alignements et orientation des niveaux excités d'une décharge haute fréquence, *J. Phys. (Paris)*, **30**, 631–642.

[277] Lombardi, M. (1976). Alignment and orientation production measurement and conversion. In: *Beam Foil Spectroscopy*, eds. I.A. Sellin and D.J. Pegg, vol. 2, 731–747 (Plenum Press, New York).

[278] Luypaert, R., Van Graen, J., Coremans, J. and De Vlieger, G. (1981). Rotational Landé factors of single rotational states in bromine, *J. Phys. B: At. Mol. Phys.*, **14**, 2575–2584.

[279] Lukomsky, N.G., Polishchuk, V.A. and Chaika, M.P. (1985). Hidden anisotropy of collisions in a low-temperature plasma, *Optika i Spektroskopiya*, **58**, 474–475. [*Opt. Spectrosc. (USSR)*, **58**, 284–285].

[280] Lukomsky, N.G., Polishchuk, V.A. and Chaika, M.P. (1985). Conversion of latent alignment into orientation in a low-pressure plasma, *Optika i Spektroskopiya*, **59**, 1008–1011. [*Opt. Spectrosc. (USSR)*, **59**, 606–608].

[281] Luntz, A.C., Kleyn, A.W. and Auerbach, D.J. (1982). Observation of rotational polarization produced in molecule–surface collisions, *Phys. Rev. B*, **25**, 4273–4275.
Kleyn, A.W., Luntz, A.C. and Auerbach, D.J. (1985). Rotational polarization in no-scattering form Ag(111), *Surf. Sci.* **152/153**, 99–105.

[282] Lyyra, A.M., Wang, H., Whang, T.-J. and Stwalley, W.C. (1991). cw all-optical triple resonance spectroscopy, *Phys. Rev. Lett.*, **66**, 2724–2727.

[283] Maltz, C., Weinstein, N.D. and Herschbach, D.R. (1972). Rotational polarization of reaction products and reprojection of angular momenta, *Mol. Phys.*, **24**, 133–150.

[284] Manabe, T., Yabuzaki, T. and Ogawa, T. (1981). Observation of collisional transfer from alignment to orientation of atoms excited by a single-mode laser, *Phys. Rev. Lett.*, **46**, 637–640.

[285] Manabe, T., Yabuzaki, T. and Ogawa, T. (1979). Theory of collisional transfer between orientation and alignment of atoms excited by a single mode laser, *Phys. Rev. A*, **20A**, 1946–1957.

[286] Marr, G.V., Morton, J.M., Holmes, R.M. and McCoy, D.G. (1979).
Angular distribution of photoelectrons from free molecules of N_2 and CO
as a function of photon energy, *J. Phys. B: At. Mol. Phys.*, **12**, 43–52.

[287] Mattheus, A., Fischer, A., Ziegler, G., Gottwald, E. and Bergmann, K.
(1986). Experimental proof of a $|\Delta m| \ll j$ propensity rule in rotationally
inelastic differential scattering, *Phys. Rev. Lett.*, **56**, 712–715.

[288] McCaffery, A.J. (1987). Dynamical stereochemistry in nonreactive
collisions, *J. Phys. Chem.*, **91**, 5451–5455.

[289] McClelland, G.M. and Herschbach, D.R. (1987). Impulsive model for
angular momentum polarization in chemical reactions, *J. Phys. Chem.*,
91, 5509–5515.

[290] McClintock, M., Demtröder, W. and Zare, R.N. (1969). Level-crossing
studies of Na_2 using laser-induced fluorescence, *J. Chem. Phys.*, **51**,
5509–5520.

[291] McCormack, J., McCaffery, A.J. and Rowe, M.D. (1980). Collision
dynamics of excited NaK. I. Reactive K–NaK and elastic He–NaK
collisions, *Chem. Phys.*, **48**, 121–130.

[292] McCormack, J. and McCaffery, A.J. (1980). Collision dynamics of
excited NaK. II. Rotationally inelastic energy transfer, *Chem. Phys.*, **51**,
405–416.

[293] Meier, W., Ahlers, G. and Zacharias, H. (1986). State selective
population of $H_2(v'' = 1, J'' = 1)$ and $D_2(v'' = 1, J'' = 2)$ and rotational
relaxation in collisions with H_2, D_2, and He, *J. Chem. Phys.*, **85**,
2599–2608.

[294] Mizushima, M. (1975). *Theory of Rotating Diatomic Molecules* (J. Wiley
and Sons, New York).

[295] Mlynek, J., Drake, K.H. and Lange W. (1979). Observation of transient
and stationary Zeeman coherence by polarization spectroscopy, *Proc. IV
Intern. Conf. Optical Sciences, Tegernsee.—Berlin: Springer*, pp.
616–618.

[296] Moruzzi, G. and Strumia F. (1991). *Hanle Effect and Level-Crossing
Spectroscopy* (Plenum Press, New York, London).

[297] Nasyrov, K.A. and Shalagin, A.M. (1981). Interaction of intense
radiation with classically rotating atoms or molecules, *Zhurnal
Eksperimental'noi i Teoreticheskoi Fiziki*, **81**, 1649–1663. [*Sov.
Phys.—JETP*, **54**, 877–883].

[298] Nesmeyanov, An.M. (1963). *Vapour Preasure of the Elements*
(Academic, New York).

[299] Novakoski, L.V. and McClelland, G.M. (1987). Orientation of CHF_3
desorbed and scattered from Ag(111): measurements using electrostatic
focusing, *Phys. Rev. Lett.*, **59**, 1259–1262.

[300] Novikov, L.N., Pokazan'ev, V.G. and Skrotskii, G.V. (1970). Coherent phenomena in systems interacting with resonant radiation, *Uspekhi Fizicheskikh Nauk*, **101**, 273–302. [*Sov. Phys.—Usp.*, **13**, 384–399].

[301] Novikov, L.N., Skrotskii, G.V. and Solomaho, G.I. (1974). The Hanle effect, *Uspekhi Fizicheskikh Nauk*, **113**, 597–625. [*Sov. Phys.—Usp.*, **17**, 542–557].

[302] Okajima, S., Saigusa, H. and Lim, E.C. (1982). Quantum beats in the fluorescence of jet-cooled diazabenzenes, *J. Chem. Phys.*, **76**, 2096–2098.

[303] Okunevich, A.I. (1981). Excited-state collisional relaxation by optical orientation of atoms with arbitrary electronic angular momentum, *Optika i Spektroskopiya*, **50**, 443–449. [*Opt. Spectrosc. (USSR)*, **50**, 239–243].

[304] Omont, A. (1977). Irreducible components of the density matrix. Application to optical pumping, *Progress in Quantum Electronics*, **5**, 69–138.

[305] (1993). Orientation and polarization effects in chemical reaction dynamics, *J. Chem. Soc. Faraday Trans.*, **89**, 1401–1590.

[306] Ottinger, Ch. and Schröder, M. (1979). Rate constants for collision-induced transitions in ground-state Li_2 from two-laser spectroscopy, *J. Phys. B: At. Mol. Phys.*, **12**, 3533–3551.

[307] Parker, D.H., Jalink, H. and Stolte, S. (1987). Dynamics of molecular stereochemistry via oriented molecular scattering, *J. Phys. Chem.*, **91**, 5427–5437.

[308] Pazyuk, E.A., Stolyarov, A.V., Tamanis, M.Ya. and Ferber, R.S. (1993). Global deperturbation analysis from energetic, magnetic and radiative measurements: Application to Te_2, *J. Chem. Phys.*, **99**, 7873–7887.

[309] Petrashen', A.G., Rebane, V.N. and Rebane, T.K. (1973). Relaxation of electron multipole moments and collisional depolarization of resonance fluorescence of an atomic ensemble in the state with an inner quantum number $j = 2$, *Optika i Spektroskopiya*, **35**, 408–416. [*Opt. Spectrosc. (USSR)*, **35**, 240–244].

[310] Petrashen', A.G., Rebane, V.N. and Rebane, T.K. (1990). Relaxation of collisional self-alignment of drifting ions in a magnetic field, *Optika i Spektroskopiya* **69**, 259–265. [*Opt. Spectrosc. (USSR)* **69**, 582–585].

[311] Pibel, C.D. and Moore, C.B. (1990). Molecular angular momentum reorientation of electronically excited hydrogen ($B^1\Sigma_u^+$), *J. Chem. Phys.*, **93**, 4804–4811.

[312] Pirags, I.Ya., Tamanis, M.Ya. and Ferber, R.S. (1986). Determination of the total cross section for deactivation of $K_2(B^1\Pi_u, v', J')$ in collisions with potassium and inert-gas atoms, *Opt. Spectrosc. (USSR)*, **61**, 18–20.

[313] Podgoretskii, M.I. and Khrustal'ev, O.A. (1963). Interference effects in quantum transitions, *Uspekhi Fizicheskikh Nauk*, **81**, 217–247. [*Sov. Phys.—Usp.*, **6**, 682–700].

[314] Pokazan'ev, V.G. and Skrotskii, G.V. (1972). Crossing and anticrossing of atomic levels and their use in atomic spectroscopy, *Uspekhi Fizicheskikh Nauk*, **107**, 623–656. [*Sov. Phys.—Usp.*, **15**, 452–470].

[315] Poliakoff, E.D., Dehmer, J.L., Dill Dan, Parr, A.C., Jakson, K.H. and Zare, R.N. (1981). Polarization of fluorescence following molecular photoionization, *Phys. Rev. Lett.*, **46**, 907–910.

[316] Pomerantsev, N.M., Rizhkov, V.M. and Skrotskii, G.V. (1972). *Physical Basis of Quantum Magnetometry* (*in Russian*) (Nauka, Moscow).

[317] Pullman, D.P. and Herschbach, D.R. (1989). Alignment of I_2 molecules seeded in a supersonic beam, *J. Chem. Phys.*, **90**, 3881–3883.

[318] Pullman, D.P., Friedrich, B. and Herschbach, D.R. (1990). Facile alignment of molecular rotation in supersonic beams, *J. Chem. Phys.*, **93**, 3224–3236.

[319] Ramsay, N.F. (1955). *Molecular Beams* (Oxford University Press, London).

[320] Rautian, S.G., Smirnov, G.I. and Shalagin, A.M. (1979). *Nonlinear Resonances in Atomic and Molecular Spectra* (*in Russian*) (Nauka, Novosibirsk).

[321] Rebane, T.K. (1986). Magnetic properties of molecules with closed electronic shells, *Up To Day Problems of Quantum Chemistry*, Ed. by Veselov M.G., Nauka, Moscow, pp. 165–211.

[322] Rebane, T.K. (1988). Effect of the momentum of inertia of an electron shell on rotational *g*-factors of a molecule, *Optika i Spektroskopiya*, **64**, 1013–1017. [*Opt. Spectrosc. (USSR)*, **64**, 603–605].

[323] Rebane, V.N. (1968). Depolarization of resonance fluorescence during anisotropic collisions, *Optika i Spektroskopiya*, **24**, 309–315. [*Opt. Spectrosc. (USSR)*, **24**, 163–166].

[324] Rebane, V.N., Rebane, T.K. and Sherstyuk, A.I. (1981). Possibility of observing the relaxation of higher polarization moments in electric-quadrupole radiation, *Optika i Spektroskopiya*, **51**, 753–755. [*Opt. Spektrosc. (USSR)*, **51**, 418–419].

[325] Rebane, V.N. and Rebane, T.K. (1992). Effect of the optical cell wall on atomic-excited-state relaxation, *Optika i Spektroskopiya*, **72**, 271–275. [*Opt. Spectrosc. (USSR)*, **72**, 148–150].

[326] Richardson, C.B., Jefferts, K.B. and Dehmelt, H.G. (1968). Alignment of H_2^+ molecular ion by selective photodissociation. II. Experiments on the radio-frequency spectrum, *Phys. Rev.*, **165**, 80–87.

[327] Rost, J.M., Griffin, J.C., Friedrich, B. and Herschbach, D.R. (1992). Pendular states and spectra of oriented linear molecules, *Phys. Rev. Lett.*, **68**, 1299–1302.

[328] Rothe, E.W., Krause, U. and Duren, R. (1980). Observation of polarization of atomic fluorescence excited by laser-induced dissociation of Na_2, *Chem. Phys. Lett.*, **72**, 100–103.

[329] Rothe, E.W., Ranjbar, F., Sinha, D. and Reck, G.P. (1981). Two-step photoionization of Na_2: dependence on alignment, *Chem. Phys. Lett.*, **78**, 16–20.

[330] Rothe, E.W., Ranjbar, F. and Sinha, D. (1981). Collision-induced dissociation of aligned ions, *Chem Phys. Lett.*, **81**, 175–178.

[331] Rowe, M.D. and McCaffery, A.J. (1978). The influence of hyperfine coherence and of elastic collisions on the circular polarization of emission from Li_2, *Chem. Phys.*, **34**, 81–94.

[332] Rowe, M.D. and McCaffery, A.J. (1979). Transfer of state multipoles in excited $A^1\Sigma_u^+ {}^7Li_2$ following rotationally inelastic collisions with He: Experiment and theory, *Chem. Phys.*, **43**, 35–54.

[333] Rubahn, H.-G. and Toennis, J.P. (1988). A molecular beam study of the potential anisotropy of laser vibrationally excited $Li_2(v = 0, 20)$ scattered from Kr, *J. Chem. Phys.*, **89**, 287–294.

[334] Sanders, W.R. and Anderson, J.B. (1984). Alignment of I_2 rotation in a seeded molecular beam, *J. Phys. Chem.*, **88**, 4479–4487.

[335] Schawlow, A.L. (1977). Laser, light, and matter, *J. Opt. Soc. Am.*, **67**, 140–148.

[336] Schawlow, A.L. (1982). Spectroscopy in a new light, *Rev. Mod. Phys.*, **54**, 697–707.

[337] Schechter, I., Levine, R.D. and Gordon, R.G. (1991). Kinematic constraints in reactive collisions, *J. Phys. Chem.*, **95**, 8201–8205.

[338] Schenk, P., Hilborn, R.C. and Metcalf, H. (1973). Time-resolved fluorescence from Ba and Ca excited by a pulsed tunable dye-laser, *Phys. Rev. Lett.*, **31**, 189–192.

[339] Schinke, R. (1980). Quantum effects in rotationally inelastic molecular scattering: $K + N_2$ and $K + CO$ collisions on simple model surfaces, *J. Chem. Phys.*, **72**, 1120–1127.

[340] Schmiedt, R., Dugan, H., Meier, W. and Welge, K.H. (1982). Laser Doppler spectroscopy of atomic hydrogen in the photodissociation of HI, *Z. Phys. A.*, **304**, 137–142.

[341] Segal, D.M. and Burnett, K. (1989). Alignment measurements in orbitally selective collision-induced fluorescence, *J. Chem. Soc., Faraday Trans. 2.*, **85**, 925–938.

[342] Seideman, T., Shapiro, M. and Brumer, P (1989). Coherent radiative control of unimolecular reactions: Selective bond breaking with picosecond pulses, *J. Chem. Phys.*, **90**, 7132–7136.

[343] Series, G.W. (1981). Thirty years of optical pumping, *Contemp. Phys.*, **22**, 487–509.

[344] Series, G.W. (1985). Studies in double resonance and optical pumping, *Ann. Phys. (Paris)*, **10**, 553–570.

[345] Serri, J.A., Morales, A., Moskowitz, W., Pritchard, D.E., Becker, C.H. and Kinsey, J.L. (1980). Observation of halos in rotationally inelastic scattering of Na_2 from Ar, *J. Chem. Phys.*, **72**, 6304–6306.

[346] Shalagin, A.M. (1977). Determination of relaxation characteristics by a polarization method in nonlinear spectroscopy, *Zhurnal Eksperimental'noi i Teoreticheskoi Fiziki*, **73**, 99–111. [*Sov. Phys.—JETP*, **46**, 50–56].

[347] Shore, B.W. (1990). *The Theory of Coherent Atomic Excitation* (Wiley, New York).

[348] Silverman, M.P., Haroche, S. and Gross, M. (1978). General theory of laser-induced quantum beats. I. Saturation effects of single laser excitation, *Phys. Rev. A*, **18**, 1507–1516.
General theory of laser-induced quantum beats. II. Sequental laser excitation; effects of external static fields, *Phys. Rev. A*, **18**, 1517–1528.

[349] Silvers, S.J., Bergeman, T.H. and Klemperer, W. (1970). Level crossing and double resonance on the $A^1\Pi$ state of CS*, *J. Chem. Phys.*, **52**, 4385–4399.

[350] Silvers S.J., Gottscho, R.A. and Field, R.W. (1981). Collisional depolarization of state selected (J, M_J) BaO $A^1\Sigma^+$ measured by optical–optical double resonance, *J. Chem. Phys.*, **74**, 6000–6008.

[351] Simons, J.P. (1987). Dynamical stereochemistry and the polarization of reaction products, *J. Phys. Chem.*, **91**, 5378–5387.

[352] Sinha, M.P., Caldwell, C.D. and Zare, R.N. (1974). Alignment of molecules in gaseous transport: Alkali dimers in supersonic nozzle beams, *J. Chem. Phys.*, **61**, 491–503.

[353] Sitz, G.O., Kummel, A.C. and Zare, R.N. (1987). Alignment and orientation of N_2 scattered from Ag(111), *J. Chem. Phys.*, **87**, 3247–3249.

[354] Sitz, G.O., Kummel, A.C. and Zare, R.N. (1988). Direct inelastic scattering of N_2 from Ag(111). I. Rotational populations and alignment, *J. Chem. Phys.*, **89**, 2558–2571.

[355] Sitz, G.O., Kummel, A.C., Zare, R.N. and Tully, J. (1988). Direct inelastic scattering of N_2 from Ag(111). II. Orientation, *J. Chem. Phys.*, **89**, 2572–2582.

[356] Skrotskii, G.V. and Izyumova, T.G. (1961). Optical orientation of atoms and its application, *Uspekhi Fizicheskikh Nauk*, **73**, 423–470. [*Sov. Phys.—Usp.*, **4**, 177–204].

[357] Smith, N., Scott, T.P. and Pritchard, D.E. (1982). Substantial velocity dependence of rotationally inelastic collision cross section in Li_2^*–Xe, *Chem. Phys. Lett.*, **90**, 461–464.

[358] Smith, N., Scott, T.P. and Pritchard, D.E. (1984). Velocity dependence of rotationally inelastic collisions: $^7Li_2^*$+ Ne, Ar, and Xe, *J. Chem. Phys.*, **81**, 1229–1247.

[359] Solomon, J. (1967). Photodissociation as studied by photolysis mapping, *J. Chem. Phys.*, **47**, 889–895.

[360] Sommerfeld, A. (1930). *Wave Mechanics* (Methuen).

[361] Sommerfeld, A. (1967). *Optics* (Academic Press, New York, London).

[362] Stenholm, S. (1984). *Foundations of Laser Spectroscopy* (A Wiley-Interscience Publication, New York, Chichester, Brisbane, Toronto, Singapore).

[363] Stock, M. and Weber, H.G. (1974). An optical pumping-experiment of ortho- and para-states of Ne_2-molecules, *Phys. Lett. A*, **50**, 343–350.

[364] Stolyarov, A.V., Pazyuk, E.A., Kuznetsova, L.A., Harya, Ya.A. and Ferber, R.S. (1990). Effects of perturbations on the term values, Landé factors of $B0_u^+$ and $A1_u$ states and on intensities of $B0_u^+$–$X1_g$ transitions of $^{130}Te_2$, *Chem. Phys. Lett.*, **166**, 290–294.

[365] Stolyarov, A.V., Klintsare, I.P., Tamanis, M.Ya., Auzinsh, M.P. and Ferber, R.S. (1992). Rotational magnetic moment of the Na_2 molecule in $A^1\Sigma_u^+$ state: Perturbation effects, *J. Chem. Phys.*, **96**, 3510–3522.

[366] Stolyarov, A.V., Klintsare, I.P., Tamanis, M.Ya. and Ferber, R.S. (1993). Observation of $A^1\Sigma_u^+ \sim b^3\Pi_u$ interaction in g factors of weakly coupled Na_2 $A^1\Sigma_u^+$ state levels, *J. Chem. Phys.*, **98**, 826–835.

[367] Sudbø, Aa.S. and Loy, M.M.T. (1982). Measurements of absolute state-to-state rate constants for collision-induced transitions between spin–orbit and rotational states of NO $(X^2\Pi, v = 2)$, *J. Chem. Phys.*, **76**, 3646–3654.

[368] Tamanis, M.Ya., Ferber, R.S. and Shmit, O.A. (1975). The influence of vapour temperatures on optical pumping of Na_2 molecules, *Izv. Akad. Nauk Latv. SSR, Ser. Fiz. Tekh. Nauk*, **4**, 33–35.

[369] Tamanis, M.Ya., Ferber, R.S. and Shmit, O.A. (1976). Optical orientation of diatomic molecules under circularly polarized laser excitation, *Opt. Spectrosc. (USSR)*, **41**, 548–550.

[370] Tamanis, M.Ya., Ferber, R.S. and Shmit, O.A. (1982). Hanle effect and collisional relaxation of the $^{130}Te_2$ ground state, *Opt. Spectrosc. (USSR)*, **53**, 449–450.

[371] Tarnovskii, A.S. (1990). Bohr–Sommerfeld quantization rule and quantum mechanics, *Uspekhi Fizicheskikh Nauk*, **160**, 155–156. [*Sov. Phys.—Usp.*, **33**, 86].

[372] Tarnovskii, A.S. (1990). A new representation of quantum mechanics, *Sov. Phys.—Usp.*, **33**, 862–866.

[373] Teets, R., Feinberg, R., Hänsch, T.W. and Schawlow, A.L. (1976). Simplification of spectra by polarization labeling, *Phys. Rev. Lett.*, **37**, 683–686.

[374] Townes, C.H. and Schawlow, A.L. (1955). *Microwawe Spectroscopy* (McGraw-Hill Publishing Company Ltd., New-York, London, Toronto).

[375] Treffers, M.A. and Korving, J. (1983). Experimental determination of the m_J distribution in inelastic scattering of Na_2 by He, *Chem. Phys. Lett.*, **97**, 342–345.
Treffers, M.A. and Korving, J. (1986). Precession induced modulation of fluorescence, *J. Chem. Phys.*, **85**, 5076–5084.
Treffers, M.A. and Korving, J. (1986). Experimental determination of scattering induced angular momentum alignment in state selected rotational transitions, *J. Chem. Phys.*, **85**, 5085–5092.

[376] van Brunt, R.J. and Zare, R.N. (1968). Polarization of atomic fluorescence excited by molecular dissociation, *J. Chem. Phys.*, **48**, 4304–4308.

[377] van Esbroeck, P.E., McLean, R.A., Gaily, T.D., Holt, R.A. and Rosner, S.D. (1985). Hyperfine structure of Na_2, *Phys. Rev. A*, **32**, 2596–2601.

[378] van Vleck, J.H. (1929). On σ-type doubling and electronic spin in the spectra of diatomic molecules, *Phys. Rev.*, **33**, 467–506.

[379] Varshalovich, D.A., Moskalev, A.N. and Khersonskii, V.K. (1988). *Quantum Theory of Angular Momentum* (World Scientific, Singapore).

[380] Vasyutinskii, O.S. (1980). Orientation of atoms during photodissociation of molecules, *Pis'ma Zh. Eksp. Teor. Fiz.*, **31**, 457–459. [*JETP Lett.* **31**, 428–430].

[381] Veseth, L. (1973). Hund's coupling case (c) in diatomic molecules. 1. Theory, *J. Phys. B: At. Mol. Phys.*, **6**, 1473–1483.

[382] Veseth, L. (1976). Theory of high precision Zeeman effect in diatomic molecules, *J. Mol. Spectr.* **63**, 180–192.

[383] Vigué, J., Beswick, J.A. and Broyer, M. (1983). Coherence effects in the polarization of the light emitted by photofragments, *J. Phys. (Paris)*, **44**, 1225–1245.

[384] Visser, A.G., Bekooy, J.P., van der Meij, L.K., De Vreugd, C. and Korving, J. (1977). Angular polarization in molecular beams of I_2 and Na_2, *Chem. Phys.*, **20**, 391–408.

[385] Waldeck, J.R., Kummel, A.C., Sitz, G.O. and Zare, R.N. (1989). Determination of population, alignment, and orientation using laser

induced fluorescence with unresolved emission. II, *J. Chem. Phys.*, **90**, 4112–4114.

[386]   Wall, L.S., Bartlett, K.G. and Edwards, D.F. (1973). Selective excitation of molecular iodine, *Chem. Phys. Lett.*, **19**, 274–277.

[387]   Wallenstein, R., Paisner, J.A. and Schawlow, A.L. (1974). Observation of Zeeman quantum beats in molecular iodine, *Phys. Rev. Lett.*, **32**, 1333–1336.

[388]   Walther, Th., Bitto, H. and Huber, J.R. (1993). High-resolution quantum beat spectroscopy in the electronic ground state of a polyatomic molecule by *IR–UV* pump–probe method, *Chem. Phys. Lett.*, **209**, 455–461.

[389]   Watanabe, H. and Tsuchlya, S. (1983). Quantum beats in the fluorescence of jet-cooled SO_2 under a weak magnetic field, *J. Phys. Chem.*, **87**, 906–908.

[390]   Wawilow, S.J. and Lewschin, W.L. (1923). Beträge zur Frage über polarisiertes Fluoreszenzlicht von Farbstofflösungen, *Z. Phys.*, **16**, 135–154.

[391]   Weber, H.G., Glass, H.-J., Huber, R., Kompitsas, M., Schmidt, G. and zu Putlitz, G. (1974). Optical pumping method for studying nuclear-spin polarization in alkali dimers, *Z. Physik*, **268**, 91–95.

[392]   Weber, H.G., Bylicki, F. and Miksch, G. (1984). Inversion of polarization by light induced stabilization, *Phys. Rev. A*, **30**, 270–279.

[393]   Weinstock, E.M. and Zare, R.N. (1973). High-field level-crossing and Stark studies of the $A^2\Sigma^+$ state of OD, *J. Chem. Phys.*, **58**, 4319–4326.

[394]   Wells, W.C. and Isler, R.C. (1970). Measurements of the lifetime of the $A^1\Pi$ state of CO by level-crossing spectroscopy, *Phys. Rev. Lett.*, **24**, 705–708.

[395]   Wick, G.C. (1948). On the magnetic field of a rotating molecule, *Phys. Rev.*, **73**, 51–57.

[396]   Wieman, C. and Hänsch, T.W. (1976). Doppler-free laser polarization spectroscopy, *Phys. Rev. Lett.*, **36**, 1170–1173.

[397]   Wood, R.W. (1908). Emission of polarized light by fluorescent gases, *Phil. Mag.*, **16**, 184–189.

[398]   Wood, R.W. and Ellet, A. (1924). Polarized resonance radiation in weak magnetic field, *Phys. Rev.*, **24**, 243–254.

[399]   Zare, R.N. (1966). Molecular level-crossing spectroscopy, *J. Chem. Phys.*, **45**, 4510–4518.

[400]   Zare, R.N. (1972). Photoejection dynamics, *Mol. Photochem.*, **4**, 1–37.

[401]   Zare, R.N. (1973). Optical pumping of molecules, *Colloques Internationaux du C. N. R. S.*, **217**, 29–40.

[402]   Zare, R.N. (1988). *Angular Momentum* (J.Wiley and Sons, New York).

[403] Zare, R.N. (1989). Photofragment angular distribution from oriented symmetric-top precursor molecules, *Chem. Phys. Lett.*, **156**, 1–6.

[404] Zewail, A.H. (1989). Femtochemistry: The role of alignment and orientation, *J. Chem. Soc., Faraday Trans. 2*, **85**, 1221–1242.

[405] Yeazell, J.A. and Stroud, C.R., Jr (1987). Rydberg-atom wave packets localized in the angular variables, *Phys. Rev. A*, **35**, 2806–2809.

Index

295

Printed in the United States
By Bookmasters